D0443761

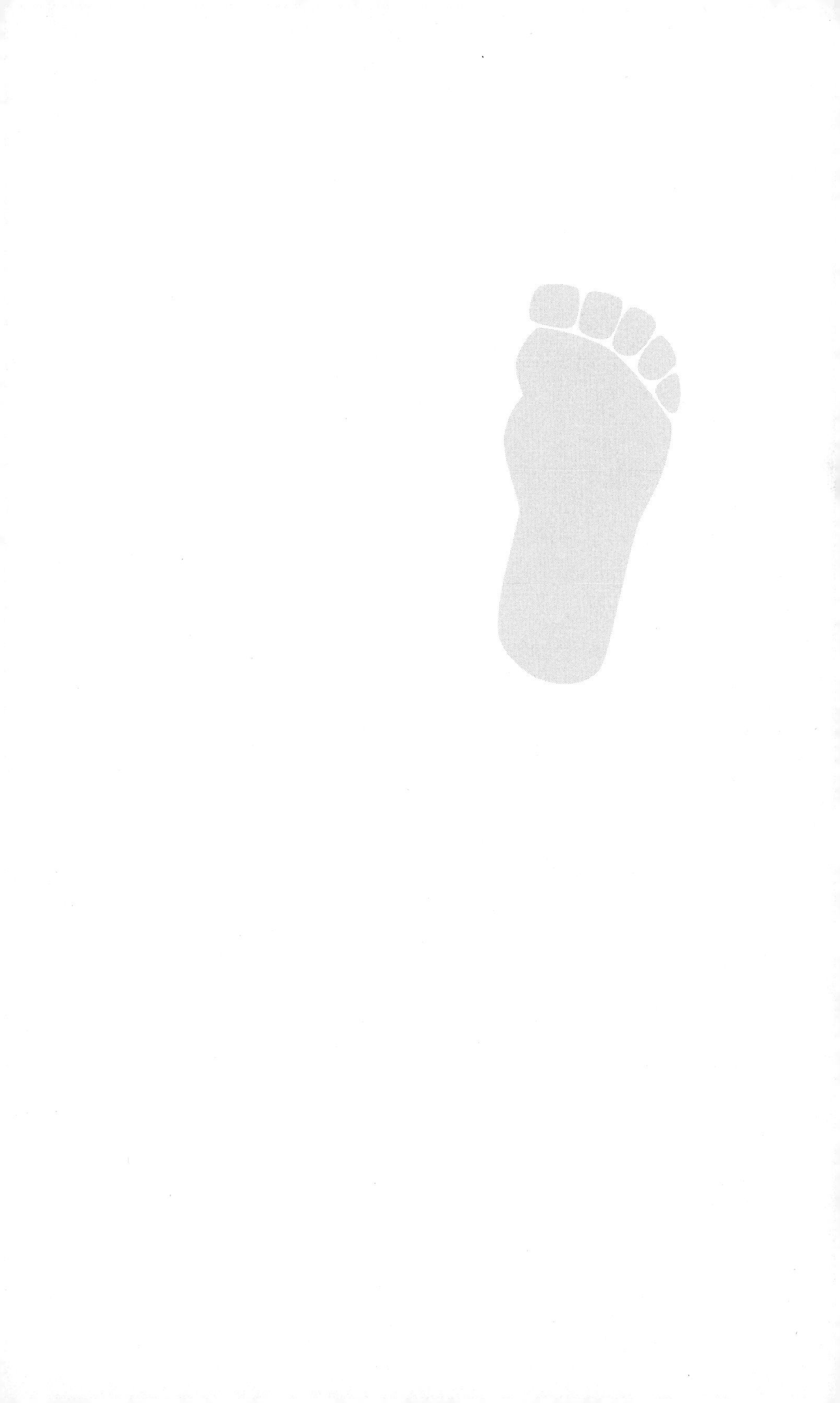

JOSHUA BLU BUHS

*The University of Chicago Press* CHICAGO AND LONDON

JOSHUA BLU BUHS is an independent scholar who studies history of science. He is the author of *The Fire Ant Wars*.

The University of Chicago Press, Chicago 60637
The University of Chicago Press, Ltd., London

Printed in the United States of America

17 16 15 14 13 12 11 10 09 1 2 3 4 5

ISBN-13: 978-0-226-07979-0 (cloth)
ISBN-10: 0-226-07979-1 (cloth)

Buhs, Joshua Blu.
Bigfoot : the life and times of a legend / Joshua Blu Buhs.
p.cm.
Includes bibliographical references and index.
ISBN-13: 978-0-226-07979-0 (cloth : alk. paper)
ISBN-10: 0-226-07979-1 (cloth : alk. paper) 1. Sasquatch. I. Title.
QL89.2S2B84 2009
001.944—dc22

2008035609

Bigfoot illustration on page 280 by Lauren Nassef.

∞ The paper used in this publication meets the minimum requirements of the American National Standard for Information Sciences—Permanence of Paper for Printed Library Materials, ANSI Z39.48-1992.

*To Kim, as always, and Morgan, too*

*What monsters men have needed to believe in they have created for themselves in words and pictures when they could not discover them in nature.*

LESLIE FIEDLER

# Contents

# *Preface*

I remember reading about Bigfoot when I was young. As a child of the 1970s, I hardly could have avoided the beast. Sasquatch didn't make much of an impression on me, though. At least, I don't think so. I was more interested in other paranormal things that were often discussed with Bigfoot—the Loch Ness monster, spontaneous human combustion, alien life forms. Like many kids, I forgot about the creature when I entered adolescence; sure, I still knew of Bigfoot, recognized the name, but I just didn't give the creature any thought. Late in 2002, though, the beast came back to me.

I was finishing my first book, *The Fire Ant Wars*, which was about the way that Americans had thought about the imported fire ant (*Solenopsis invicta*) and how differing ideas about the insect had led to different public policies for dealing with it. That book was frustrating to write. The problem was the ant. *Solenopsis invicta* was an agent in the story; it didn't have as much power as the humans who confronted it, but neither could it just be ignored. And so I had to struggle to make sense of ant biology at the same time that I had to figure out the various human motivations. I succeeded, I think, and taking account of the ant made the book better, but still, writing it was hard. I started to wonder if there was an easier way. I was still interested in human ideas about nature, how they come into being, the effect that they have, but I was being driven less by theoretical considerations than laziness. I wanted a subject that left me free to consider the intellectual history without also having to think about the ways that nature changed. Bigfoot seemed to fit the bill. Here was a creature, I imagined then, that embodied various ideas about the natural world . . . but didn't exist. So I didn't have to worry about its agency, its ability to shape the story.

A lot has happened since I started to noodle around with this book. My wife and I moved halfway around the world, from California to Japan, then back again—bringing with us our first child. These have been some of the happiest years of my life, but I also experienced some of my darkest lows. My hope that this would be a less frustrating book to write proved true—this has been an incredibly fun project: a blast, I've said more than once. But, boy, was I wrong about Bigfoot being powerless. I still don't think that Bigfoot exists—indeed, writing this book actually gave substance to what was before only a vague kind of skepticism. Nonetheless, Bigfoot took over this story and led me in directions I did not expect. Almost nothing of my original ideas about the book survived. And that, I think, is what made writing the book so much fun.

Still, even with Bigfoot as a guide, this book could not have been written without a great deal of support. For their help, I thank three anonymous reviewers for the University of Chicago Press; Ellen Alers and the staff at the Smithsonian Institution Archives; Lenny Barshack; Pete Beatty; Joan Berman and Edie Butler at the Humboldt State University special collections; Janet Bord; Yves Bosson; Matthew Burton Bowman; Jessica Brown-Velez; Sally Byers and Wiley-Blackwell Publishing Ltd.; the staff at the California State Library's California Room; Diane Coulson; Ray Crowe; Meredith Eliassen at the San Francisco State special collections; Nathan Ensmenger; Paul Fattig; Leanda Gahegan and the staff at the National Anthropological Archives; Mary Gehl; Oliver Glaizot at the Musée Cantonal de Zoologie; Roy Goodman, Earl Spamer, and the rest of the staff at the American Philosophical Society; Mary Hammer at the Washington State Archives; Marissa M. Hendriks at the University of Pennsylvania's Rare Book and Manuscript Library; Christie Henry; Don Herron; Kathy Hodges and the staff at the Idaho State Historical Society; Mort Künstler; Annette Lambert; Mark Madison at the U.S. Fish and Wildlife Service's National Conservation Training Center; Lisa Marine and the staff at the Wisconsin Historical Society; Brenda Marsh; Paula McEvoy; Lorna McIlnay; Alison Miner with the University of Pennsylvania Museum; Sherry Orth, Shawna Butler, and the staff at the Yakima County Clerk's office; Daniel Perez; Gabe Perillo; Janet Peterson with World Book, Inc.; David Price; Dave Rubert; Sarah Strong, Bob Swofford, and Teresa Meikle with the *Santa Rosa Press-Democrat*; Steven Torrington and the *Daily Mail*; Mike Van Wagenen; Robert Voelker-Morris; Ron Westrum; Joy Wheeler and the archival staff at the Royal Geographical Society; Charles Winkler and the staff at the *Times-Standard*; and Morgan Yates.

I owe a special debt of gratitude to my family, who supported me over the many years I've been at work on this project. Thank you! Sasquatch himself is not big enough to carry the burden that my wife, Kim McIlnay, has borne while I spent my time in the office scribbling away at this book. She has been a constant source of support, inspiration, and humor. I love you. And Morgan, my own little Bigfoot, came into the world as this book was being written and enriched my life in ways I couldn't have imagined possible. Kim and Morgan, this book's as much yours as it is mine.

# *Dramatis Personae*

DON ABBOTT: British Columbia anthropologist who studied Bigfoot tracks and Roger Patterson's movie.

GEORGE AGOGINO: Anthropologist who acted as a consultant to Tom Slick's expeditions in search of the Yeti and Bigfoot.

BETTY ALLEN: Journalist who investigated Bigfoot sightings in northern California.

P. T. BARNUM: Nineteenth-century showman who exhibited, among other things, a succession of wildmen.

DMITRI BAYANOV: Russian wildman aficionado who succeeded Boris Porshnev as the leading hominologist in that country.

FRED BECK: Miner who claimed that Bigfoot attacked his party in 1924.

FRANK BEEBE: Illustrator who studied Roger Patterson's movie for the Royal British Columbia Museum.

JESS AND CORALIE BEMIS: California residents who helped spread rumors about Bigfoot in 1958. Coralie eventually contacted Andrew Genzoli, helping bring the monster to the attention of the press.

LESLIE BREAZEALE: Along with Ray Kerr, he claimed to have seen Bigfoot after being hired by Ray Wallace to hunt the monster.

JOHN BURNS: Teacher on the Chehalis Reservation in Canada who translated First Nations legends of wildmen into stories about Sasquatch.

PETER BYRNE: Irish big-game hunter who lead several expeditions in search of the Yeti and Bigfoot.

CLIFFORD CARL: Head of British Columbia's natural history museum.

JEANNIE AND GEORGE CHAPMAN: First Nations couple who in 1941 saw what they thought was a Sasquatch. Their tale became a classic reference among Bigfoot hunters.

LOREN COLEMAN: Fortean who became interested in wildmen late in the 1950s. He has written extensively on Bigfoot and other unexplained phenomena.

CARLETON COON: Anthropologist and another of Tom Slick's scientific consultants.

JERRY CREW: Member of the Wallace brothers' construction crew who found huge tracks around his bulldozer. He helped to publicize Bigfoot during the late 1950s.

TERRY CULLEN: Led Ivan Sanderson and Bernard Heuvelmans to Frank Hansen.

RENÉ DAHINDEN: Swiss immigrant to Canada who became one of the most well-known Sasquatch hunters.

AL DEATLEY: Roger Patterson's brother-in-law. He developed the movie of Bigfoot that Patterson made and helped to promote it.

MICHAEL DENNETT: Skeptic who revealed that Grover Krantz had been hoaxed.

JOHN DIENHART: Director of the *World Book Encyclopedia*'s publicity. He arranged for the encyclopedia to sponsor Edmund Hillary's hunt for the Yeti.

DESMOND DOIG: Creator of *Bing, the Abominable Snow-Baby* and chronicler of Edmund Hillary's expedition in search of the Yeti.

FRANK EDWARDS: Author of books on paranormal topics who wrote about Bigfoot.

CHARLES FORT: Writer of the early twentieth century who compiled four books documenting a host of unexplained phenomena. His followers became known as Forteans.

ANDREW GENZOLI: Columnist for the *Humboldt Times* who helped bring the story of Bigfoot to the world's attention.

BOB GIMLIN: Roger Patterson's friend who was on the trip when Patterson supposedly filmed a Sasquatch.

JOHN GREEN: The Dean of Sasquatchery. Green was newspaper editor who became intrigued by the tales told about Sasquatch. He wrote a number of books on the subject.

J. RICHARD GREENWELL: Secretary of the International Society of Cryptozoology.

DONALD GRIEVE: British anatomist who studied Roger Patterson's movie.

GEORGE HAAS: Fortean who created the *Bigfoot Bulletin*.

MARJORIE HALPIN: Anthropologist who organized the first scientific conference on Bigfoot.

FRANK HANSEN: Carnival showman who exhibited a wildman beginning in the late 1960s that some Bigfooters came to think was real.

BERNARD HEUVELMANS: French writer whose work focused on undiscovered animals, such as the Yeti. He was the first president of the International Society of Cryptozoology.

EDMUND HILLARY: Mountaineer who, along with Tenzing Norgay, was the first to summit Mt. Everest. He later went in search of the Yeti and concluded the beast did not exist.

AL HODGSON: Proprietor of the general store in Willow Creek, California. He helped John Green, René Dahinden, and Betty Allen, among others, in their investigations.

JOHN HUNT: Leader of the expedition that conquered Everest and a believer in the Yeti.

RALPH IZZARD: Journalist who organized an expedition in search of the Abominable Snowman under the aegis of the London *Daily Mail*.

RAY KERR: Along with Leslie Breazeale, he claimed to have seen Bigfoot after being hired by Ray Wallace to hunt the monster.

ED KILLAM: Bigfoot hunter who won grant money to search for the monster in the early 1970s. He was also involved with Robert Morgan's expedition.

GROVER KRANTZ: Anthropologist who studied Bigfoot, arguing that the creature existed.

PAUL KURTZ: Philosophy professor and founder of the Committee for the Scientific Investigation of Claims of the Paranormal.

IAN MACTAGGART: Zoologist whom John Green consulted about Sasquatch tracks.

VLADIMIR MARKOTIC: Anthropologist who became intrigued with wildman reports. He worked closely with Grover Krantz putting out a book supporting Bigfoot's existence.

IVAN MARX: Bear hunter who was part of Tom Slick's California hunt for Bigfoot and later was involved with numerous hoaxes, including a prominent one in Bossburg, Washington.

JIM MCCLARIN: Bigfoot hunter who dropped out of Humboldt State to chase the beast. He carved a huge statue of the monster for Willow Creek.

JEFF MELDRUM: Grover Krantz's successor. He is an anthropologist who believed in Bigfoot's existence and inherited much of Krantz's material after Krantz died.

JOE METLOW: Prospector who was involved with hoaxes in Bossburg, Washington.

ROBERT MORGAN: Bigfoot hunter active in the 1970s and again in the 1990s. Peter Byrne accused him of faking evidence.

RANT MULLENS: Creator of numerous fake feet, and supposed perpetrator of the hoax that convinced Fred Beck and his men that they were being attacked by Sasquatches.

JOHN NAPIER: Primate anatomist and anthropologist who became interested in Bigfoot during the late 1960s, studying Roger Patterson's movie, trackways, and Frank Hansen's wildman. He is the author of *Bigfoot*.

HENRY NEWMAN: Journalist with the *Calcutta Statesman* who coined the phrase "Abominable Snowman."

RON OLSON: Filmmaker who introduced Roger Patterson and Al DeAtley to four-walling and later made his own Bigfoot movie.

W. C. OSMAN HILL: British primatologist who served as one of Tom Slick's scientific advisors. He later studied Roger Patterson's movie for Ivan Sanderson.

ALBERT OSTMAN: Prospector who claimed to have been abducted by a family of Sasquatches in 1924.

ROGER PATTERSON: A sometime-rodeo rider who, with Bob Gimlin, claimed to have filmed a Bigfoot in 1967.

DANNY PEREZ: Bigfooter who dropped out of Humboldt State to chase the monster. He compiled a huge bibliography on the beast and also wrote a history of Roger Patterson's film.

MARLIN PERKINS: Naturalist on Edmund Hillary's expedition that debunked the Yeti's existence.

MARIAN PLACE: Journalist and children's author who chronicled Bigfoot's early history and also wrote several books about the creature for kids.

BORIS PORSHNEV: Russian historian who led a small group of researchers there who believed that wildmen still inhabited the Russian wilds. He cowrote a book with Bernard Heuvelmans arguing that Neanderthals still existed.

ROBERT MICHAEL PYLE: Naturalist who wrote *Where Bigfoot Walks*, arguing that Bigfoot was an incarnation "of nature, the earth, and all that is green and contrary to control."

ROBIN RIDINGTON: Anthropologist who studied Roger Patterson's film.

WILLIAM ROE: Construction worker who saw Sasquatch.

GERALD RUSSELL: Explorer and animal collector who was on both Ralph Izzard's and Tom Slick's hunts for the Yeti.

IVAN SANDERSON: Naturalist and Fortean who wrote a number of articles and a huge book, *Abominable Snowman: Legend Come to Life*, arguing that a number of undiscovered wildman species exist.

ERIC SHIPTON: Mountaineer who found Yeti tracks in 1951.

BARBARA ANNE SLATE: UFOlogist who claimed that Bigfoot was associated with flying saucers.

TOM SLICK: Texas oilman and millionaire who sponsored hunts for the Yeti in the Himalayas and the Sasquatch in North America.

FRANK SMYTHE: British mountaineer who claimed that the Abominable Snowman was a bear.

LAWRENCE SWAN: Biologist who was on the Hillary expedition that debunked the Yeti's existence.

H. W. TILMAN: Mountaineer who thought that the Yeti existed.

BOB TITMUS: Taxidermist who became involved with the hunt for Bigfoot.

LEE TRIPPETT: Tried to contact Bigfoot with extrasensory perception.

MARIO TRUZZI: Sociologist who helped to found Committee for the Scientific Investigation of Claims of the Paranormal before becoming disillusioned and leaving the organization.

WLADIMIR TSCHERNEZKY: Anatomist who reconstructed the Yeti's foot from tracks.

RAY WALLACE: One of three Wallace brothers who ran a construction company in northern California during the late 1950s. He was accused of faking the tracks that Jerry Crew found, later admitting to that hoax and several others. His brothers were Leslie and Wilbur (known as Shorty).

WHAT-IS-IT: Wildman exhibit created by Barnum. In its first incarnation, Edmund Leech played the What-Is-It. A second, unknown actor reprised the character in the 1860s. William Johnson played the role into the 1920s.

ZADIG: Titular character in a story by Voltaire who was able to reconstruct a dog's appearance from its tracks.

Bigfoot has had a complicated relationship with its many fans. They loved the creature, but wanted to kill it, too—or at least some did. They idolized it and feared it. They saw the monster as hope for a better world, but could only gain access to it through the stuff of the world as it was. They prized it as the epitome of authenticity, but had to make do with replicas. They wanted to be it, and—often, anyway—wanted it to remain free, away from them. Here, Bigfoot seeker Loren Coleman stands with a replica of the wildman. (Photo courtesy of Loren Coleman.)

CHAPTER ONE

# *Wildmen*

Robert Hatfield heard the dogs howling. At least that's what he claimed later. Hatfield was a logger, down from Crescent City, in the far northwestern corner of California, visiting his sister and brother-in-law at Fort Bragg, along the coast. It was Wednesday night, February 7, 1962. Hatfield went outside to see what was bothering the dogs. He saw a huge creature, a beast, Hatfield said, "much bigger than a bear, covered with fur, with a flat, hairless face and perfectly round eyes." It stood "chest and shoulders above a six-foot-high fence." Hatfield went back into the house and woke Bud Jenkins, his brother-in-law, telling him to come out and see the largest bear he'd ever see.[1]

When Jenkins and Hatfield returned to the yard, the beast was no longer there. Jenkins went back inside to get his gun; Hatfield started scouting. He rounded the corner of the house—and bumped into the beast. It knocked him down so hard that his arm and shoulders were "sore for the next three days." Hatfield scrambled back inside, yelling a warning to Jenkins: there was a "half-man, half-beast" monster after them! Once Hatfield was in the house, he and Bud tried to slam the door shut, but the beast caught it and pushed back. "Let it in and I'll get it!" Jenkins shouted, holding his shotgun. The two men let go, but before Jenkins could fire off a round, the beast turned and left. Good thing—it turned out the gun wasn't loaded.[2]

The sheriff's department came out to investigate the disturbance. There were few clues. A bad odor wafted heavily in the air. There was also a muddy

**1.** Hector Lee, "This Is the 'Big Foot' Edition," *From the Sourdough Crock: A Newsletter from the California Folklore Society* 1, no. 4 (1962): 1–2 (all quotations); Frank Edwards, *Strange World* (New York: Lyle Stuart, 1964), 53–55; John Green, *The Sasquatch File* (Agassiz, BC: Cheam, 1973), 29.

**2.** Lee, "This Is the 'Big Foot' Edition," 1–2 (all quotations); Edwards, *Strange World*, 53–55; Green, *The Sasquatch File*, 29.

handprint on the door, eleven-and-a-half inches long with stubby fingers. Journalists for the local Santa Rosa *Press-Democrat* did some probing into the matter, too. And six Fort Bragg men formed a hunting party. For the past several years, newspapers had been reporting on a monstrous, manlike beast said to range from British Columbia down into northern California. It was called Bigfoot. Four of the men in the hunting party were convinced that Hatfield and the Jenkinses had seen Bigfoot. As evidence, they pointed to broken branches along a path about a mile from the Jenkinses' property, some tracks, and some dung. The two other members of the party were skeptics. The branches proved nothing, they said; deer could have broken them. Bears could have made the tracks. The dung was horse manure—indeed, the whole matter seemed to be not much more than manure. Maybe that accounted for the lingering odor.[3]

The newspaper reports caught the eye of Hector Lee at nearby Sonoma State College. Lee was an important folklorist, having done seminal work on Mormon folktales, and was building an archive of California folklore. Apparently he had not paid any attention to Bigfoot prior to the creature being sighted in Mendocino County. But now he started gathering information about the beast. It was part of a small spurt of interest in Bigfoot among folklorists and those interested in folklore. Unlike the sheriffs, journalists, or hunters, the folklorists were not interested in whether Hatfield actually saw a half-man, half-beast monster. Most likely, Hatfield saw no such thing, but that didn't mean his stories, and other tales about the creature, weren't important, or didn't reveal something about the human condition.[4]

That is the main contention of this book. Maybe there is no Bigfoot walking the forests of the American Pacific Northwest, but the creature is still *real*—it is part of the American cultural landscape, something about which people can, and do, talk, something that most everyone recognizes and knows. Understanding the monster helps to explain some of twentieth-century America. Tracing the creature's fortunes as it passed through America in the second half of the twentieth century sheds light on how knowledge moves through society, on the intersection of class, technology, science, and

**3.** "Skeptics and Enthusiasts Clash on Bigfoot Search," *Santa Rosa (CA) Press-Democrat*, February 18, 1962; Lee, "This Is the 'Big Foot' Edition," 1–2; Green, *The Sasquatch File*, 29.

**4.** Lynwood Carranco, "Three Legends of Northwestern California," *Western Folklore* 22, no. 3 (1963): 179–85; idem, "It Wasn't Bigfoot After All," *Western Folklore* 23, no. 4 (1964): 271–72; Daniel Hoffman, *Paul Bunyan: Last of the Frontier Demigods* (Lincoln: University of Nebraska Press, 1983), 12–21.

belief. It is a worthy reason to write a biography of a legend, a way of showing that what seems trivial and ridiculous is not.[5]

## Wildmen through History

As the folklorist Bacil Kirtley pointed out a few years after Hatfield's night of terror, Bigfoot is a contemporary example of a well-known folkloric character, the wildman—a hairy, sometimes giant, humanlike beast that lives on the outskirts of civilization. If wildmen are not a universal myth, then they are close. Peoples as diverse as the Maya, English, Chinese, and Melanesians have stories about them. *The Epic of Gilgamesh*, one of the first written documents—it dates to almost two thousand years before the birth of Christ—features Enkidu, a wildman with "hair that sprouted like grain." Enkidu ate with the gazelles, drank at a watering hole with other animals, and protected beasts from hunters. Genesis 6:4 says, "The giants were on the earth in those days—and also afterward—when the sons of God went to the daughters of men and had children by them. They were the heroes of old, men of renown." Esau, Jacob's brother, was a hairy hunter. The ancient Greeks' fervid imaginations populated the earth with all sorts of wildmen and wildwomen: Amazons and centaurs and cyclopes and fauns and giants and mænads and satyrs and sileni and titans. For modern readers the most famous wildman—although long since domesticated—is Santa Claus: hairy and strange, often depicted garlanded in holly and other vegetation, he visits civilization only around the winter solstice, bringing punishment, reward, and the promise that the days will lengthen again.[6]

Throughout history, stories about wildmen have provided a way of thinking about what it means to be human: the contradictions, difficulties, limits, and the glorious wonder of it all. Sometimes, wildmen were stand-ins for other, distant—strange and frightening—peoples. In his *Natural History*, the Roman writer Pliny the Elder claimed that a number of monstrous races

**5.** David J. Daegling, *Bigfoot Exposed: An Anthropologist Examines America's Enduring Legend* (Walnut Creek, CA: Altamira Press, 2004), 4, 259.

**6.** Bacil F. Kirtley, "Unknown Hominids and New World Legends," *Western Folklore* 23 (1964): 77–90; Richard Bernheimer, *Wild Men in the Middle Ages: A Study in Art, Sentiment, and Demonology* (New York: Octagon, 1970); Myra Shackley, *Still Living? Yeti, Sasquatch and the Neanderthal Enigma* (London: Thames & Hudson, 1983), 16–22; Roger Bartra, *Wild Men in the Looking Glass: The Mythic Origins of European Otherness* (Ann Arbor: University of Michigan Press, 1994); Phyllis Siefker, *Santa Claus, Last of the Wild Men: The Origins and Evolution of Saint Nicholas, Spanning 50,000 Years* (New York: McFarland & Co., 2006).

FIGURE 1. Bigfoot is a modern example of a well-known mythological archetype, the wildman. (Martin Schongauer, *Wild Man Holding a Shield with a Greyhound*, ca. 1480/1490. Rosenwald Collection. Courtesy of the Board of Trustees, National Gallery of Art, Washington, D.C.)

wandered across the world. He wrote about tribes of people with hairy tails, with dog heads, with horse hooves. Bestiaries after Pliny's *Natural History* continued to include wildmen, all the way through the works of Carl von Linné, who wrote what are considered to be the first scientific taxonomies. Wildmen could be disgusting, cannibals or man-eaters, the things that the civilized defined themselves against. They were, at times, interpreted as signs of God's wrath: their horrible disfigurement was a warning of what would happen should God withdraw His blessing. They were also proof of God's magnificence or nature's munificence: that the world was filled with every possible form of life, every gap between different kinds of creatures filled by some beast. Some wildmen were thought to have special faculties, abilities lost to civilized humans: the power to command animals, to call up winds and storms, to survive without society. In some myths, the wizard Merlin was a wildman.[7]

For many centuries, learned Europeans took reports of such creatures seriously. As explorers extended the boundaries of the known, tramping through terra incognita, they found bizarre creatures—the platypus with its crazy-quilt patchwork of bird, reptile, and mammalian parts was probably the most notorious—and among these wonders were wildmen. Orangutans were discovered in Southeast Asia—the name literally means "man of the forest"—and chimpanzees in Africa. Other peoples were sometimes classified as wildmen—Native Americans, for instance, forcing the Pope to decree that they were, indeed, full-fledged humans, children of God. During the Renaissance, savants were drawn to the cases of feral children, boys and girls raised by animals or who otherwise grew up in isolation and so were wild.[8]

Over time, belief in wildmen fell into disfavor. The ones that had been found were less than advertised; upon inspection, they could be classified as either human or animal, not something in between. And the great majority of the monstrous races that had been cataloged in old bestiaries did not exist. One thirteenth-century European visitor to the Tartars, for example, remarked, "I asked about the monsters, or monstrous men, about which Pliny and Solinus wrote. They told me they had never seen such creatures, which led me to wonder greatly if it were true." There were no dog-headed men, no humans with eyes in their chests, none with hooves.[9]

By the nineteenth century, the category of wildmen had been carved up by various scientific disciplines and explained away. After Linné, bestiaries no longer included wildmen; taxonomists asserted that the globe housed no more wildmen than those apes already cataloged. Physiologists demonstrated that it was impossible for a race of giant humans ever to have existed

**7.** Shackley, *Still Living?* 16–22; Richard Nash, *Wild Enlightenment: The Borders of Human Identity in the Eighteenth Century* (Charlottesville: University of Virginia Press, 2003).

**8.** Lee Eldridge Huddleston, *Origins of the American Indians: European Concepts, 1492–1729* (Austin: University of Texas Press, 1967), 15; Shackley, *Still Living?* 16–22; Harriet Ritvo, *The Platypus and the Mermaid and Other Figments of the Classifying Imagination* (Cambridge, MA: Harvard University Press, 1997); Birute Mary Galdikas, *Orangutan Odyssey* (New York: Harry N. Abrams, 1999), 8; Julia V. Douthwaite, *The Wild Girl, Natural Man, and the Monster: Dangerous Experiments in the Age of Enlightenment* (Chicago: University of Chicago Press, 2002); Michael Newton, *Savage Girls and Wild Boys: A History of Feral Children* (New York: Picador, 2004).

**9.** Lorraine Daston and Katherine Park, *Wonders and the Order of Nature, 1150–1750* (New York: Zone Books, 2001), 63–64 (quotation).

and offered naturalistic explanations for oddities such as people who were covered entirely with hair. Geologists and evolutionists pushed wildmen into the past, making them into humanity's ancestors, rather than neighbors: they once had existed, but no longer did. Sigmund Freud built on this evolutionary idea and also twisted it, so that wildmen still existed, but only *inside* the human psyche. The wildman was the residue of our evolutionary past, the animalistic, untamed, uncivilized part of us. "In modern times," wrote the scholar Hayden White, "the notion of a 'wild man' has become almost exclusively a psychological category rather than an anthropological one." Continued reports of wildmen were dismissed as superstitions of the ignorant or as misidentifications.[10]

## What-Is-It

Just because science banished wildmen from nature did not mean that interest in the monsters ended. In the United States and elsewhere, fascination with wildmen continued throughout the nineteenth century—maybe in part *because* of the hardening opinion among scientists that such creatures did not exist: official denunciations made the creatures seem rare and wonderful and uncanny. Literature was rife with wildmen, from Sir Walter Scott's Wandering Willie to Tarzan the Ape Man. Newspapers, too, were filled with reports of wildmen haunting the darkness beyond the edge of town. At circuses, audiences paid their pennies to gawk at geeks, wildmen so savage that they ate live animals. The actor Hervey Leech made something of a career playing apes and baboons in theatrical performances. P. T. Barnum brought a wildman to London in 1846; it was supposed to have been captured in the "wilds of California," where it had been living with a tribe of Indians. The wildman growled and, like any good geek, ate disgusting things—in this case, raw meat.[11]

**10.** Sigmund Freud, *Totem and Taboo: Some Points of Agreement Between the Mental Lives of Savages and Neurotics* (New York: W. W. Norton, 1962); Hayden White, "The Forms of Wildness: Archaeology of an Idea," in *The Wild Man Within: An Image in Western Thought from the Renaissance to Romanticism*, ed. Edward Dudley and Maximillan E. Novak (Pittsburgh, PA: University of Pittsburgh Press, 1972), 3–38 (quotation, 34); Henri F. Ellenberger, *The Discovery of the Unconscious* (New York: Basic Books, 1981); Walter Stephens, *Giants in Those Days: Folklore, Ancient History, and Nationalism* (Lincoln: University of Nebraska Press, 1989), 2.

**11.** Phillips Verner Bradford and Harvey Blume, *Ota Benga: The Pygmy in the Zoo* (New York: St. Martin's, 1992); James W. Cook, *The Arts of Deception: Playing with Fraud in the Age of Barnum* (Cambridge, MA: Harvard University Press, 2001), 133 (quotation); Fred Nadis, *Wonder*

In 1869 came a report that the petrified body of a giant had been found in Cardiff, New York, proof that the Bible was right, physiologists wrong: that there had indeed been giants in those days. A journalist reporting on the stream of visitors to the Cardiff giant "noticed on the faces of all a momentary spasm of awe, a short involuntary holding of breath, as their gaze fell upon what they firmly believed to be the stony remains of an American Goliath." That same year Mark Twain published a mock-interview with a wildman for the *Buffalo Express*. "There has been so much talk about the mysterious 'wild man' out there in the West for some time, that I finally felt it was my duty to go out and interview him," he wrote. "I felt that the story of his life must be a sad one—a story of suffering, disappointment, and exile—a story of man's inhumanity to man in some shape or other—and I longed to persuade the secret from him." Barnum re-created his wildman exhibit in the 1860s, claiming this time that the creature had been captured in Africa, and again in the 1870s. In the mid-1880s he also began displaying Krao the Human Monkey and Jo-Jo the Dog-Faced Boy, a man with a rare condition known as hypertrichosis but who was rumored to be the offspring of a Russian woman and a bear—a real, live wildman.[12]

Although these wildmen were associated with leisure-time entertainments, they provoked serious questions about what it meant to be human, to be civilized. These were questions much on the mind of Americans during the middle of the nineteenth century—after the publication of Charles Darwin's *Origin of Species* blurred the line between humans and beasts and as America first teetered on the brink of civil war, and then fell into it, in large part because of questions about the humanity of blacks. "What-Is-It," Barnum titled his second wildman exhibit, daring his audience to grapple with the thorny scientific and political issues of the day. Some whispered that the wildman was "an advanced chimpanzee," others that it resulted from

*Shows: Performing Science, Magic, and Religion in America* (New Brunswick, NJ: Rutgers University Press, 2005); Chad Arment, *The Historical Bigfoot* (Landisville, PA: Coachwhip Publications, 2006).

**12.** Andrea Stulman Dennet, *Weird and Wonderful: The Dime Museum in America* (New York: New York University Press, 1997), 69–85; Jan Bondeson, *The Two-Headed Boy, and Other Medical Marvels* (Ithaca, NY: Cornell University Press, 2000), 18–25; Cook, *The Arts of Deception*, 126–150; Mark Twain, *The Curious Republic of Gondour and Other Whimsical Sketches* (Whitefish, MT: Kessinger Publishing, 2004), 22–24 (second quotation); Michael Pettit, "'The Joy in Believing': The Cardiff Giant, Commercial Deceptions, and Styles of Observation in Gilded Age America," *Isis* 97 (2006): 659–77 (first quotation, 667).

FIGURE 2. Wildmen were common in nineteenth-century American popular culture, appearing in stories and newspapers, as well as on the stage. This picture shows P. T. Barnum's What-Is-It from the 1860s. The wildman was clearly an actor in a costume—widely, though apparently wrongly, assumed to be William Henry Johnson, who was actually Barnum's third What-Is-It—and yet the exhibit still stirred up controversies over the exact nature of the creature. (*What Is It?* National Portrait Gallery, Smithsonian Institution, Frederick Hill Meserve Collection.)

a "cross between a nigger and a baboon." Deciding exactly what was on the stage, where the line was drawn between human and animal, between white and black, was important, desperately so, to understanding how the social order should be built, the morality that should be instilled.[13]

Others who came out to see Barnum's What-Is-It were convinced that the series of so-called wildmen were just actors in costume. Of course, they were correct. Barnum's first What-Is-It display lasted less than half an hour, ending when a competitor recognized the wildman as Hervey Leech dressed in a hair shirt, his skin stained. According to one report, the rival entered the cage, stripped the shirt from Leech, and offered to buy him a *cooked* steak, enough of that raw meat. And what was true of Barnum's exhibits applied to all the wildmen of the day. The Cardiff giant was the invention of George Hull, a tobacconist who wanted to expose the gullibility of Christians and so bought Montana rock, employed Chicago sculptors to carve a giant from it, weathered the statue himself with water and sand, then secretly shipped the humbug to an in-law's property. Louis T. Stone, a journalist, fabricated the adventures of a wildman for his newspaper in Connecticut; H. L. Mencken did, too, when he was city editor of the *Baltimore Herald*, inventing new reports every Sunday for a month, piquing the curiosity of his readers and giving himself something to write about on slow nights. Twain admitted at the end of his article that the wildman was only a "sensation"—something to stir up interest and sell papers, not something real.[14]

Fakery, though, did not detract from the displays and news stories—quite the opposite. Barnum noted, "The public appears disposed to be amused even when they are conscious of being deceived." He shrugged off his exposure in London, going on to become the most famous man in nineteenth-century America (his biography was likely the second-most-read book, behind only the Bible). Twain and Mencken also went on to fame and fortune. The possibility—the near certainty—that the wildmen were frauds was one of the reasons that they were so popular. Humbuggery raised a whole other set of

**13.** Eric Lott, *Love & Theft: Blackface Minstrelsy and the American Working Class* (New York: Oxford University Press, 1993); Cook, *The Arts of Deception*, 119–62 (quotations, 139); Benjamin Reiss, *The Showman and the Slave: Race, Death, and Memory in Barnum's America* (Cambridge, MA: Harvard University Press, 2001).

**14.** H. L. Mencken, *Newspaper Days, 1899–1906* (New York: Alfred A. Knopf, 1941), 136–38; Curtis D. MacDougall, *Hoaxes* (New York: Dover, 1958), 3; Cook, *The Arts of Deception*, 27–28; Shelley Streeby, *American Sensations: Class, Empire, and the Production of Popular Culture* (Berkeley: University of California Press, 2002); Twain, *The Curious Republic Of Gondour and Other Whimsical Sketches*, 22–24; Pettit, "'The Joy in Believing,'" 659–77.

pressing questions, as serious and important in their way as the questions about the morality of slavery and the limits of human nature.[15]

In the second third of the nineteenth century, the American middle class became acutely concerned with fraud and authenticity. The country was growing, its economy switching from agrarian to industrial. Everyday transactions were increasingly conducted between strangers—raising the specter of fraud since personal reputations no longer warrantied goods—and new technologies made possible the creation of fake things: furniture that looked as though it were from the colonial period, pictures that seemed to reproduce moments otherwise lost to time. For the American middle class, these anxieties were exacerbated by their own rise to respectability. Nineteenth-century America valorized self-made men, Horatio Algers who lifted themselves up by their own bootstraps. The problem with self-made men, though, was that they lacked the traditional measures of refinement—learned manners, old money, reputable family names—and so the line between a respectable man and a con man was vanishingly thin. Nineteenth-century wildmen, the geeks, What-Is-Its, and petrified giants played on these anxieties. It was a time when the very notion of reality seemed to warp and become unmoored, when determining truth was difficult, if not impossible, when superstition could become science and science come to seem nothing more than the barking of a carnie: *What* is it? What *is* it? What is *it?*[16]

## The Abominable Snowman

At the end of the nineteenth century, European mountaineers began to explore the Himalayas, returning with stories of a wildman, often called the Yeti. The first mention came in 1832 when B. H. Hodgson, the British Resident in the court of Nepal, reported that the natives whom he employed to hunt the local fauna had seen a furred, upright, tailless demon. (Hodgson thought it an orangutan.) In 1899, L. A. Waddell, a major in the Indian army medical corps and Fellow of the Linnaean Society, told of finding what his porters insisted were Yeti tracks in northwestern Sikkim. (He dismissed it as bear spoor.) More than a decade later, William Hugh Knight, a member of

**15.** Neil Harris, *Humbug: The Art of P.T. Barnum* (Boston: Little Brown, 1973); Cook, *The Arts of Deception*, 16 (quotation); Reiss, *The Showman and the Slave*, 3–4.

**16.** Karen Halttunen, *Confidence Men and Painted Women: A Study of Middle-Class Culture in America, 1830–1870* (New Haven, CT: Yale University Press, 1986); Miles Orvell, *The Real Thing: Imitation and Authenticity in American Culture, 1880–1940* (Chapel Hill: University of North Carolina Press, 1989).

the Royal Societies Club, saw a wildman in Tibet—not its print, but the animal itself. "He was a little under 6 ft. high, almost stark naked in that bitter cold—it was the month of November. He was a kind of pale yellow all over, about the colour of a Chinaman, a shock of matted hair on his head, a little hair on his face, highly splayed feet, and large, formidable hands." In 1915, J. R. P. Gent, a British forestry officer, reported hearing from locals of gargantuan tracks, also in Sikkim.[17]

Like most wildmen, these Himalayan monsters were confounding. Part of the difficulty was linguistic. "Almost everything about the Yeti is controversial," said anthropologist Myra Shackley, "even the name." In Nepal, *yeti* (or *yeh-teh*) is apparently a generic term, referring to several different animals. One is the *dzu-teh*, a large, ferocious creature, quite possibly a bear. There are also reports of a small, monkey-like wildman, sometimes called *thelma*, or *teh-lma*, and sometimes, confusingly—possibly wrongly—called *yeh-teh*. And then there is the *meh-teh*, the classic wildman—half-man, half-beast—like the one that Robert Hatfield saw in Fort Bragg that February night. Surrounding regions—indeed, most of montane Asia—have their own, differently named, wildmen.[18]

The other difficulty in making sense of the Himalayan wildman was metaphysical. Many of the stories about wildmen were quotidian, reflecting an agricultural people's familiarity with local fauna: a herder saw a Yeti in the evening while rounding up a stray yak, for instance, or a man saw one while on the edge of town. "This is what makes Yeti stories so convincing," said journalist Desmond Doig, "the casual, matter-of-fact way in which they are told." The wildmen were treated as though they were as real, as substantial—as unremarkable—as bears and pika. But there were also fantastic tales. In Bhutan, Sikkim, and Nepal, there were stories of a time, long ago, when wildmen plagued villages until the creatures were made drunk and then tricked into killing themselves with fire or swords. Wildmen were said to have their feet on backward; said to abduct women; said to chase children—who escaped only by running downhill so that the wildmen's long hair fell over their eyes and obscured their vision; and said to be harbingers of bad luck:

**17.** John Napier, *Bigfoot: The Yeti and Sasquatch in Myth and Reality* (New York: E. P. Dutton, 1973), 36–38; Shackley, *Still Living?* 54 (quotation).

**18.** Ralph Izzard, *The Abominable Snowman* (New York: Doubleday & Company, 1955), 100; Charles Stonor, *The Sherpa and the Snowman* (London: Hollis & Carter, 1955), 63–64; William C. Osman Hill, "Abominable Snowmen: The Present Position," *Oryx* 6 (1961): 87; Edmund Hillary and Desmond Doig, *High in the Thin Cold Air* (Garden City, NJ: Doubleday, 1962), 31–32; Shackley, *Still Living?* 52 (quotation).

to see one was to be cursed, certain to die. Wildmen body parts—scalp, skin, hands—were kept as holy relics by religious leaders in many Sherpa towns, the beast revered as well as feared. The scalp kept at the monastery in Pangboche, Nepal, was supposed to have come from a wildman that cared for the monastery's founding lama when he meditated alone high in the mountains, bringing him food and water, until the day that the lama found the wildman dead in a cave.[19]

In purifying science of popular delusions, wonders, and prodigies, elite scholars were sometimes too diligent; it wasn't just wildmen and rains of frogs and multiple suns that were banished, but true things, too, like meteors, which for a long time cognoscenti considered mere superstition—reports that the sky rained rocks were no more believable than reports that it rained blood. Such mistakes raised the question: perhaps wildmen *did* exist, not in circuses or zoos or the pages of newspapers, but in nature, just in far away places where scientists rarely traveled. Perhaps the Yeti was real, a mundane creature that had inspired curious legends, the way the rhinoceros may have been distorted into the unicorn, or the way rabbits gained human traits in Beatrix Potter's stories. Perhaps the trouble was not the Yeti at all, but that conceptual categories in Tibet and Nepal and surrounding regions did not match those in the West, combining as they did the real and unreal in obscure ways.[20]

Exploration of the Himalayas became more serious later in the twentieth century—and the reports of wildmen more frequent. The major world powers were competing to assert dominance over central Asia, a continuation and reworking of the so-called Tournament of Shadows that had played out between Britain and Russia for control of Afghanistan and neighboring areas in earlier decades. Mountaineering became intertwined with geopolitics, climbing a way of proclaiming mastery over a region. Britain had demonstrated its power trigonometrically in 1852 by calculating that Himalayan

**19.** C. K. Howard-Bury, *Mount Everest: The Reconnaissance, 1921* (London: Edward Arnold, 1922), 141; Hillary and Doig, *High in the Thin Cold Air*, 77, 116–17 (quotation, 117); Sherry Ortner, *Sherpas through Their Rituals* (Cambridge: Cambridge University Press, 1978), 10–33; Shackley, *Still Living?* 59–62.

**20.** Ronald Westrum, "Social Intelligence About Anomalies: The Case of Meteorites," *Social Studies of Science* 8 (1978): 461–93; Lawrence W. Swan, *Tales of the Himalaya: Adventures of a Naturalist* (La Crescenta, CA: Mountain Air Books, 2000), 42; Sherry Ortner, *Life and Death on Mt. Everest: Sherpas and Himalayan Mountaineering* (Princeton, NJ: Princeton University Press, 2001), 113; Richard Conniff and Harry Marshall, "In the Realm of Virtual Reality," in *The Best American Science and Nature Writing 2002*, ed. Natalie Angier (New York: Houghton Mifflin, 2002), 24–33.

Peak XV was the world's tallest mountain and naming it in honor of a countryman, Sir George Everest, former surveyor-general of India. After the Great War, European countries and Japan raced to prove their mastery with more than mathematics—by putting boots on the roof of the world and planting their nation's flag. To climb Everest was to conquer it.[21]

In 1921, Lieutenant Colonel C. K. Howard-Bury led Britain's first reconnaissance of the path to Everest—leaving from Tibet and walking south. On Lhakpa La, more than 20,000 feet up, the expedition "came across tracks in the snow" that seemed humanlike. The porters, Howard-Bury wrote, "at once jumped to the conclusion that this must be 'The Wild Man of the Snows.'" Howard-Bury thought the porters' claim preposterous. Probably a wolf, he decided, "a large 'loping' grey wolf, which in the soft snow formed double tracks rather like those of a barefooted man." In his opinion, the wildman was only a make-believe monster, a fairy tale that—like the legend of Santa Claus—was used to scare children into behaving. Still, Howard-Bury sent word of the discovery to the press. Henry Newman, a columnist at the *Calcutta Statesman*, an English-language newspaper in India, did some follow-up reporting, talking with some Tibetans; "the whole story," he thought, "seemed a joyous creation" and so he wrote an article on the subject. In the process, however, he added to the linguistic confusion. The beast had been referred to as *metoh kangmi*, which meant "man-like wild creature." Newman recognized *kangmi* as meaning snowman, but he garbled the translation of the first word, confusing *metoh* for the Tibetan word *metch*, which meant filthy or dirty. And so Newman gave the beast the English name "Abominable Snowman."[22]

"The result was like the explosion of an atom bomb," wrote naturalist Ivan Sanderson. Previous Yeti reports often went little noticed beyond a small circle of people, mountaineers and zoologists and connoisseurs of those subjects. Newman apparently expected the same to be true of his story, sending it to one or two newspapers. Instead, word of the beast rocketed around

**21.** James Ramsey Ullman, *The Age of Mountaineering* (Philadelphia, PA: J. B. Lippincott Company, 1954), 234–35; Peter Hopkirk, *The Great Game: The Struggle for Empire in Central Asia* (New York: Kodansha America, 1994); Karl Ernest Meyer and Shareen Blair Brysac, *Tournament of Shadows: The Great Game and the Race for Empire in Central Asia* (New York: Counterpoint Press, 2000).

**22.** Henry Newman, letter, July 27, 1937 (final quotation), HWT 27/1/17a, H. W. Tilman papers, Royal Geographical Society, London; C. K. Howard-Bury, *Mount Everest: The Reconnaissance, 1921* (London: Edward Arnold, 1922), 141 (first quotations); Osman Hill, "Abominable Snowmen," 87.

the world, from Tibet to India to England to Atlanta. "It was seized upon by cartoonists," Newman remembered, "who drew abominable and grotesque figures and put on them the heads of well-known people." William Henry Johnson, the man who played the part of Barnum's final What-Is-It, died in 1926, performing almost to the very end; the Abominable Snowman took What-Is-It's place in the popular consciousness. For a time, "Abominable Snowman" became a generic phrase referring to any wildman—as What-Is-It had once been. Questions that had been asked of circus geeks were now asked of the confounding Yeti: What Is It? Was it real or an invention? Did it belong to the music hall or mammalogy? Maybe it was all a joke. But then legends do not show up against the snow, and superstitions do not leave tracks. Maybe there was a wildman, out there, in nature. Maybe the scientific consensus was wrong.[23]

Over the next fifteen years or so, Europeans exploring the mountains of southern Asia continued to send back reports of the Abominable Snowman. In 1925, N. A. Tombazi, a photographer and Fellow of the Royal Geographical Society, saw something like an Abominable Snowman near the Zemu Glacier. "Unquestionably," he wrote afterwards, "the figure in outline was exactly like a human being, walking upright and stopping occasionally to uproot or pull at some dwarf rhododendron bushes. It showed up dark against the snow and, as far as I could make out, wore no clothes." Tombazi hiked to where the creature had been and found manlike tracks; they were about six inches long. Mountaineers also came across purported Yeti scalps in lamaseries. But analyses of them were either inconclusive or indicated that the hair had come not from a primate, but an ungulate.[24]

A report by mountaineer Ronald Kaulback in July 1937 prompted a volley of letters to the *London Times*, mountaineers and biologists speculating on the trackmaker's identity. Langurs? Bears? Pandas? A romp of otters jump-

**23.** Henry Newman, letter, July 27, 1937 (final quotation), HWT 27/1/17a, H. W. Tilman papers; Ralph Izzard, "Why the 'Snowman' Is Called 'Abominable,' *Daily Mail* (London), December 9, 1953, Abominable Snowman clippings file, *Daily Mail* office, London; untitled article, *Atlanta Constitution*, December 21, 1921; "The Snowman," *Los Angeles Times*, January 3, 1922; Ivan T. Sanderson, *Abominable Snowmen: Legend Come to Life*, rev. abridgment (New York: Pyramid, 1968), 52 (first quotation).

**24.** "Himalaya 'Scalp' Studied as Clue," *New York Times*, December 20, 1954; "The 'Scalp' of an Abominable Snowman? A Yeti-Hide Head-Dress," *Illustrated London News*, March 27, 1954, 477; Bernard Heuvelmans, *On the Track of Unknown Animals*, trans. Richard Garnett (New York: Hill & Wang, 1958), 134; Osman Hill, "Abominable Snowmen," 93; John Napier, *Bigfoot: The Yeti and Sasquatch in Myth and Reality* (New York: E. P. Dutton, 1973), 39–40 (quotation, 39).

ing through the snow, each one landing in the hole left by the one before it, until the holes were widened to resemble a primate's footfall? The Yeti? Most likely, mountaineer Frank Smythe wrote in November, summarizing the opinion of those skeptical that an Abominable Snowman existed, bears left the tracks, and the stories of wildmen had been promulgated by superstitious natives who, although nominally Buddhist, continued to practice a primitive nature worship. Reports of the Abominable Snowman, in short, were no more credible than reports of dog-faced men. They were a species of that disreputable genre, the traveler's tale.[25]

But while the Abominable Snowman raised questions similar to those provoked by the What-Is-It, the context in which those questions were asked was quite a bit different. Barnum had been master of nineteenth-century American popular culture; the Yeti was preeminently a product of mass culture. Popular culture, according to historian Michael Kammen, is social, public, and interactive, allowing the audience to participate in its own entertainment: Barnum's wildman was directly in front of the crowd. It could be talked to, touched, perhaps undressed. Mass culture, by contrast, is passive and private. Products of mass culture—movies, paperbacks, magazines, television shows, newspapers—reach a huge audience, but the audience is dispersed: in their homes, on the train, away from one another. Mass culture is also intimately tied to another development in American society: the rise of consumerism. Buying is the fuel that keeps mass culture expanding. In a mass culture, advertising exploits its audience's insecurities, finds its secret hopes, and offers products—consumption—as a way to repair those inadequacies, achieve those dreams. According to Kammen, mass culture superseded, but did not replace, popular culture, growing through the early part of the twentieth century and exploding after World War II, when consumerism also became dominant.[26]

These cultural shifts altered what it meant to be a person, changed the way that Americans thought about themselves. Nineteenth-century America

**25.** Ronald Kaulback, "20 Months in Tibet. II. Along the Salween," *Times* (London), July 13, 1937; F. S. Smythe, "Abominable Snowmen—Pursuit in the Himalayas—A Mystery Explained," *Times* (London), November 10, 1937; F. S. Smythe, "The Abominable Snowmen," *Times* (London), November 16, 1937.

**26.** Roland Marchand, *Advertising the American Dream: Making Way for Modernity, 1920–1940* (Berkeley: University of California Press, 1986); Jackson Lears, *Fables of Abundance: A Cultural History of Advertising in America* (New York: Basic Books, 1995); Michael Kammen, *American Culture, American Tastes: Social Change and the Twentieth Century* (New York: Basic Books, 1999), 18–26, 162–64.

had celebrated what historian Warren Susman called a "culture of character." Character was associated with citizenship and duty, work, manners, and integrity. "The stress was clearly moral and the interest was almost always in some sort of higher moral law," Susman noted. A man with strong character—and character was tightly coupled with the notion of manhood—controlled his desires, his impulses. In the twentieth century, focus shifted away from character toward personality. Personality was not something one possessed, but something one built. Personality unfettered desires, allowed them to be sated; mass culture provided the material from which personalities were created—advertisements, movies, magazines entering Americans homes to tell them what they needed to succeed, to invent and present their true selves.[27]

The possible existence of the Yeti raised questions about this cultural arrangement, about what counted as truth and authenticity in an era when there was no essential self, but only various performances. Was the Abominable Snowman just another media creation? Something to sell newspapers? Or was the creature something more—proof that there was an obdurate world, a real and true one, behind the blur of mass culture? Did the world yet hold a wonder—a creature that was authentic? That was the view of mountaineer H. W. Tilman. He dismissed the brouhaha in the *London Times* as so much hot air—the usual cacophony of mass culture. The following year, climbing with Eric Shipton, he came across tracks that his porters told him had been made by a Yeti, a man-eating Yeti. Tilman was a joker and joshed with the Sherpas that as no one had been in the area for thirty years, the Yeti must have been "devilish hungry." But he also respected the Sherpas and their lore, romantically valuing it above the opinion of scientists—he never had much taste for science anyway. Tilman followed the tracks for a mile. He was not sure what had left the prints—not a bear, of that he was certain—but was disposed to think it had been the Abominable Snowman. The bloviating in London, Tilman said, proved nothing. "The Abominable Snowman re-

**27.** Warren I. Susman, "'Personality' and the Making of Twentieth-Century Culture," in *Culture as History: The Transformation of American Society in the Twentieth Century* (New York: Pantheon, 1984), 271–85 (quotations, 273–74); Loren Baritz, *The Good Life: The Meaning of Success for the American Middle Class* (New York: Perennial Library, 1990), 212; Philip Cushman, *Constructing the Self, Constructing America: A Cultural History of Psychotherapy* (Reading, MA: Addison-Wesley, 1995), 210–78; Tom Pendergast, *Creating the Modern Man: American Magazines and Consumer Culture, 1900–1950* (Columbia: University of Missouri Press, 2000), 8–13, 209.

mained to continue his evasive, mysterious, terrifying existence, unruffled as the snows he treads, unmoved as the mountains amongst which he dwells, uncaught, unspecified, and not unhonoured."[28]

## The Life and Times of Bigfoot

The Bigfoot that supposedly lumbered across the Jenkinses' property in Fort Bragg and scared Robert Hatfield so much was the lineal descendant of medieval wildmen, Barnum's What-Is-It, and the Yeti. Indeed, the six men who investigated the beast in February 1962 bonded themselves into what they called an ABSMism Club—an Abominable Snowmanism Club. When the folklorists turned their attention to the matter, they had a chance to explore why Bigfoot caused such a stir and probe the connections between the beast's career in the mass culture and the evolution of that very same mass culture. Many communities have their own monsters. Bigfoot, however, was "America's Number One Monster," as the title of one book about the creature had it. Why? What made the creature so popular among such a wide swath of people? That seems like an interesting question, worthy of professional folklorists. But they didn't bother with such questions. They were content to note that stories about Bigfoot were modern manifestations of the classic wildman story. And after the spurt of interest in the early 1960s, few folklorists bothered with the subject. Nor did scholars investigate the popularity of the Abominable Snowman. The rare scholarly investigations that were undertaken usually reiterated the point that Bigfoot was just a reworked example of a creature known from the Middle Ages, focused only on whether the stories about wildmen were true, or offered belief in wildmen as a regrettable example of a popular delusion.[29]

**28.** Tilman, *Mount Everest 1938*, 127–37; Eric Shipton, *Eric Shipton: The Six Mountain-Travel Books* (Seattle, WA: Mountaineers Books, 1997), 247 (quotations); Peter Steele, *Eric Shipton: Everest & Beyond* (Seattle, WA: Mountaineers Books, 1999), 84.

**29.** "Skeptics and Enthusiasts Clash on Bigfoot Search"; Alan R. Velie, "The Dragon Killer, the Wild Man and Hal," *Fabula: Zeitschrift für Erzählforschung* 17 (1976): 269–74; David Hufford, "Humanoids and Anomalous Lights: Taxonomic and Epistemological Problems," *Fabula: Zeitschrift für Erzählforschung* 18 (1977): 234–41; Daniel Cohen, *Bigfoot: America's Number One Monster* (New York: Pocket Books, 1982); Manfred F. R. Kets de Vries, "Abominable Snowman or Bigfoot: A Psychoanalytic Search for the Origin of Yeti and Sasquatch Tales," *Fabula: Zeitschrift für Erzählforschung* 23 (1982): 246–61. An exception to the generalization is Linda Milligan, "The 'Truth' about the Bigfoot Legend," *Western Folklore* 49 (1990): 83–98.

FIGURE 3. Bigfoot reward poster. This early image of Bigfoot makes its genealogy clear: like the medieval wildman and Barnum's What-Is-It, Bigfoot is more human than animal. But unlike earlier examples of the type, this Bigfoot is also rather silly looking—not something to be taken seriously. (By permission of the *Santa Rosa Press Democrat* [February 18, 1962].)

While it is difficult to explain why something did not occur, it's possible to offer informed speculation. The most likely reason that folklorists ignored Bigfoot and other twentieth-century wildmen is exactly because they were popular. During the first half of the twentieth century, folklorists made it their duty to seek out stories shared among members of small, isolated groups. They were interested in authentic legends and authentic folk communities. Mass culture and its products seemed the antithesis of authenticity: mass culture was fake, an attempt to appeal to the lowest common denominator and separate rubes from their dollars. Once folklore seeped into newspapers and other mass media, it became what the dean of American folklorists, Richard Dorson, called "fakelore." Certainly there were elements of the Fort Bragg story that seemed to validate these biases. The *Santa Rosa Press-Democrat* jokingly offered a $25 reward for Bigfoot's capture, illustrating the wanted poster with the drawing of an oafish looking giant. "Meanwhile," the paper noted, "in Fort Bragg itself a shoe store ha[d] a giant boot in its window. A sign offers to sell Bigfoot a pair." And a "used-car dealer" advertised "a Bigfoot Auto Sale." Not long after the Fort Bragg monster made the papers, it found its way into the fanciful writing of Frank Edwards, a radio personality and author who collected stories of things that were, as he said, "stranger than science." For folklorists, stories of wildmen were just more examples of the unreality of mass culture; authenticity had to be found elsewhere.[30]

Much has changed in the interpretation of culture since the early 1960s, and now mass culture is not seen as worthless, as fake, but, rather, as another venue in which historical changes can be teased out. All sorts of topics that once seemed trivial are now grist for the academic mill, from comics to advertisements to romance novels to sitcoms to gangster rap. And that's where this book comes in. It picks up where the folklorists stopped, trying to understand how Bigfoot became prominent in American culture, why some people believed the creature existed, the function that such belief served, and how the debate over the existence of wildmen fit into twentieth-century American culture.[31]

First, this book shows how the modern myth of Bigfoot emerged out of, and diverged from, traditional wildman tales. The Yeti's career continued after World War II, eventually becoming entwined with two other legends,

**30.** Richard M. Dorson, "Folklore and Fakelore," *American Mercury*, March 1950, 335–43; "Skeptics and Enthusiasts Clash on Bigfoot Search" (quotations); Edwards, *Strange World*, 46–55; Sam G. Riley, "A Search for the Cultural Bigfoot: Folklore or Fakelore?" *Journal of Popular Culture* 10 (1976): 377–87; Richard M. Dorson, *Man and Beast in American Comic Legend* (Bloomington: Indiana University Press, 1982), 75; Regina Bendix, *In Search of Authenticity: The Formation of Folklore Studies* (Madison: University of Wisconsin Press, 1997).
**31.** Jan Harold Brunvand, *The Vanishing Hitchhiker: American Urban Legends and Their Meanings* (New York: W. W. Norton, 1989); Linda Deagh, *American Folklore and the Mass Media* (Bloomington: Indiana University Press, 1994); John Storey, *Cultural Studies and the Study of Popular Culture* (Athens: University of Georgia Press, 2003).

that of the Canadian wildman called Sasquatch and a California beast known as Big Foot—the creature that visited Hatfield and the Jenkinses. Each of these monsters gained notoriety during the middle of the twentieth century, as mass media promoted what were local legends into international sensations. These various traditions were interwoven into a single legend, a legend that passed into obscurity during the middle of the 1960s before returning to prominence at the end of that decade.

Second, this book connects these modern tales of wildmen to concerns over the maturation of mass culture and consumerism. For many people, a world filled with plastic gewgaws, a world in which shopping was the engine of economic growth, a world that was seen mostly through the blue light of a TV screen seemed increasingly unreal, fake. Bigfoot—although itself most likely fake—seemed to stand against this trend, representative of the really real, the world beyond the facade, a world of life and death and vital things. Bigfoot was the last representative of the old order, symbolic of the values that had been lost. And this was especially true among white, working-class men—who gave Bigfoot its warmest reception.

Speaking generally, these men were bothered by the irruption of mass culture—much as folklorists were—by the American emphasis on consumerism, and the change from a culture of character to a culture of personality. Stories about Bigfoot were a way for these men to confront and work through their anxieties, to try to resist the changes, just as Barnum's What-Is-It allowed middle-class Americans to grapple with their concerns in the nineteenth century. For some, Bigfoot was also a way to assert their dignity. By claiming that they knew—from their studies, from their hunting, from their investigations—that Bigfoot existed, that the elite consensus was wrong, they made themselves feel powerful. They understood reality, its workings, better than scientists. To proclaim Bigfoot's existence was to insist upon one's dignity against a world that either denied it, or, worse, went on spinning about its axis as though dignity did not even *matter*, as though the world was nothing but gewgaws and shopping and TV.

Bigfoot, though, was an unreliable champion for resisting mass culture. Bigfoot did not lead its working-class fans away from mass culture and consumerism, but into it. The hunt for the creature did not prove one's dignity, but destroyed it. The beast was not a representative of the really real, the authentic, genuine, true world beneath the facade, but was itself a creation of mass culture and consumerism. By the end of the twentieth century, Bigfoot was an advertising icon, used to sell everything from ketchup to TVs; its most ardent defenders were dead or defeated. And so, in one sense, this story is a tragic one, of hunters who hoped to find a better world and never did, of working-class men who bet their dignity on a beast that never existed. In another sense, however, the story is about the triumph of Bigfoot. In an American society so taken by consumerism, advertising helps to create reality. Maybe Bigfoot didn't exist in nature, but it was more than a circus geek: in important respects, Bigfoot became real, part of the cultural landscape that every American knew and recognized.

CHAPTER TWO

# Yeti 1951–1959

In September 1951, after the monsoon season ended and the snowfall lightened, Eric Shipton, at the time probably Britain's most esteemed mountain climber, led Michael Ward, Bill Murray, Tom Bourdillon, ten Sherpas, and two New Zealanders—H. E. Riddiford and Edmund Hillary—north from the Nepal town Namche Bazaar into the high Himalayas, establishing a base camp on the Khumbu Glacier. The team was scouting a new path to Everest—the one taken by prior expeditions such as Howard-Bury's had been blocked when Communist China took over Tibet. Ward, a doctor and one of his generation's best mountain climbers, had noticed this route while studying aerial photographs of the region. His discovery had been tantalizing enough to convince the British Himalayan Committee to ask Nepal for permission to send the team; unexpectedly, Nepal approved the request, and Shipton's was the first full-scale exploration of the path to Everest since the conclusion of World War II.[1]

Toward the end of September, Shipton and Hillary left their base camp and climbed 20,000 feet up a ridge of Mount Pumori to get a view of the course ahead. A steep-walled basin at Everest's base—what mountaineers call a cwm, or a cirque—rose gently to a saddle that connected Everest to its sister, Mount Lhotse. The approach from the saddle to Everest's peak seemed clear. It appeared that if the team could reach the cwm, the rest of the climb would be straightforward. Theirs would not be a reconnaissance; they would conquer Everest. They would be the first humans to stand on the roof of the world.[2]

**1.** Peter Steele, *Eric Shipton: Everest & Beyond* (Seattle, WA: Mountaineers Books, 1999), 146–51; Margalit Fox, "Michael Ward, 80, Doctor on '53 Everest Climb, Dies," *New York Times*, October 25, 2005.

**2.** James Ramsey Ullman, *The Age of Mountaineering* (Philadelphia, PA: J. B. Lippincott Company, 1954), 260–64; Steele, *Eric Shipton*, 145.

FIGURE 4. Members of the 1951 Mount Everest expedition in Nepal. Wildmen came back into vogue during the 1950s thanks to the discovery of a humanlike footprint in the snows of the Himalayas by mountaineers Eric Shipton and Michael Ward. Shipton is on the far left of the back row. From left to right are Bill Murray, Tom Bourdillon, and H. E. Riddiford. Michael Ward (on the left) and Edmund Hillary are sitting in front of them. (Image S0001295. By permission of the Royal Geographical Society.)

The approach to the cwm, however, was treacherous: its entrance was at the end of a 2,000-foot icefall riven by canyons and ridges and crevasses that reminded Shipton of a "bombed-out area in London during the war." The team carefully picked its way along the icefall, eventually reaching its edge, so close—yet so far: a wide crevasse separated the icefall and cwm. The team had no way to cross and so would have to climb down one side and up the other. It would be slow, difficult, potentially deadly work, requiring the porters to repeatedly cart heavy loads over the last hundred feet of the icefall, an especially dangerous stretch prone to avalanches and collapse. Hillary later wrote, "Over the next few days, we discussed the problem of the Icefall and there was much talk about 'unjustifiable risk' and 'unsafe for porters.' But I think we all realised that these were attitudes from the past, that nobody was going to get up Everest without taking a few risks, that the Icefall would never be a place for the cautious or the faint-hearted." Shipton, though—"ever a cautious leader," according to historian James Ullman, "to whom human lives were more important than victory"—judged the path too hazardous.[3]

The team fell back to Namche Bazaar at the end of October, spending several nights in the town, relaxing after the hard work of climbing. It was festival season. There was much drinking of the local alcohol, *chang*, and "non-stop dancing throughout most of the night." Shipton and his Sherpa hosts laughed at Hillary in the midst of the bacchanal, "tall and loose-limbed, supported by two stocky Sherpa wenches, an expression of powerful concentration on his bearded face as he strove to catch the complicated rhythm." The team was supposed to leave again on Saturday, but Shipton could not entice the Sherpas out of the village until Sunday, November 4. The Kiwis took a direct route to India, needing to return home to their businesses, Hillary to selling honey, Riddiford to law. Meanwhile, the Brits wanted to explore the west. Murray and Bourdillon trekked along the Nangpa La, a trade route between Nepal and Tibet that carried heavy traffic despite being 19,000 feet above sea level; Shipton, Ward, and six Sherpas headed into the Pangbuk Valley, near the border between Tibet and Nepal.[4]

"To the west and south-west," Shipton wrote, "there was a score of high mountains whose position in relation to the watershed we could only conjecture." Shipton loved exploration even without glory, loved the mountains so much that he named some after girlfriends, loved "to arrange the peaks and valleys and glaciers in their true perspective, and gradually learn to know them with an intimacy and understanding." And so he and Ward and the Sherpas explored the tangle. The weather was fine, the sun bright and hot. Early Thursday morning, November 7, Shipton, Ward, and the Sherpa Sen Tensing hiked away from their base camp carrying a tent and a week's worth of food, leaving the other Sherpas to await Murray and Bourdillon and point them in the right direction. The team had cut steps along the route the day before, which compensated for the heavy loads they carried. By 2:00 that afternoon, Shipton, Ward, and Tensing reached the top of a saddle, 20,000 feet above sea level. Over the next hour and a half, they descended 1,000 feet along

**3.** Ullman, *The Age of Mountaineering*, 257 (first and last quotations); Edmund Hillary, *Nothing Venture, Nothing Win* (New York: Coronet, 1977), 161 (second quotation); Steele, *Eric Shipton*, 153–57.

**4.** Eric Shipton, "Footprints of the 'Abominable Snowman,'" *Times* (London), December 6, 1951 (all quotations); Steele, *Eric Shipton*, 157.

an easy icefall to the head of the Menlung Glacier. The going became more difficult for a time, the snow knee-deep.[5]

Around 4:00, the three men came across "some strange tracks in the snow." Shipton's first thought was that Murray and Bourdillon had passed this way already. But he had spied out the landscape over the past few days and knew that there was no other approach from Nangpa La than the one that he, Ward, and Tensing had just walked. Something else had made the tracks, two animals, it seemed, and not too long ago—the balls of snow that they had kicked up remained, not yet melted "despite the warm sun which had been shining all day." The tracks had not survived the heat so well, most "distorted by melting into oval impressions, slightly longer and a good deal broader than those made by our large mountain boots." But, Shipton noted, "where the snow covering the ice was thin," there was a "well preserved impression of the creature's foot. It showed three broad 'toes' and a broad 'thumb.'" The tracks crossed a crevasse and "one could see quite clearly where the creature had jumped and used its toes to secure purchase on the other side." Shipton and Ward followed the footprints for about a mile, but lost them on moraine-covered ice. Shipton took several photographs of the prints, posing Ward, Ward's boot, and an ice pick next to them for scale; and then they continued further along the glacier until darkness forced them to camp. That night, Shipton was haunted by an "eerie feeling . . . that somewhere in that moon-lit silence the strange creatures that had preceded us down the glacier were lurking." Murray and Bourdillon found the tracks a few days later, nearly obliterated from melting, and followed them for two miles before losing the trail.[6]

Sen Tensing said that Yetis had made the tracks—not bears nor monkeys, both of which he knew well, but a Yeti, what the British called an Abominable Snowman. He had seen a Yeti himself two years before, in the town of Thyangboche, not far from Namche Bazaar. The creature was "half man half beast, about five feet six inches tall, covered with reddish brown hair but with a hairless face." Shipton "became convinced," he said, "particularly by the unmistakable evidence of the toes, of the existence of a large, apelike

**5.** Shipton, "Footprints of the 'Abominable Snowman'" (first quotation); idem, *That Untravelled World: An Autobiography* (New York: Charles Scribner's Sons, 1969), 195 (second quotation); idem, *Eric Shipton: The Six Mountain-Travel Books* (Seattle, WA: Mountaineers Books, 1997), 260; Steele, *Eric Shipton*, 158.
**6.** Shipton, "Footprints of the 'Abominable Snowman'" (remainder of quotations); idem, *That Untravelled World*, 196 (second quotation), 197 (last quotation); Steele, *Eric Shipton*, 160.

FIGURE 5. This is one of the Yeti footprints that Eric Shipton discovered, with an ice ax placed next to it for scale. (Image S0001202. By permission of the Royal Geographical Society.)

creature, either quite unknown to science or at least not included in the known fauna of Central Asia." Murray and Bourdillon were also convinced that the footprints belonged to the Yeti. "The Abominable Snowman is not a myth," Bourdillon wrote in a letter home.[7]

## Sensation

Shipton was chronicling his reconnaissance in the pages of the *London Times*; word of his discovery appeared in the paper on December 6, 1951, a photograph the day after. As before, the mass media carried word of the Abominable Snowman around the world. *The Illustrated London News*, *Life* magazine, and *Newsweek* each published stories on the tracks and their maker before the year was out. The Yeti was a sensation: it became a star. A number of movies featuring the beast debuted during the 1950s, *Snow Creature* and *Man Beast* and *Half-Human* and *The Abominable Snowman of the Himalayas*. So popular was the beast that Nepal began to sell hunting licenses for the creature; ac-

7. Shipton, "Footprints of the 'Abominable Snowman'" (first quotation); idem, *That Untravelled World*, 197 (second quotation); Steele, *Eric Shipton*, 160 (final quotation); W. H. Murray, *W. H. Murray: Evidence of Things Not Seen: A Mountaineer's Tale* (London: Baton Wicks, 2002).

cording to *Newsweek*, the goal was to raise money for disaster relief. "The Abominable Snowman is serious business," noted *Sports Illustrated*'s Don Connery. "Those who are skeptical keep their secret doubts to themselves. As good patriots, why should they undermine what has become an important source of income and just about the only thing guaranteed to land this obscure, backward little nation in newspapers"?[8]

Debate over the Yeti's existence, which had flagged during the war, reignited. What had made those tracks? Was it, as Tensing said, the Abominable Snowman? Or, as Smythe had it, a bear? "Something must have made the Shipton footprint," wrote anthropologist and wildman aficionado John Napier. "Like Mount Everest, it is there, and needs explaining." Less than a week after the article appeared in the newspaper, the British Museum (Natural History) mounted in its central hall a display about the footprints. The museum's exhibit tried to show that langurs had left the tracks. The *London Times'* correspondent found the claim convincing, despite langurs being much too small to leave prints of the size that Shipton had found. Others claimed that the tracks had been left by ascetic monks who lived high in the Himalayas, so mentally disciplined that they could walk through the snow nearly naked; or by snowshoes with fronts so worn that the wearer's toes left imprints in the snow. Still others claimed that the beast was real. Slavomir Rawicz published a book recounting his supposed escape from a Soviet gulag by walking from Siberia to India. In the course of his trek, Rawicz claimed, he watched a pair of Abominable Snowmen for a couple of hours. And American George Moore published in *Sports Afield* an account of his encounter with the terrible beasts.[9]

In 1955, the Indian scholar and Fellow of the Royal Geographical Society Swami Pranavananda argued that the mystery could be solved if only the linguistic muddle made by Newman and Howard-Bury could be clarified. A

**8.** Don Connery to Earl Burton, June 20, 1958 (quotation), General Correspondence A–B 1958 file, Carleton S. Coon papers, National Anthropological Archives, Smithsonian Institution, Suitland, MD; "Monkey or Bear? The Abominable Snowman's Footprints Compared with Impressions Taken from Zoo Animals," *Illustrated London News*, December 15, 1951, 973–75; "Abominable Snowman," *Newsweek*, December 17, 1951, 33; "Abominable Himalayan," *Life*, December 31, 1951, 88; "Flight and Prayers," *Newsweek*, June 16, 1958, 51; Loren Coleman, *Tom Slick and the Search for Yeti* (Boston: Faber & Faber, 1989), 141.

**9.** "Tracks on Everest," *Times* (London), December 13, 1951; George Moore, "I Met the Abominable Snowman," *Sports Afield*, May 1957, 41–45, 91–93; "You Call This Abominable?" *Newsweek*, December 29, 1958, 63; John Napier, *Bigfoot: The Yeti and Sasquatch in Myth and Reality* (New York: E. P. Dutton, 1973), 36–37, 49–50, 141 (quotation); Slavomir Rawicz, *The Long Walk: The True Story of a Trek to Freedom* (Guilford, CT: Lyons Press, 1997), 223–32.

correct understanding of the language proved that there was no such thing as an Abominable Snowman. The term *mi-te* (*metoh* or *meh-teh*) did not mean filthy, but man-bear, he said, and referred to the red bear. *Kangmi*—snowman—was just a colloquial way of referring to the same animal. The Sherpas had not told Howard-Bury of a wildman, but of a bear that is like a man because it sometimes rises onto its hind legs. William L. Strauss, an anthropologist at Johns Hopkins University, heavily promoted Pranavananda's philological argument. "The matter, of course, cannot be conclusively settled until a specimen of undoubted 'snowman' is secured for study," he admitted. "On the basis of the best evidence," however, "the 'Abominable Snowman' would seem . . . to be no other than the Himalayan red bear."[10]

Biologist Lawrence Swan objected that the linguistic argument made no biological sense: red bears inhabited the western Himalayas, while the Yeti tracks had been found in the eastern parts of the mountain range. "There is a fairly striking faunal difference between these two regions," Swan wrote, "and it is not legitimate, nor is it good zoogeography, to attempt to discredit the legend on the evidence obtained from the western Himalayas or the plateau of Tibet. The Abominable Snowman, presumably, has no business in these parts." Swan also pointed out that Tibetans, Nepalese, and Sherpas did not describe the Yeti as bearlike but as a "bipedal ape." Could apes inhabit the harsh Himalayas? Swan did not make the point, but yes: langurs do. (One name for the region, *Mahalangur Himal*—in which *langur* means ape and *himal* mountain—translates as *mountains of the great apes*.) And there are thick rhododendron forests in the valleys. Apes could live in the lush foliage, occasionally crossing the snow, leaving their marks.[11]

Meanwhile, other evidence came to light that seemed to simplify the Yeti's metaphysical status. Czech anthropologists working in Mongolia found books of Tibetan natural history from the eighteenth and nineteenth centuries that included pictures and descriptions of a wildman that lived in the

**10.** Swami Pranavananda, "Abominable Snowman," *Indian Geographical Journal* 30 (1955): 99–104; idem, "The Abominable Snowman," *Alpine Journal* 61 (1956): 110–17; William L. Straus Jr., "Abominable Snowman," *Science* 123 (1956): 1024–25 (quotation, 1025); idem, "Again the Abominable Snowman," *Science* 124 (1956): 22; Pranavananda, "The Abominable Snowman," *Journal of the Bombay Natural History Society* 54 (1957): 358–64; idem, "Footprints of Snowman," *Journal of the Bombay Natural History Society* 54 (1957): 448–450; William L. Straus Jr., "Abominable Snowman," *Science* 126 (1957): 858.

**11.** Lawrence W. Swan, "Abominable Snowman," *Science* 127 (1958): 882–84 (quotations, 883); Edmund Hillary and Desmond Doig, *High in the Thin Cold Air* (Garden City, NJ: Doubleday, 1962), 27; Ivan T. Sanderson, *Abominable Snowman: Legend Come to Life* (Philadelphia, PA: Chilton, 1961), 398–420, 487.

mountains. "The authenticity of these illustrations of the wild man," one of the scientists wrote, "is supported by the fact that among tens of illustrations of animals of various classes . . . there is not a single case of fantastic or mythological animal. . . . The creatures mentioned here are actually living animals observed in nature." The Abominable Snowman was remarkable only because it had not yet been captured, named, and classified by Western scientists. "The people who claim the whole thing to be hogwash had better look into the facts," Swan said.[12]

As during the interwar years, the Yeti also raised questions about society more generally, particularly the spread of mass culture and consumerism. Anthropologist John Napier complained that the 1950s was when Tibetan and Nepalese folklore about the Yeti "started to deteriorate into fakelore." In a strict sense, he was correct: Rawicz's story was most likely fake; his book's geography was so confused that Shipton, among others, doubted its veracity. Moore's account also seemed nothing more than an invention to cash in on interest in the Himalayan monsters.[13]

But interest in the Yeti was not just driven by greed, even when the interest manifested itself as movies and fabulous memoirs. That was just another example of the mistake that folklorists made about twentieth-century wildmen. Fascination with the Yeti grew out of and reflected important themes in British culture. It was more than fakelore—it was folklore for an industrial age, spread on the currents of mass media. In the Abominable Snowman, many British, especially those of a romantic cast, saw both an affirmation and critique of their national character. Often, the Yeti was portrayed as a repository of ancient, hard-won wisdom—the kind of wisdom possessed by Britain after centuries of imperial rule. "I'm wondering, wondering how old that face is," one character says of a Yeti in the movie *The Abominable Snowman of the Himalayas*. "It's seen a long life."[14]

The Yeti, though, was untouched by the materialism of modern life: it preserved a purity of motive that the British themselves had lost, or at least

**12.** Tom Benet, "Bay Man Ready for Snowman," *San Francisco Chronicle*, Lawrence Swan clipping file, San Francisco State University, San Francisco, CA (last quotation); Emanuel Vlcek, "Old Literary Evidence for the Existence of the 'Snow Man' in Tibet and Mongolia," *Man* 59 (1959): 133–34 (quotation, 134); Emanuel Vlcek, "Diagnosis of the 'Wild Man' According to Buddhist Literary Sources from Tibet, Mongolia, and China," *Man* 60 (1960): 153–55.

**13.** Napier, *Bigfoot*, 41–47 (quotation, 47).

**14.** *The Abominable Snowman of the Himalayas*, DVD, directed by Val Guest (Twentieth Century Fox Film Corp.; Anchor Bay Entertainment, 1999); Donald S. Lopez Jr., *Prisoners of Shangri-La: Tibetan Buddhism and the West* (Chicago: University of Chicago Press, 1999).

buried—a criticism, then, but also an affirmation that there was, somewhere still, a thing that embodied Britain's true greatness. That was attractive, drawing the British to the monster, but also worrying. When Britain first decided to conquer Everest, there were many who bemoaned that, as the *London Evening News* put it, "Some of the last mystery of the world will pass when the last secret place, the naked peak of Everest, shall be trodden by those trespassers." There was a similar reluctance about capturing the Yeti and enmeshing it in the corrupt world. Napier's complaint about fakelore reflected some of this worry: that what had been pure in Tibet became corrupted in the West. John Masters, a former officer in the British Army, offered a starker example of this anxiety. He suggested that instead of catching the Abominable Snowman, the Yetis "put us in cages, labeled *Loathsome Earthman (self-extinguishing)* ♂♀, take away our nasty toys, return to their caves and rocks, and live happily ever after."[15]

## The Yeti, Science, and Zadig's Method

The sensation that followed in the wake of Shipton's discovery was more than an expression of British ambivalence about their national character, however, and more than an occasion for history to repeat itself, farce following farce. In the 1930s, the matter had been thrashed out in the *London Times*. This time, debate over the Yeti's existence took place not only in newspapers, magazines, and movie theaters, but also in scientific journals. Pranavananda published in the *Indian Geographical Journal* and the *Journal of the Bombay Natural History Society*. Strauss published in America's most renowned scientific journal, *Science*, as did Swan. Britain's leading scientific journal, *Nature*, reported on the beast. "There is no doubt," John Napier said, "that the footprint on the Menlung Glacier gave the whole business of the Himalayan Bigfoot an air of scientific respectability."[16]

It's true that there was no body of a Yeti in a museum or laboratory to study, that the best evidence for its existence was a footprint. But science deals with many entities that have never been observed directly. No one has seen a black hole. No one has seen a boson. These things are known by their effects, by the traces that they leave: by bending light, by tracks left in par-

**15.** John Masters, "The Abominable Snowman," *Harper's*, January 1959, 30–34 (second quotation, 34); Henrika Kuklick, *The Savage Within: The Social History of British Anthropology, 1885–1945* (Cambridge: Cambridge University Press, 1993); Walt Unsworth, *Everest: The Mountaineering History* (Seattle, WA: Mountaineers Books, 2000), 24 (first quotation).
**16.** Napier, *Bigfoot*, 49.

ticle detectors. Unseen things can be scrutinized scientifically, using what the famed nineteenth-century scientist Thomas Henry Huxley called Zadig's method.[17]

Zadig was the perspicacious titular character in a 1748 novel by Voltaire. In the course of the story, Zadig was asked if he had seen the queen's lost dog. Although he had not, he knew that it had long ears, a limp, and was a bitch that had recently given birth—because he had seen its tracks and knew how to read them: faint marks outside the prints of the forepaws, he reasoned, were made by the dog's ears; the track of one paw was shallower than the others because she favored it; and furrows between the footprints were made by her dragging teats. Huxley, a vociferous champion of Darwin's evolutionary theories, argued that Zadig's logic was common in science. Archaeologists rely on Zadig's method. So do astronomers and geologists and historians and paleontologists. Georges Cuvier, the inventor of modern paleontology, wrote in 1834, "Today, someone who sees the print of a cloven hoof can conclude that the animal which left the print was a ruminative one, and this conclusion is as certain as any that can be made in physics or moral philosophy. This single track therefore tells the observer about the kind of teeth, the kind of jaws, the haunches, the shoulder, and the pelvis of the animal which has passed: it is more certain evidence than all Zadig's clues." Zadig's method is also the method of police detection: Sherlock Holmes and his literary descendants use Zadig's form of ratiocination. From signs no one else notices they can deduce the workings of the world, reconstruct a crime scene, identify the murderer. As a fingerprint or Cuvier's hoof print, the track that Shipton, Ward, and Tensing found was a clue that could be used to identify its maker. "If fingerprints can hang a man, as they frequently do," H. W. Tilman wrote, "surely footprints may be allowed to establish the existence of one."[18]

Bernard Heuvelmans, a French biologist—also popular writer, jazz artist, and, according to the press, "The Sherlock Holmes of Zoology"—applied Zadig's method to Yeti prints in his compendious *On the Track of Unknown Animals* (1955; translated into English, 1958). Snow leopards or wolves (or otters!) had not made the prints, he said. That was obvious from the shape of

**17.** Carlo Ginzburg, "Morelli, Freud and Sherlock Holmes: Clues and Scientific Method," *History Workshop*, no. 9–10 (1980): 5–36; Louis Liebenberg, *The Art of Tracking: The Origin of Science* (Cape Town: New Africa Books, 1995).

**18.** Ralph Izzard, "Abominable Snowmen," *Daily Mail*, June 5, 1953 (last quotation), Abominable Snowman clippings file, *Daily Mail* office, London; Georges Cuvier, *Recherches sur les Ossements Fossiles*, vol. 1 (Paris: 1834), 185 (quotation); Thomas Huxley, *Science and Culture* (London, 1881), 128–48; Ginzburg, "Morelli, Freud and Sherlock Holmes," 5–36.

FIGURE 6. Studying wildmen required the skills of a detective. Bernard Heuvelmans, author of *On the Track of Unknown Animals* and an early advocate for the existence of the Yeti, shown here, was known as the "Sherlock Holmes of Zoology." (Image 05955. © Musée de Zoologie—Lausanne/Agence Martienne.)

the tracks. Nor had langurs: the distance between prints was too long, the tracks too large, whatever the British Museum (Natural History) asserted. The only way a langur could have left such marks was by leaping across the snowfield, all four of its feet landing in the same spot. But the tracks were too clear, toes and heels too obvious, to have been made by a leaping monkey. A bear was more likely. There were no claw marks, but Heuvelmans expected as much: bears, he wrote, usually walk so that their claws do not press into the ground. If it was a bear, then Shipton's most famous, clearest photograph showed a hind foot, Heuvelmans deduced from the tracks. What was more, based on the position of the toes, he reasoned that it had to show the right hind foot. Had Smythe been right? Was the Abominable Snowman just a bear?[19]

When he first examined the photograph that Shipton took, Heuvelmans said, he thought that the bear "theory seem[ed] very satisfactory." But he

**19.** Bernard Heuvelmans, *On the Track of Unknown Animals*, trans. Richard Garnett (New York: Hill & Wang, 1958), 140–48; "Bernard Heuvelmans," *Times* (London), September 8, 2001.

changed his mind. One of Shipton's photographs showed a line of tracks that—according to press reports—had been made by a Yeti. The angle at which that photograph was taken obscured the details of the prints, but the shape of the gait could be determined—and it was not a trail that could have been left by a bear; the steps were in the wrong place, Heuvelmans concluded. He was right, although for the wrong reason. Almost twenty years later, John Napier, a British anthropologist who did seminal research on primate anatomy, learned that Shipton had taken the photograph of the trail earlier in the day; a goat had left it. The negative had been misfiled, the photograph mislabeled, and the error repeated because the detail was so poor. Heuvelmans did not know of the mistake. He only knew that no animal known to inhabit the Himalayas could have made both the track and the trail—so Shipton had found evidence of an unknown animal. Indeed, combining the (misinterpreted) picture of the trail and the photograph of the print, Heuvelmans concluded that the animal was a biped, a large biped, and that meant some kind of ape, just as Sen Tensing had described the Yeti. Heuvelmans was convinced that Shipton had found the track of the Abominable Snowman.[20]

At about the same time that Heuvelmans was writing, Wladimir Tschernezky, with the zoology department at Queen Mary College in London, was studying the tracks that Shipton had found. The detail, he thought, ruled out the possibility that the footprint was the melted remains of a smaller track; and the pattern of the snow kicked up by the walker was characteristic of a human's gait. Using plaster, Tschernezky re-created the foot from the prints. The plaster reconstruction had several notable characteristics—a large heel, a powerful hallux, and a long second toe—that were clues to the trackmaker's identity. The large heel indicated that the beast combined the humanoid tendency toward long tarsal bones and the gorilloid tendency toward wide heels. The hallux, or big toe, was probably adapted for grasping; its shape would make walking upright difficult, but the long, thick second toe compensated, helping the creature to balance. Adding these traits together, Tschernezky concluded that the tracks could not have been made by anything but a huge, heavy, bipedal primate.[21]

**20.** Heuvelmans, *On the Track of Unknown Animals*, 145–51 (quotation, 145); Napier, *Bigfoot*, 49.

**21.** W. Tschernezky, "A Reconstruction of the Foot of the 'Abominable Snowman,'" *Nature* 186 (1960): 496–97; W. Tschernezky and C. R. Cooke, "Unpublished Tracks of Snowman or Yeti," *Mankind Quarterly* 15(1975): 163–77.

Back in the 1930s, G. H. Ralph von Koenigswald, a Dutch paleontologist, found large primate teeth at a Chinese pharmacy in Hong Kong. Large teeth meant a large jaw, which meant a large body. They were the teeth of a giant. The Japanese captured Von Koenigswald during World War II, and the teeth spent the duration of the hostilities in a milk bottle buried in a friend's backyard. By the mid-1950s, scientists had established that the molars belonged to a creature they named *Gigantopithecus*, a huge ape that existed as recently as five hundred thousand years ago. Heuvelmans fingered *Gigantopithecus* as a relative of the Abominable Snowman, if not the Yeti itself—*Gigantopithecus* having escaped extinction and found sanctuary from humans in the high Himalayas. Working independently, Tschernezky came to a similar conclusion. Zadig's method seemed to function nicely: three lines of evidence—the errors in one not yet recognized—inescapably led to the scientific conclusion that a massive ape lived in southern Asia. The Yeti was more than fakelore, more even than folklore. The Abominable Snowman existed.[22]

## Britain Hunts the Yeti

As Heuvelmans and Tschernezky analyzed the purported Yeti tracks, the race to conquer Everest intensified. Shipton's team came close but failed in 1951; Swiss teams failed twice in 1952. (On one ascent, the climbers did find Abominable Snowman tracks, however, and Norman Dyhrenfurth claimed in *Argosy* magazine that a Yeti had shaken his tent.) Nepal granted the British permission to try again in 1953, the Swiss in 1954, and the French in 1955. There were also rumors that the Soviets were joining the competition. It seemed certain that the mountain would be conquered soon; pressure to succeed was great. "We've got this one chance," said a British climber, "and if we don't get it right we've had it. And dammit, it's our mountain." Already Britain had sent eight expeditions to climb Everest, and none had made it to the top.[23]

The Himalayan Committee, which coordinated the British assaults, tapped Shipton to lead the 1953 expedition. He had more experience in the Himalayas than anyone else, and he was in prime physical condition. But

**22.** Heuvelmans, *On the Track of Unknown Animals*, 151–58; Tschernezky, "A Reconstruction of the Foot of the 'Abominable Snowman,'" 497; Russell Ciochon, John Olsen, and Jamie James, *Other Origins: The Search for the Giant Ape in Human Prehistory* (New York: Bantam Books, 1990); Russell Ciochon, "The Ape That Was," *Natural History*, November 1991, 54–62.
**23.** Ullman, *The Age of Mountaineering*, 260–73; Ralph Izzard, *The Abominable Snowman* (New York: Doubleday & Company, 1955), 30–31; Steele, *Eric Shipton*, 187 (quotation).

there was some animosity toward him from the other climbers. He was perceived as too tentative, and this was no time for timidity. There could be no repeat of 1951. He preferred small expeditions, but the situation's desperation called for a huge effort. And he was not as motivated by nationalism as others, though national pride was at stake. The disgruntled rumblings proved persuasive. Late in 1952, the committee replaced Shipton with the army colonel John Hunt. The decision upset Bourdillon and Hillary, both slated to rejoin Shipton for the 1953 climb. Bourdillon withdrew, and Hillary considered pulling out as well. He respected Shipton a great deal and had never heard of John Hunt. Shipton, although stung by the decision, convinced Bourdillon to rescind his withdrawal. Hillary also decided to follow Hunt into the Himalayas, as did Michael Ward.[24]

This expedition made it through the icefall, across the cwm, and sent two of its members to the highest point on earth. At 11:30 in the morning on May 29, 1953, Edmund Hillary and Tenzing Norgay stood together atop Mount Everest, more than 29,000 feet above sea level. "My initial feelings," Hillary later said, "were of relief. Relief that there were no more steps to cut, no more ridges to traverse, and no more humps to tantalize us with hopes of success." News of the triumph reached London on the eve of Queen Elizabeth II's coronation.[25]

With Everest conquered—all the stairs cut, the ridges traversed, the humps in the background, Hillary and Norgay knighted, the queen honored—attention shifted toward the still-mysterious creature said to live in Everest's shadow. In late 1953, journalist Ralph Izzard asked the *London Daily Mail* to sponsor another expedition to the Himalayas, this time to hunt the Yeti. Izzard and the *Daily Mail* had a long-standing relationship—and a long-standing interest in mythical beasts. The paper had sent Izzard with Hunt's team to chronicle the ascent of Everest and before that had sponsored his search for the Buru, a large, aquatic reptile said to inhabit the marshy Himalayan Apa Tani valley. Those joint ventures must have worked out well enough for the *Daily Mail*, because it agreed to pay for Izzard's Abominable Snowman hunt.[26]

The involvement of the mass media raised concerns about the integrity of the hunt. Was this all just a bid for publicity? How could such an expedition

**24.** Hillary, *Nothing Venture, Nothing Win*, 169–70; Steele, *Eric Shipton*, 185–202; Peter Gillman, "How the Sacking of Britain's Top Mountaineer Sparked a Mutiny," *Times* (London), May 14, 2003.

**25.** Ullman, *The Age of Mountaineering*, 282–84 (quotation, 282).

**26.** Ralph Izzard, *The Hunt for the Buru* (London: Hodder & Stoughton, 1951).

prove anything real when it seemed to be in the business of only selling newspapers and the memoirs of its participants? Obviously seeing the hunt as only a joke, someone cabled Izzard, "Your announcement intolerable interference. Referring the matter to UN (signed) The Abominable Snowman," and then followed it by another, "Come up and see me sometime (signed) The Abominable Snowwoman." *Punch* magazine published a satirical poem about the expedition titled "Nothing Definite Yeti" that winked at the crass commercialism involved in the hunt for the Abominable Snowman:

> There are fascinating footprints in the snows of Katmandu
> On a slightly less than super-human scale:
> There are numerous conjectures on the owner of the shoe
> And the money it has cost the *Daily Mail*.[27]

Maybe, as the *New York Times* reported, the Yeti's popular (and ridiculous-sounding) name, Abominable Snowman, "saddled all discussions with comedy," but Izzard's hunt could not be easily dismissed as showmanship. "It may well have been true that Ralph Izzard's ambitious project, sponsored as it was by a great daily newspaper, was regarded by many as a stunt. Yet nothing could have been farther from the facts," Lord John Hunt said. The expedition, he noted, was "fraught with immense difficulties": "the remoteness and vastness of the area chosen combined with its rugged character, the lack of resources, the altitude, and the rigorous climate." Those willing to slog through snow, risk avalanches, and endure bone-chilling temperatures to look for a beast could not be ignored. The beast might exist. Lord Hunt was inclined to think so. "I believe in the Yeti," he said. "I have seen his tracks, heard his yelping call, listened to first-hand experiences of reputable people." How could one dispute the opinion of those who had been there? Of the man who had led the team that conquered Everest? Surely he knew a great deal about the region—had tested his knowledge against reality and been proven correct.[28]

Izzard recruited eight others to accompany him into the Himalayas for sixteen weeks, including John Jackson and Tom Stobart, also members of Hunt's successful expedition; Biswamoy Biswas, a scientist with the Zoo-

**27.** Izzard, *The Abominable Snowman*, 68 (first quotations); Myra Shackley, *Still Living? Yeti, Sasquatch and the Neanderthal Enigma* (London: Thames & Hudson, 1983), 58 (final quotation).
**28.** John Hunt, "No Joke Seeking the Yeti," *Daily Mail*, September 8, 1955 (second, third, and fourth quotations); Izzard, *The Abominable Snowman*, 68; Charles Stonor, *The Sherpa and the Snowman* (London: Hollis & Carter, 1955), vii (fifth quotation); A. M. Rosenthal, "Texan Will Lead 'Snowman' Hunt," *New York Times*, February 5, 1957 (first quotation); Bruce Hevly, "The Heroic Science of Glacier Motion," *Osiris* 11 (1996): 66–86.

logical Survey of India; and Gerald Russell, an American living in New Delhi who had been with the first Western expedition to capture a live panda. In Izzard's opinion, Russell was a master of Zadig's method. "A layman would possibly be astonished at the picture a man such as Mr. Russell . . . can construct from a few scant traces," Izzard said. Twelve Sherpas and two hundred porters were also part of the expedition, each porter carrying sixty pounds of gear. It was "possibly the best equipped scientific expedition ever sent to the Himalaya [*sic*]," Izzard boasted.[29]

In December, Dr. Charles Stonor set out in advance of the rest of the team to scout out the area, talk with locals about the Yeti, and investigate a putative Yeti scalp that had been found the previous autumn at a monastery in Pangboche, near Thyangboche. Stonor was a biologist, formerly with the London Zoo, and a romantic—the trip for him was a way of coming in touch with the purity of the English past. Walking through Nepal, he felt as though he had "been transported back into medieval England, with its mingling of harsh reality and colourful pageantry; an atmosphere so far untarnished by the dire slavery of materialism that has robbed and cheated us of so much that is best in our heritage." Early on, Stonor met Pasang Nima, who told him he had seen a Yeti only three months before. It was "a small, squat beast, the size of a teenage boy, covered with stiff red-brown and black hair, a flat face like a monkey, no tail, and normally walking on two legs." Unlike Smythe twenty-five years before, Stonor did not doubt the native reports. "The Sherpas," he said, "are a highly-intelligent, hard-headed race of realists." In January 1954, Stonor found what appeared to be Yeti tracks. He reported, "I am, shall we say, 95 per cent certain the snowman exists." The scalp was a less helpful clue—Stonor was unable to identify the animal from which it came, although he was convinced that it was authentic. So he sent home a few strands of hairs for further analysis. Later negotiations to borrow the scalp for more intensive study failed.[30]

The rest of the team hiked out of Katmandu at the tail end of the month. They carried with them an azure and blue flag emblazoned with two footprints. Between the tracks was the whiskered face of Bing, the Abominable

**29.** Ralph Izzard, "The Snowman Team Moves to Katmandu," *Daily Mail*, January 18, 1954 (quotations), Abominable Snowman clippings file.

**30.** Ralph Izzard, "Stonor Talks with Man Who Saw the Snowman," *Daily Mail*, January 19, 1954 (second quotation); idem, "Yeti Track Discovered; Like Human Foot," *Daily Mail*, January 26, 1954 (fourth quotation); Charles Stonor, "I'm Fully Convinced the Yeti Is There," *Daily Mail*, July 12, 1954 (third quotation), all in Abominable Snowman clippings file; Izzard, *The Abominable Snowman*, 99–115; Stonor, *The Sherpa and the Snowman*, 10–11, 38–39, 65, 117–18, 204 (first quotation).

FIGURE 7. After Edmund Hillary and Tenzing Norgay reached the summit of Mount Everest, British attention turned to tracking down the Yeti. In 1954, the *Daily Mail* sent this rather large expedition in search of the creature. At center is Bing, The Abominable Snow-Baby, the star character of a comic strip and the expedition's mascot. (© Associated Newspapers Ltd.)

Snow-Baby, the title character in a comic strip that journalist Desmond Doig drew for the *Calcutta Statesmen*. Surveying the team as it marched under the Bing flag, Izzard noted, "the total colour effect is, happily, that of the United Nations flag, which, in our case, can be taken to be symbolic; for counting our paid helpers, we represent five nations, all working together in perfect harmony." After 170 miles of walking, the team joined Stonor at Namche Bazaar on February 11. With the porters paid and sent home, the expedition divided into three groups of three explorers and some Sherpas.[31]

One team headed up the Chola Kola Valley, near where the Swiss had found tracks in 1952 and where the Yeti had been sighted six weeks earlier, according to what locals told Stonor. The second entered the Imba Kola Valley and investigated an area that was the scene of a supposed Yeti attack in 1949. The

**31.** Ralph Izzard, "Snowman Team Waits for Walkie-Talkie Radios," *Daily Mail*, January 25, 1954 (quotation); idem, "Three-Prong Hunt for the Snowman Has Begun," *Daily Mail*, March 1, 1954, both in Abominable Snowman clippings file.

third followed the Dudh Kosi Gorge to Mah, a yak-grazing village where Yetis were reported to be active, calling, leaving tracks, and killing two yaks. John Jackson found tracks. Russell analyzed alleged Yeti droppings, declaring that he had could see pica hairs in the feces—although the British zoologist W. C. Osman Hill thought that such a determination almost impossible to make in the field, without a microscope. Stonor saw a bear, which he thought was probably the inspiration for stories about the *dzu-teh*—but certainly not for all Yeti reports.[32]

In March, Izzard and Russell found what they believed to be the Yeti tracks while searching "among the towering peaks, rugged cliffs, glaciers, and ice-falls of the Upper Dudh Kosi Valley." They followed the trail for eight miles. "A fascinating picture unfolded itself of a shy, timid creature which uses man-made paths with the utmost caution," Izzard wrote. The tracks crossed a slope and approached a plateau, where they became jumbled. Izzard and Russell thought that this was where the creature had dropped to all fours, cautiously peering over the edge "to ensure the coast was clear before advancing." After detouring around a village, the tracks joined up with another pair; the two Yetis crossed a river and headed up a steep climb to Lang Boma Lake. They slid down another slope on their rumps and continued on across the country.[33]

The hunt ended in May, the results ambiguous. Izzard and Russell were certain that the Abominable Snowman existed; they had followed its tracks, reconstructed its behavior. They knew it intimately, thought that they understood it. Charles Stonor upgraded his earlier assessment. "I for one am completely convinced" that an unknown primate lives in the Himalayas, he said. He was conflicted only about whether it should be caught: "I rather favoured the possible quixotic ending to the story," he wrote. "The opened door of the cage; the Abominable Snowman taking a last look round his quarters and then shuffling off into the distance in imitation of the fade-out of a Charlie Chaplin film." Stobart, however, was less certain. The evidence was meager—some tracks, a few hairs that proved inconclusive on further analysis. No one had seen a Yeti, not its silhouette in the distance, not its hairless face.

**32.** Izzard, "Three-Prong Hunt for the Snowman Has Begun," *Daily Mail*, March 1, 1954; idem, "Stobart Glimpses a 'Dzu-Teh,'" *Daily Mail*, May 25, 1954, both in Abominable Snowman clippings file; idem, *The Abominable Snowman*, 127–28, 137–44, 150; Stonor, *The Sherpa and the Snowman*, 140–75; William C. Osman Hill, "Abominable Snowmen: The Present Position," *Oryx* 6 (1961): 95.

**33.** Ralph Izzard, "'A Shy, Timid Creature,'" *Daily Mail*, March 23, 1954, Abominable Snowman clippings file.

No one had photographed an Abominable Snowman, touched one, or killed one. "We can still give no more than a guess," he said.[34]

## America Hunts the Yeti

After the *Daily Mail* expedition ended, the quest for the Yeti continued. The Abominable Snowman remained enigmatic, but the evidence for its existence was good enough, the reward for its capture lucrative enough to entice adventurers. In 1955, Peter Byrne, an Irish-born big-game hunter, announced that he was putting together an expedition to seek the monster. He had seen tracks in 1948 and now wanted to see the creature. Japan sent parties to investigate. The Soviets established a "Snowman Commission" and sent a number of expeditions into the Pamirs looking for the wildman of Russian legend. Swiss-American climber Norman Dyhrenfurth—whose tent supposedly had been rattled by a wildman in 1952—promised that the Swiss team he was leading up Mount Lhotse, Everest's southern sister, would spend time investigating the mystery that surrounded the Yeti. And French teams looked for the Yeti during their explorations of Mount Makalu.[35]

In 1956, American Tom Slick, scion of a wealthy oil family, visited India and started plotting a hunt for the Yeti using helicopters and dogs; *Life* magazine had offered him $25,000 for the first photograph of a Yeti. Slick was a larger-than-life character: Texan, rancher, millionaire, cofounder of Slick Airways, friend of Howard Hughes, international playboy, peace advocate, and scientific philanthropist. He endowed the Southwest Research Institute, at its founding the second largest research center in the United States, and the Foundation for Applied Research, later called the Southwest Foundation for Biomedical Research.[36]

For all he did to support scientific research, Slick had a dilettante's simplistic view of science. He believed that science in America was too cautious.

**34.** Ralph Izzard, "The Snowman Hunters Sift Their Evidence," June 10, 1954, *Daily Mail* (third quotation); Charles Stonor, "I'm Fully Convinced the Yeti Is There," *Daily Mail*, July 12, 1954 (first quotation), both in Abominable Snowman clippings file; Shackley, *Still Living?* 57 (second quotation).

**35.** Christian Bordet, "Traces de Yeti dans l'Himalaya," *Paris, Museum National d'Histoire Naturelle Bulletin* 27 (1955): 433–39; "To Climb Lhotse," *New York Times*, February 18, 1955; "Japanese Hunting for Abominable Snowmen," *Los Angeles Times*, November 27, 1959; "New Snowman Clues," *Life*, February 15, 1960, 83–84; Napier, *Bigfoot*, 51, 66–72; Peter Byrne, *The Search for Big Foot: Monster, Myth or Man?* (New York: Pocket Books, 1976), 97; Dmitri Bayanov, *In the Footsteps of the Russian Snowman* (Surrey, BC: Hancock House, 1996).

**36.** Coleman, *Tom Slick and the Search for Yeti*, 1–32.

He wanted a revolutionary science. So he sought out undiscovered animals throughout the world, thinking he could prod a leap that way. The Yeti, he said, was the missing link, the discovery of which would revolutionize anthropology. It was a desire born of hope: that great mysteries remained yet unsolved, that the world had not yet been fully explored, that monsters existed unknown to science, their discovery promising huge leaps in knowledge. It was, as well, a desire born of ignorance. The term "missing link" was in disrepute among biologists and anthropologists, as they had already discovered a number of transitional forms between apes and humans, and, anyway, the phrase was certainly not applicable to the Yeti, which was clearly an ape, *Gigantopithecus*, perhaps, or a relative—but not a human ancestor.[37]

Slick's initial forays into Nepal were tied up in red tape. But he did learn of Byrne; the two swapped correspondence and made plans for a hunt. Byrne had been with the Royal Air Force before joining a British tea company, but then went into big game hunting, "the life of a tea planter" being, he said, "less attractive after Indian independence." Meanwhile, the American anthropologist Carleton Coon was visiting India on his "Faces of Asia" tour, taking pictures of the peoples of Asia for the U.S. Air Force so that downed pilots could learn to identify their position based on physiognomies of the locals—a way of asserting control over an area not so different from measuring the mountains of the Himalayas or conquering Everest; Americans had been slow to come to the area but were making up for lost time. Coon was also a consultant for *Life* on matters anthropological. Someone in Henry Luce's Time-Life Corporation had reconsidered the tender for the Yeti photograph and asked Coon for his opinion of Slick and his chances for success. Coon investigated and, although he thought that the Yeti probably existed, advised the company to rethink its offer. Slick was "a very nice guy," he thought, but his expedition was "inadequately staffed" and his plans "impractical." The magazine eventually withdrew its offer. Coon, however, impressed upon Slick how very much Byrne wanted to capture a Yeti—a recommendation that Byrne credited with cementing his and Slick's relationship.[38]

**37.** Peter J. Bowler, *Evolution: The History of an Idea* (Berkeley: University of California Press, 1989), 322–25; Coleman, *Tom Slick and the Search for Yeti*, 1–32.

**38.** Liz Barker to Coon, October 29, 1957; Byrne to Coon, June 10, 1957, both in General Correspondence A–E 1957 file, Carleton S. Coon papers; Byrne, *The Search for Big Foot*, 97–98; Carleton Coon, *Adventures and Discoveries: The Autobiography of Carleton S. Coon* (Englewood Cliffs, NJ: Prentice Hall, 1981), 231, 298, 317–18 (quotations on 318); Coleman, *Tom Slick and the Search for Yeti*, 53–58; David Price, "Interlopers and Invited Guests: On Anthropology's

FIGURE 8. Discoveries such as those by Eric Shipton convinced some academics that wildmen were worth studying seriously. Carleton Coon, shown here in his University of Pennsylvania office, was one of the first anthropologists to suggest that the Abominable Snowman existed. (Image S4-54942. By permission of the University of Pennsylvania Museum.)

Despite the loss of *Life*'s support, Slick, Byrne, and N. D. "Andy" Bachkheti (superintendent of the Delhi Zoological Park) spent a month in the relatively unexplored Arun Valley looking for the Yeti. Slick interviewed fifteen people who claimed to have seen the Abominable Snowman and showed them photographs of various animals that were similar to the Yeti. They all chose the gorilla as most closely resembling the Yeti, followed by an artist's rendering of *Australopithecus* and an orangutan. They recognized both bears and langurs and stated categorically that these were not Yetis. The three men also found tracks and what they took to be Abominable Snowman dung.[39]

The evidence was enough to convince Slick that a bigger expedition was warranted. He corralled fellow Texas oilmen and big-game enthusiasts

Witting and Unwitting Links to Intelligence Agencies," *Anthropology Today* 18, no. 6 (2002): 16–21.

**39.** Coleman, *Tom Slick and the Search for Yeti*, 59–67.

F. Kirk Johnson Sr. and Jr. to contribute funds to the hunt; he also put together a team of scientific consultants—coordinated by the American anthropologist George Agogino and including Heuvelmans and Coon—to evaluate whatever evidence his hunters found. And he made sure that the events stayed in the news, arranging for William Randolph Hearst's *New York Journal-American* to publish weekly articles by himself and Byrne about the expedition. Slick was confident, betting a friend $1,000 that the creature would be caught by the end of the year. He did not go into the field this time, however. His biographer, Loren Coleman, supposed that Slick did not join the team because his strong-willed mother, worried about his safety, forbade him from going back into the Himalayas. His absence also might have been a reflection of his leadership style, developed while he was in the navy during World War II: "I start things, then turn them over to someone else to run," he told the *Houston Post*.[40]

Early in February 1958, Gerald Russell, veteran of the *Daily Mail* expedition, led Byrne, Byrne's brother Bryan, fifteen Sherpas, seventy-five porters, and a government liaison into the Arun Valley, where a large species of Abominable Snowman was reported to live, a monster that grew to over seven feet tall and terrorized the locals; the mountaineer Norman Dyhrenfurth and Gerald Holton, an American photographer, caught up with the team later. In the field, the expedition divided into small parties. According to Byrne, Russell wanted the white hunters to disguise themselves as natives. "We will wear rough woolen Sherpa vests, woolen hats, and felt Tibetan knee boots," Byrne told *Journal-American* readers. "Our faces will be stained brown. And, while hunting, we will all try to act like the wandering yak herders or Sherpa berry pickers." On the *Daily Mail* expedition, Russell had noted that wolves approached the Sherpas but fled when they saw whites and reasoned that the disguises would "make all the watchful creatures of the Himalayas—including the Yeti—take [the hunters] to be no more than locals" and allow them to get close enough for photography.[41]

As they wandered through the Arun Valley and its tributaries, the hunters found what they took to be abundant evidence that the Abominable Snowman existed. Dyhrenfurth and the Sherpa Ang Dawa discovered a cave lined

**40.** Jim Mousner, "Tom Slick Proves Independence of Research, Science," *Houston Post*, n.d. (quotation), Bigfoot file 7, Andrew Genzoli papers, Humboldt State University, Arcata, CA; Tom Slick, "The Search for the Abominable Snowman," *New York Journal-American*, April 20, 1958; Loren Coleman, *Tom Slick: True Life Encounters in Cryptozoology* (Fresno, CA: Craven Street Books, 2002), 85–86, 95, 109–15, 204–5.

**41.** Peter Byrne, "The Search for the Abominable Snowman," *New York Journal-American*, April 27, 1958 (quotations); Coleman, *Tom Slick*, 101–4.

with junipers and filled with droppings. "You'd have to be immensely strong to pull those juniper branches out of the ground," Dyhrenfurth said later. "We tried. We couldn't. The Yeti must be stronger than a man." Meanwhile, Russell and Da Temba, the Sherpa working with him, found what looked to be small Yeti tracks. Some locals testified that Yeti were common at the edges of rivers, where they scrounged for frogs.[42]

In April, Russell met a man who claimed to have seen a Yeti only the night before. Russell had Da Temba and the witness scout the area the next night—when frog hunting was best—while he rested for his early morning watch. "After patrolling up stream then down stream without seeing anything, and about to turn off from the stream to the camp about 30 yards away [Da Temba and the local] noticed a wet footprint on a stone and soon after saw a small Yeti in the torch light 10 yards away," Russell recounted in a report to Slick. "The Yeti took one step towards them whereupon they ran and spent the night in a small settlement a few hundred yards away."[43]

Russell, Da Temba, and the man who had first seen the Yeti stayed up the next night, watching. They saw nothing but found tracks the next morning. Russell and Da Temba continued their vigil the following two nights, while Peter and Bryan Byrne hightailed it to Russell's camp. "We are making a forced march to be—we hope—in on a capture of the Snowman," Peter Byrne wrote in the *Journal-American*. Was it a small variety of Yeti? the Byrnes wondered. Or was it the young of the large variety? The surrounding mountain villages, Peter Byrne had learned, were "full of accounts of the creature's strength and habit of killing and mutilating men." Was that why Da Temba had run? Because he was frightened? The Byrnes arrived a few days later; shepherds had settled their flocks in the area, but Russell and Da Temba were still finding tracks, although they had not yet glimpsed their maker.[44]

Byrne's article about his march to Russell's camp did not appear in the newspaper until May 18, 1958. That same day, the *Journal-American* reported that Slick had just received a coded message from the expedition indicating that Byrne had pressing information. "We have a code fitting all important situations, including the capture of a Snowman and evidence that it is a missing link," he told the paper. But he did not divulge the nature of the message.

**42.** Peter Byrne, "The Abominable Snowman," *New York Journal-American*, May 11, 1958; Gardner Soule, *Trail of the Abominable Snowman* (New York: G. P. Putnam's Sons, 1966), 35 (quotations).

**43.** Coleman, *Tom Slick*, 105.

**44.** Peter Byrne, "Searching for the Abominable Snowman," *New York Journal-American*, May 18, 1958 (quotations); Coleman, *Tom Slick*, 105.

Had his team captured an Abominable Snowman? Was the creature real—not Sherpa folklore, not the creation of greedy journalists, but a living, breathing, eating, shitting zoological specimen? "Next Sunday," the *Journal-American* promised, "another fascinating report from the Slick, Johnson Expedition."[45]

There was an article the following week, but it was anticlimactic. Peter Byrne claimed that the sheep and goats had obliterated the Yeti tracks and scared off the little Yetis. The night of their arrival, the Byrne brothers sat up with Russell and Da Temba, but they saw nothing, no Yeti, no footprints. Russell left the next day, headed for home. Peter spent the following night "huddled in the hollow of a big rock," watching. "The rain was cold and persistent and the roar of the waterfall drowned out all sounds. The moonlight came through the rain clouds only fitfully." He never saw the Abominable Snowman.[46]

Despite the disappointment, Slick told the hunters to stay in the field as long as possible. The Byrnes remained on the Choyang River. A few weeks after Russell's departure, they found tracks again. These ran along the river "to a flat rock on which were the remains of a half-eaten frog. Toe prints were clearly visible in the sand." Nearby rocks had been overturned—maybe by the Yeti looking for food, the hunters speculated. "Some were so large that it took two of us to move," Peter Byrne wrote in the *Journal-American*. That night, and others, the Byrnes baited rocks with frogs and spied on the river from blinds, but they never caught the creature, never even saw it. At the end of the month, they met Dyhrenfurth, who had found what he thought was Yeti fur, and together visited monasteries in Thyangboche and Pangboche, where they photographed Yeti scalps and saw, as well, a purported Yeti hand.[47]

The expedition ended in June. As others before them, Slick's hunters had not caught the Yeti, but they were satisfied with their results. Based on what he had seen and what the locals had told him, Russell surmised that as many as four thousand small Yetis inhabited the Himalayas. Dyhrenfurth was also convinced that the larger, fiercer species of Abominable Snowman existed. The evidence was persuasive enough to Slick—and the Kirk Johnsons—that they tapped the Byrne brothers to go into the field again, just the two of them, living off the land, no tents, no food. Shortly before Christmas, Bryan and Peter left behind—as Peter said—"the delights of a Kathmandu hotel and the

**45.** "Big News Expected!" *New York Journal-American*, May 18, 1958.

**46.** Peter Byrne, "Bad Breaks Slow up Search for Brute," *New York Journal-American*, May 25, 1958 (quotations); Coleman, *Tom Slick*, 105.

**47.** Peter Byrne, "Frogs Lure Abominable Snowman," *New York Journal-American*, June 5, 1958 (quotations); "Big News Expected!"; Coleman, *Tom Slick*, 103–8.

fleshpots of an eastern city and its swinging international set" for a Spartan existence and $100 per month each because they were "very keen young men, dedicated to a task to which [they] were prepared to devote [their] lives, to what Tom Slick . . . called the Ultimate Quest." So strong was their conviction that they stayed in the field throughout the winter. So strong was Slick's that over the years he sunk $100,000 into the hunt, according to his own accounting, and the Kirk Johnsons supposedly contributed another $100,000. "The Abominable Snowman exists," Slick wrote in the *Journal-American*. "Someday, and soon, he will be found."[48]

## What the Evidence Told

By the middle of the twentieth century, the global empire that the British had commanded for centuries was crumbling. India, Pakistan, Burma, and Ceylon were each granted sovereignty before Shipton found those tracks. But the British still saw themselves as key players in the geopolitical game. Broadly speaking, they thought that centuries of imperial rule had made them wise in the ways of the world, sagacious about the honeypots that tempted colonial powers, the intractability of many social problems, the complex and often dark motives of people, the varieties of human experience. That was the point of Izzard praising the color scheme of his expedition. It was a hopeful image: that Britain still led the world.[49]

In contrast, the British saw Americans as superficial and garrulous, naive and concerned only with making money; the United States' growing international clout—taking pictures of everyone in Asia, competing in the Himalayas, involved in Vietnam and elsewhere—was well intentioned but bumbling. The British empire of enlightened rule was being replaced by an American empire that used mass media to turn the world into consumers of the goods that Americans produced. Americans were the antithesis of the Yeti, representing not authenticity but the plasticity of the contemporary age. The point was illustrated well in the British movie *The Abominable Snowman of the Himalayas*, which pitted a sensitive British botanist against an American, Tom Friend, who was, as one character said, "nothing but a cheap fairground trick-

**48.** Betty Allen, "Pacific Northwest Scientific Expedition," 1959, Bigfoot file 7, Andrew Genzoli papers; Tom Slick, "Elusive Snowman Still in the Hills," *New York Journal-American*, June 22, 1958 (penultimate and final quotations); Byrne, *The Search for Big Foot*, 101 (quotations); Coleman, *Tom Slick*, 106–7, 109, 116–17.

**49.** Denis Judd, *Empire: The British Imperial Experience, from 1765 to the Present* (New York: Basic Books, 1998), 13–15, 297–433.

ster," desiring a Yeti only so that he could display it. Driven by motives alien to British gentlemen, possessed of an almost willful ignorance, Friend brought grief to the expedition: he was responsible for the death of two expedition members, had a Yeti killed, and caused his own death, leaving the botanist alone to mourn the death of the magnificent Yeti.[50]

Some have speculated that Tom Friend was modeled on Tom Slick. They shared a Christian name and monosyllabic surnames that were also common words. But a direct connection was hardly necessary: Slick, like Friend, embodied postwar Britain's stereotype of newly powerful Americans. Slick burst onto the scene with his dogs and helicopters and guns and millions, promising to solve a long-standing mystery, pledging to capture a creature that the British had hunted for decades. But Slick had no experience with the Abominable Snowman; America had no tradition in the Himalayas. Unlike the Brits who had preceded him, Slick did not even go into the mountains himself, did not pit his ideals against the harsh environment. Izzard, who met Slick, thought him "a bit of a showman"—a devastating critique from Izzard, who thought, "It is impossible to combine sincere scientific investigation with circus showmanship." Ignoring that his own expedition had been about British pride, about making money, and that his book on the expedition had made reference to an Abominable Snowwoman, Izzard complained that Slick seemed "anxious to promote a big American prestige success. He seemed particularly interested in the possible existence of an Abominable Snow-woman. It was the first time I had even considered such a possibility but it dawned on me at once that a big breasted woman would arouse much more public interest than a man." So it is no wonder that, as Slick's biographer notes, the millionaire "received a good deal of bad press about his very Texan approach to the Yeti hunt, especially from British writers in England and India."[51]

But one need not be influenced by British chauvinism to conclude that Slick's expedition lacked scientific integrity. Some of the stories published in the *Journal-American* made no sense except as attempts to garner publicity. Why would Russell send Da Temba to hunt the Yeti instead of going himself?

**50.** Ariel Dorfman, *The Empire's Old Clothes: What the Lone Ranger, Babar, and Other Innocent Heroes Do to our Minds* (New York: Pantheon Books, 1983); Guest, *The Abominable Snowman of the Himalayas*; Matthew Fraser, *Weapons of Mass Distraction: Soft Power and American Empire* (New York: Thomas Dunne Books, 2003); Graham Greene, *The Quiet American* (New York: Penguin Classics, 2004).

**51.** Izzard, *The Abominable Snowman*, 68 (second quotation); "It's Abominable Snow Woman!" *Chicago Tribune*, April 25, 1957; Coleman, *Tom Slick*, 84 (final quotation), 104 (first and third quotations).

Why would Da Temba run hundreds of yards to a village instead of thirty yards to get Russell? Slick's announcement that he had received important news appeared in the newspaper the same day as the report of Da Temba's sighting—but that was a week after the Byrnes had arrived in Russell's camp and been disappointed. Since it took runners only four days to reach Katmandu from the expedition's camps, more than enough time had passed for Slick to learn what had happened. So what was the news? Nothing ever followed. Don Connery, a writer for *Sports Illustrated*—another in Henry Luce's stable of magazines, along with *Life*—thought that the whole episode a little too fortunate for the expedition's contract to provide the newspaper with good copy.[52]

Slick's consultants also doubted his probity. "I don't trust him," George Agogino told Carelton Coon. Slick insisted that all of his consultants keep their studies confidential, even from other researchers. If Slick's later actions can be used to explain this insistence, he wanted to keep the matter out of the press—except on his own terms—so that he could sell the story of the wildman's capture later. (According to Connery, Dhyrenfurth came down from the mountains claiming "he had in his possession 'proofs which will convince scientists of existence of human like creature of Himalayas' but refused to produce same being honor bound by" Slick's contract with the newspaper.) Agogino worried that publicity concerns would trump scientific analysis, worried that Slick would go to the press and claim that his scientific consultants said something that they had not. He kept a file of all his work on the Yeti in case he needed to disprove something Slick said publicly, he had the other consultants carefully check statements that Slick attributed to them, and he bucked Slick's confidentiality rules when he thought it necessary. For example, he had Coon make a plaster cast of a footprint so that he could study it at his leisure although he knew that Slick would fire him if he found out that Agogino "had duplicates of anything." Agogino only—and barely—tolerated the restrictions because he thought that he could bring integrity to the study of the Abominable Snowman and keep it out of the "hands of crackpots and publicity seekers."[53]

The evidence gathered during the expedition did not do much for Slick's reputation, nor did it make the case for the Abominable Snowman's existence

**52.** Don Connery to Earl Burton, June 20, 1958 (first quotation), General Correspondence A–B 1958 file, Carleton S. Coon papers.

**53.** Agogino to Coon, March 7, 1959 (second quotation); Agogino to Coon, May 15, 1959 (first and third quotations); Agogino to Coon, September 22, 1959; Agogino to Coon, October 2, 1959; all in General Correspondence, A–E 1959 file, Carleton S. Coon papers; Betty Allen, "Pacific Northwest Scientific Expedition," 1959, Bigfoot file 7, Andrew Genzoli papers.

any stronger than had Izzard's hunt. One purported Yeti hand turned out to be the paw of snow leopard. Analysis of the Yeti fur was inconclusive (and possibly fabricated; years later, the primatologist William Montagna said that he had studied supposed snowman fur provided by Slick and had concluded that the sample was not fur at all but fibers). Examination of the stool samples was also inconclusive, although a number of Slick's scientific consultants thought it unlikely that the droppings came from any kind of primate. The stool had parasite eggs, one of which—according to a French scientist—was from a parasitic worm unknown to science. From this evidence, Heuvelmans concluded that Slick had found the feces of a mysterious beast: "Since each species of mammal has its own parasites, this indicates that the host animal is equally an unknown animal." But Heuvelmans's application of Zadig's method was poor: there is not a one-to-one correspondence between parasites and their hosts. Animals have many parasites and even a known animal might harbor some parasites that have not yet been discovered by scientists, classified, and named. The parasite eggs proved nothing, except that Heuvelmans may not have been as unbiased in his examination of evidence as he claimed.[54]

Initially, pictures of a Yeti hand from the monastery in Pangboche excited observers. W. C. Osman Hill, a British zoologist, had studied the photographs and thought the relic suggestive of a "an unknown anthropoid." If he could see the hand itself, he could be more certain—the hand, like Cuvier's hoof print, a certain clue. Early in 1959, Peter Byrne visited the Pangboche monastery and while studying the hand replaced a phalanx and thumb with bones from a human, pocketing the originals and packing them out of Nepal. In Calcutta, the Byrne brothers met the actor Jimmy Stewart and his wife Gloria; they were friends of the Kirk Johnsons and happened to be traveling through the area. The Stewarts agreed to smuggle out what Byrne later called "the grisly trophy" and passed it to Osman Hill when they returned to London. Upon seeing the bones, Osman Hill vacillated but ultimately changed his mind: he told Slick that they were human bones, not from an unknown anthropoid after all. Agogino disagreed; the metatarsals were flat—a characteristic of apes, not humans—and large. Coon thought that while the bones were bizarrely shaped, they were within the range of normal human variation, as did Fred Ullmer, a mammalogist with the Philadelphia Zoo. (Unfortunately, chemical analyses of the skin were inconclusive and so could not settle the controversy.)[55]

**54.** Osman Hill, "Abominable Snowmen," 95; William Montagna, "From the Director's Desk," *Primate News* 14, no. 8 (1976): 7–9; Coleman, *Tom Slick*, 116–25 (quotation, 118).

**55.** Agogino to Coon, March 7, 1959; Agogino to Coon, May 15, 1959; Agogino to Coon, May 28, 1959, all in General Correspondence A–E 1959 file, Carleton S. Coon papers; Osman Hill,

In the spring of 1959, Slick sent Agogino the cast of a footprint that had been made on the 1957 reconnaissance. The cast had dirt in it, suggesting that it had been made on a riverbank, as Slick claimed, and between the toes Agogino could see impressions left by tufts of hair. One of Slick's consultants, Adolf Schultz, a German scientist, whom Coon thought knew "more about primate feet than anyone else in the world," concluded that the track had been left by a panda—maybe a new species of panda, but still a panda, not a primate. Agogino recognized that some parts of the cast were "bear-like," but thought that Schultz had been mislead by some idiosyncrasies of the track. Others also doubted Schultz's conclusion—but no one knew what had made the track. Damning with faint praise, Coon called the cast "the first really concrete piece of evidence turned up" by Slick's expedition.[56]

Osman Hill, Agogino, and Coon continued to believe that the Abominable Snowman might exist, but the evidence was only suggestive at best, not conclusive. "The goodies brought back by [Slick's] expedition[s]—the mummified paw of a snow leopard, a mummified human hand and a footprint or two—add up to nothing at all," said John Napier, accepting Osman Hill's description of the Pangboche relic as human. "No single item contributed one jot or tittle of proof." The wildman of the Himalayas was no longer a local legend—British and American mass media had promoted it to international stardom. But what, exactly, the creature was remained a mystery. It was a fairy tale, but also endorsed by men hardheaded enough to pit their ideals against the world's most unforgiving landscape. It left traces, but those traces did not quite form themselves into a coherent image. The Abominable Snowman was Barnum's What-Is-It for a new generation, offspring of a mass media that could bring the world to everyone's doorstep, but also made that world a little harder to gauge, a little more plastic and unreal.[57]

"Abominable Snowmen," 95; Soule, *Trail of the Abominable Snowman*, 37; Byrne, *The Search for Big Foot*, 101–2 (second quotation, 102); Coleman, *Tom Slick*, 121–24 (first quotation, 124).

**56.** Agogino to Coon, May 15, 1959 (second quotation); Agogino to Coon, May 28, 1959, both in General Correspondence A–E 1959 file; Coon to Adolf Schultz, May 27, 1959 (first and third quotations), General Correspondence, S–Z 1959 file; Agogino to Coon, January 5, 1960, General Correspondence A–D 1960 file, all in Carleton S. Coon papers; Coleman, *Tom Slick*, 119–20.

**57.** Osman Hill, "Abominable Snowmen," 97; Napier, *Bigfoot*, 52 (quotation); George Agogino, "An Overview of the Yeti-Sasquatch Investigations and Some Thoughts on Their Outcome," *Anthropological Journal of Canada* 16 (1978): 11; Coon, *Adventures and Discoveries*, 318; John P. Jackson, "'In Ways Unacademical': The Reception of Carleton S. Coon's *The Origin of Races*," *Journal of the History of Biology* 34 (2001): 253.

CHAPTER THREE

# Sasquatch 1929–1958

About the same time that word of the Abominable Snowman first raced around the globe on newswires, another wildman was gaining some attention in Canada, albeit on a much smaller scale. John Burns, a teacher on the Chehalis Indian Reservation, near the resort town of Harrison Hot Springs in British Columbia's Fraser River Valley, was asking his First Nations students, their families, and their friends about a wildman said to haunt the region. He had heard about the creature from local anthropologist Charles Hill-Trout and, intrigued, started gathering all the stories that he could. In 1929, he compiled some of the tales into an article for the Canadian magazine *Maclean's*. Burns called the wildman Sasquatch, an Anglicization of the giant's name in the Stalo dialect of Halkomelem language, sɛ́sq̓əc.[1]

Superficially similar stories about wildmen could be heard from Native American tribes up and down the Pacific coast—such creatures are, after all, almost universal archetypes. These legendary giants were complex figures that nestled into the specific worldview of particular tribes, playing different roles in different tribes' mythologies. Most often the wildmen and ogresses were reported to be nocturnal, but not always. Some were completely covered in fur; others were no more hairy than a human. Some abducted women and children; some aided ostracized young men. Some talked; some only whistled. Some had legs that could not bend, so that they could run only downhill, while others had spiked toes.[2]

**1.** J. W. Burns, "Introducing B.C.'s Hairy Giants," *Maclean's*, April 1, 1929, 9, 61–62; Wayne Suttles, "On the Cultural Track of the Sasquatch," *Northwest Anthropological Research Notes* 6 (1972): 65–90; John Green, *Bigfoot: On the Track of the Sasquatch* (New York: Ballantine, 1973), 4–5; Christopher L. Murphy, *Meet the Sasquatch* (Surrey, BC: Hancock House, 2004), 31–33.
**2.** Franz Boas, *Kwakiutl Tales*, vol. 2 (New York: AMS Press, 1969), 117–22; Bruce Rigsby, "Some Pacific Northwest Native Language Names for the Sasquatch Phenomenon," *Northwest*

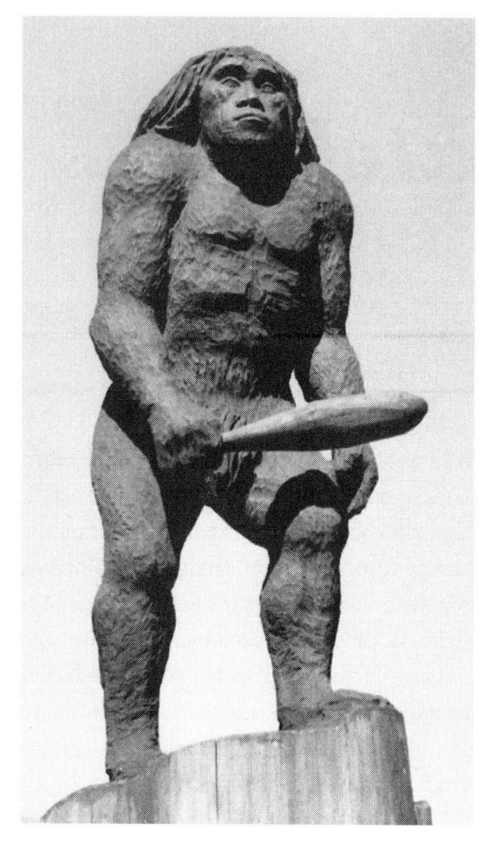

FIGURE 9. Around the same time that British expeditions generated interest in the Abominable Snowman, tales were being told in the Canadian press of a local wildman, called Sasquatch, by J. W. Burns, a teacher on the Chehalis Indian Reservation. This statue shows Sasquatches as Burns imagined them: a race of uncivilized Indians, bestial and dangerous, but still human. (Image I-51761. Courtesy of Royal British Columbia Museum, BC Archives.)

Burns pared down this diversity. In his rendering, Sasquatches were a race of giants living in the high mountains. Their appearance was uniform: "Eight feet tall, covered head to toe with black wooly hair," their faces, according to one of Burns's informants, "negro black." The giants were uncivilized, but they could speak. And they possessed magical powers. One whose baby was accidentally shot by a successful hunter, for instance, cursed the man so that he could never kill a bear again. The First Nations people whom Burns interviewed lived in fear of the giants. Sasquatches warred with them, hectoring them with volleys of rocks, coveting their women. Burns spoke with a

*Anthropological Research Notes* 5 (1971): 153–56; Wayne Suttles, "Sasquatch: The Testimony of Tradition," in *Manlike Monsters on Trial*, ed. Marjorie Halpin and Michael M. Ames (Vancouver: University of British Columbia Press, 1980), 245–54; Claude Lévi-Strauss, *The Way of the Masks* (Vancouver: Douglas & McIntyre Ltd., 1982).

woman who claimed to have been abducted by Sasquatch. She lived with him and his elderly parents for twelve months. Finally set free, she returned home and gave birth to a child. It "lived but for a few hours," the woman told Burns, "for which I was glad. I hope that I never again shall see a Sasquatch."[3]

Over the next fifteen years or so, Burns published several more articles about the wildman, and what had been a local legend became something more—a minor celebrity, although not one that could rival the Yeti's popularity. All through the middle of the twentieth century, stories about Sasquatch appeared in Canadian and American media. In 1934, for instance, two brothers from the University of California's medical school announced that they were going to hunt the wildman for the school's anthropology department. The area around the Chehalis Reservation seized on the Sasquatch tales as a way to entice tourists. Along Highway 7 there was a Sasquatch Inn, and, in 1938, Harrison Hot Springs hosted Sasquatch Days, featuring a First Nations ceremony.[4]

Burns eventually left the area, but the legend he had helped to create and popularize persisted, growing in ways that Burns could no longer control. Burns apparently believed that Sasquatch existed, and offered his accounts of the beast as a bit of secret knowledge, information passed to him from natives because they respected him. The rest of the world, however, took Sasquatch to be a creature of fable. It probably didn't help that the *Maclean's* article appeared on April Fool's Day, 1929, making the stories seem a hoax, an Indian legend. The tales traded about the beast in British Columbia often took the form of jokes and ribald stories. There's even some doubt whether the Californian medical students meant their expedition to be anything more than a joke—they said that they were going to lasso a Sasquatch, hardly a serious proposition, and anthropologists at the University of California said that they neither knew of the Sasquatch or the brothers.[5]

Sasquatch might have faded away into obscurity—fated to become a small-time legend of only local interest—were it not for the Swiss émigré

**3.** Burns, "Introducing B.C.'s Hairy Giants," 9, 61 (second quotation); Green, *Bigfoot*, 4–5; Murphy, *Meet the Sasquatch*, 31–34 (first quotation, 31; third, 33).

**4.** "Heavens! That Horrible Old Sasquatch Is at Large Again," *Washington Post*, April 9, 1934; "Giant Wildmen Reported Seen," *Washington Post*, June 18, 1934; "'Whoopee' with Grass Skirts and Totem Poles in British Columbia," *Los Angeles Times*, June 19, 1938; Green, *Bigfoot*, 4–5, 35.

**5.** Yvonne H. Stevenson to University of California, 1943, and Edward Winslow Gifford to Yvonne H. Stevenson, 1943, both in Stevenson file, box 140, Anthropology Department Records, CU-23, Bancroft Library, University of California, Berkeley; Burns, "Introducing B.C.'s Hairy Giants," 9, 61–62; "Luck to Sasquatch," *Washington Post*, April 10, 1934; "Tales from Tall Timber Heard in Northwest," *Washington Post*, April 14, 1949.

René Dahinden and the Canadian journalist John Green. In the late 1950s, the two men started investigating Sasquatch. They never saw the wildman—only gathered tales from others and studied tracks—but they were confident in their ability to ferret out the truth. And the truth was, they claimed, that Sasquatch was real. Coming at a time when the Abominable Snowman was making international headlines, Dahinden and Green's conclusions thrust Sasquatch back under the klieg lights of modern mass media and made the Canadian wildman into a celebrity.

## The Great Sasquatch Hunt

Born in Lucerne, Switzerland, on August 23, 1930, René Dahinden lived a Dickensian childhood, into and out of orphanages and foster homes, taken back and rejected by his mother twice, put to work on a farm where he was "five steps lower than a dog," but where he was also made tough. "Later I wrote and thanked those people," he said. "Compared to there, everything I met was a joke." After graduating from the farm, Dahinden wandered across Europe before emigrating to Canada in October 1953 and settling on another farm, that of Wilbur Willick in Calgary. Two months later, he heard of the *Daily Mail*'s expedition to catch the Abominable Snowman. "Something clicked in me then," he said later, "and, looking back, it seemed that maybe I'd been searching all my life for a chance like that, a chance to really accomplish something." He said to Willick, "Now wouldn't that be something; to be on the hunt for that thing?"[6]

Willick told him that he didn't have to travel all the way to the Himalayas for a crack at a wildman. One was reported to live on the West Coast, he said, probably remembering Burns's articles. Dahinden thought that Willick was joshing him—pulling one over on the immigrant. But he couldn't let the subject go. Barbara Wasson, one of Dahinden's closest friends and a clinical psychologist, said she thought that the immigrant saw Sasquatch as an "ambiguous father image"—the search for the hairy wildman, at least initially, was a response to being born a bastard, rejected time and again, looking always for that animal that was hidden, elusive, and wary of entanglements.[7]

**6.** James Halpin, "Sasquatch," *Seattle*, August 1970, 31–34 (second quotation, 34), 58–59; Don Hunter and René Dahinden, *Sasquatch* (Toronto: McClelland & Stewart, 1973), 74–78 (remainder of quotations).

**7.** Barbara Wasson, *Sasquatch Apparitions* (Bend, OR: privately printed, 1979), 31.

Dahinden dogged his employer with questions. Finally convinced that he wasn't having his chain yanked, Dahinden moved to British Columbia in the spring of 1954, taking up a series of jobs that consumed most of his time. "But the issue kept floating around in my head," he later recalled. High spirited, compact, and muscled, Dahinden shrugged off all those who said the Sasquatch was only an Indian legend, sneers and rolling eyes being nothing compared to what he had endured as a child. His independent streak was legendary. Once, on a radio program, an audience member told him that two hundred million people thought that he was wrong, Sasquatch didn't exist. Without missing a beat, Dahinden responded, no, there were two hundred *and twenty* million people against him, and every one of *them* was wrong.[8]

In 1956, Dahinden's search for information on Sasquatch took him to the office of the *Agassiz Advance*, where he met John Green. Lanky, with a rubbery face, Green was the son of a politician, a graduate of the prestigious Columbia School of Journalism, and the editor of the *Advance*. Like many, probably most, residents of the area, Green thought that Sasquatch was a legend, an imaginary bogey—the previous April Fool's Day he had run a prank story about a Sasquatch that made off with a nubile guest of the Harrison Hot Springs Hotel—and he told Dahinden so. The immigrant's "visit made a good story," Green later wrote, "but I felt rather sorry for him."[9]

Green found himself reconsidering his opinion of Sasquatch the following year. In 1957, British Columbia began gearing up for its centennial anniversary. The provincial government offered grants for local communities to create memorial projects. Harrison Hot Springs was small, and so was slotted to receive a small grant, only $600. The village council settled on the idea of using the money for a Sasquatch hunt. The provincial government eventually rejected that idea, and instead the money went toward the purchase of a furnace for the community hall. But before the village's suggestion was denied, word of the hunt reached the press. Coming amid the repeated attempts to capture the Abominable Snowman, Harrison Hot Springs' proposal attracted a lot of notice, locally, nationally, and internationally. "Newspaper and radio reporters flocked around," Green wrote. "Papers all over Canada played the story on the front page. There were numerous offers

**8.** Hunter and Dahinden, *Sasquatch* (Toronto: McClelland & Stewart, 1973), 74–78; "The Bigfoot Debate—Part II," *The ISC Newsletter* 10, no. 4 (1991): 2.
**9.** Green, *Bigfoot*, 5, 35; Hunter and Dahinden, *Sasquatch*, 77–78; John Green, *Sasquatch: The Apes Among Us* (Seattle, WA: Hancock House, 1978), 49 (quotation).

from would-be Sasquatch hunters, even from young ladies prepared to act as 'bait.'" Dahinden returned to lead the expedition.[10]

John Burns, now living in San Francisco, was irritated by the hunt. He didn't appreciate people referring to Sasquatches as monsters or preparing to track them down with dogs and expose them "to the gaping gaze of the curious." Sasquatches, he insisted, were "harmless people of the wilderness." That they were being treated like circus freaks raised the old question: Who is the real monster—the freak, or the one who pays to gawk at the geek? "Our veneer of civilization," he reported sadly, "does not hesitate to even use monsters for commercial purposes."[11]

Many locals had no such objections. The hunt for Sasquatch was supposed to be a circus. Gaping looky-loos were what the village council wanted. The Fraser River Valley was changing, its reliance on timber and farming giving way to tourism and a service economy. Reports of a Sasquatch hunt and the attendant publicity was good for business. *Vancouver Providence* columnist Eric Nicol wrote a comic article suggesting that the hunt wouldn't result in the capture the wildman, but would turn the hunters wild: they'd get drunk and get frisky with the local women—or the bait, as it were. "After all," he wrote, "that tangled jungle behind Harrison Lake can do strange things to a white man"—like turn mild-mannered men into consumers.[12]

But the transmutation of the unreal into money was not this circus' best trick—nor was it the celebration of a downtrodden people's mythology or the transformation of white men into savages. Better than all that, the hunt for a legendary monster brought out evidence that the monster wasn't legendary at all, but real. "I was quickly exposed to the fact that there were local people who took the Sasquatch very seriously indeed, and not all of them were Indians," Green said, and that exposure started him thinking, his perceptions changing, as though he was in a hall of mirrors and everything looked different: Sasquatch, journalism, investigative skills, Dahinden, the role of scientists in society, truth.[13]

**10.** Green, *Bigfoot*, 5–6 (quotations); Hunter and Dahinden, *Sasquatch*, 77–83; Green, *Sasquatch*, 49–64.

**11.** Alex MacGillivray, "'Shouldn't Be Captured,'" *Vancouver Sun*, May 25, 1957, http://www.bigfootencounters.com/articles/jwburns2.htm (accessed February 9, 2008).

**12.** John Green, *The Sasquatch File* (Agassiz, BC: Cheam, 1973), 5; Hunter and Dahinden, *Sasquatch*, 79 (Nicol quotation); John A. Cherrington, *The Fraser River Valley: A History* (Madeira Park, BC: Harbour Publishing, 1992), 321.

**13.** Green, *Sasquatch*, 51.

Amid the excitement over the hunt, a game guide mentioned to Green that Jeannie Chapman, a First Nations woman from Ruby Creek—about twelve miles away—had seen a Sasquatch back in 1941. Also during the preparations for the hunt, a printer working for Green heard the same story from Esse Tyfting, the janitor at Agassiz's high school and someone whom Green considered estimable. "Thoroughly intrigued," Green later wrote, he spoke with Tyfting, who admitted that he had not seen the creature, only its tracks. Tyfting had been a railroad maintenance worker at the time and had been called to the Chapman house after the Sasquatch had gone. He sketched for Green what he had seen, a footprint sixteen inches long, four inches across the heel, and eight inches across the ball of the foot. The creature had stepped over a "fence more than four feet high without breaking stride." Green was a newsman, and this must have seemed a good story; he didn't have to believe in the beast, but other people were admitting that they did, and that was worth some ink. Along with Dahinden, he went to talk with Jeannie Chapman.[14]

The event had frightened her so much that the family had moved, and she was still scared when she spoke with them. She thought that Sasquatch was an omen of impending death. Jeannie Chapman told Green and Dahinden that on the day in question her child—in some accounts a daughter, in some a son—had returned home from playing outside saying that a "big cow" was making its way to their house; Chapman looked out and saw a giant two-legged creature. When it entered the family's storage lean-to, she gathered her children and ran to Ruby Creek station. Along the way, she crossed paths with her husband, George, also a railroad maintenance worker. He and some other men inspected the house, finding tracks and a broken barrel of salted salmon. Later, a deputy sheriff from Bellingham, Washington, investigated, bringing Tyfting along with him. In addition to interviewing Mrs. Chapman (twice), Green and Dahinden visited the Chapmans' old house, talked with a handful of other people, including the deputy sheriff's son, who showed them the file his late father had opened on the case. It included a sketch of a footprint that to Green's eyes looked like the one Tyfting had provided.[15]

The Chapmans' experience had made the papers at the time, John Burns covering it for the *Vancouver Province*—except that he didn't call the creature Sasquatch. He reported that it was a bear, "one of the largest bears ever known in the vicinity." (Apparently, Burns revised this opinion later.) After talking

**14.** Green, *Bigfoot*, 8–15 (first quotation, 11); idem, *Sasquatch*, 51–52 (second quotation, 52).
**15.** Green, *Bigfoot*, 8–15; Hunter and Dahinden, *Sasquatch*, 60–63; Green, *Sasquatch*, 51–52; Murphy, *Meet the Sasquatch* (Surrey, BC: Hancock House, 2004), 35.

with the witnesses, Green doubted this explanation. Only one of the people whom he interviewed thought that the beast was a bear, but that person admitted it had walked on two legs. Otherwise, the creature seemed like an ape or wildman. It was not the only matter on which Green thought that Burns had been wrong. He suspected that the Sasquatches were not a race of giant, hairy Indians, but animals, and so brought the evidence that he collected to the Provincial Museum and showed it to Ian MacTaggart, a zoologist.[16]

Green was not a scientist. He was not a mountaineer. And the Ruby Creek Incident—as it came to be called—was told to him by Indians, not the most trusted of witnesses by the standards of 1950s British Columbia. But Green had a way of making the stories seem more than legends—solid and real. The credibility came from Green's journalistic skills. Green thought that he had what Hemingway reportedly said any good journalist needed: a built-in bullshit detector. "Interviewing people and gathering facts is my regular occupation," he said, "and if I were being fooled very often my readers would be bringing it to my attention."[17]

The stories Green heard about the Ruby Creek Incident "rang true," he wrote, and that was revelatory: these weren't just stories, weren't just First Nations myths. They were stories that Green could test, could investigate, and they withstood his probing. Certainly, there were discrepancies in the stories, and Mrs. Chapman's account changed a number of times, but Green's stint as a court reporter had taught him that eyewitnesses often contradicted one another. That didn't invalidate the case; it just meant that memories were fallible and had to be tested. That the stories varied in some details made them seem more authentic, not less. The Ruby Creek Incident "forced me to begin taking the subject seriously," Green said, as though he had no choice. And so, confident in his new judgment, Green reached out to Dahinden, hiring him on to the staff of the *Advance* in December 1957. Together, they spent increasing amounts of time investigating Sasquatch.[18]

## The Kidnapping of Albert Ostman

Green and Dahinden contacted Bruce McKelvie, a British Columbian journalist who continued to keep tabs on Sasquatch after Burns's departure. "An admirable hack," according to one biography, McKelvie "was seldom averse

**16.** "Sasquatch Return Frightens Indians in British Columbia," *Long Beach (CA) Independent*, November 28, 1941; Green, *Bigfoot*, 9 (quotation), 39.

**17.** Green, *Bigfoot*, 8–15 (quotation, 13); idem, *Sasquatch*, 51–52.

**18.** Green, *Bigfoot*, 8 (quotation); Hunter and Dahinden, *Sasquatch*, 74–78.

to sensationalizing the past, based on fanciful research, if it meant getting more people to pay attention." The journalist told Green and Dahinden that he knew a man who had killed a Sasquatch but was sworn to secrecy so could say no more. He did, however, have another lead, putting them on the trail of Jacko.[19]

At the British Columbia archives, Green and Dahinden found an article from an 1884 edition of the *Vancouver Daily Colonist* detailing the capture of what appeared to be a young Sasquatch during railroad construction along the Fraser River. "Something of a gorilla type," according to the article, the "half man and half beast" creature was more than four and a half feet tall and weighed 127 pounds. It resembled a human being except that its entire body was covered with inch-long glossy fur. The railroad crew initially thought that it was a "crazy Indian" but then decided that they had found something altogether more interesting. They dubbed it "Jacko"—probably slang for monkey—and one of the men planned to bring it to London for exhibition. In the meantime, they fed Jacko berries and milk and withheld meat for fear he would turn savage.[20]

The story of Jacko wasn't as convincing as the Chapmans' tale, and there was little investigating to be done, but the article could still be tested. McKelvie said that he verified that everyone mentioned in the report had actually been living in the area at the time. And Green met a man who said that he had been a child when Jacko was caught. He had never gone to catch a peek of the beast but remembered the hullabaloo that greeted its capture. Additionally, the story added to Green's suspicions that the Sasquatch was a real animal—not a legendary giant—because the account did not rely on Native Americans at all. Apparently, Sasquatch had been seen by people of some credibility, people who could not be accused of being confused by the haze of First Nations lore.[21]

Another of those people was William Roe, who was inspired by Harrison Hot Springs' proposed hunt to write the *Advance* confessing his encounter with a Sasquatch. In October 1955, he said, while taking a break from his work on the highway near Tête Jaune Cache, at the confluence of the Fraser and Robson rivers in British Columbia, Roe wandered into the mountains and happened across a female Sasquatch, six feet tall, hairy, with large breasts and a head "that somewhat resembled a Negro's." She stripped leaves from a

**19.** "McKelvie, B. A." *ABC Bookworld*, 2003, http://www.abcbookworld.com/view_author.php?id=857 (accessed April 9, 2008); Green, *Bigfoot*, 35–38.
**20.** Green, *Bigfoot*, 35–38.
**21.** Ibid., 7, 37–38.

bush with her teeth. Roe almost shot the beast but thought better of it—she looked too human. Later, he followed her spoor to where she slept. Green learned that zoologists had corresponded with Roe about bison and considered him a creditable observer of wildlife.[22]

Green published Roe's account in his newspaper and a little while later, Albert Ostman, a retired logger, wrote to Roe, saying that he too had seen a Sasquatch—actually four—but had kept the story to himself for more than thirty years because he feared being ridiculed. Roe set up an interview between Ostman and another journalist. As preparation, Ostman gathered items from his past to help jog his memory and then wrote out a long, involved, and curious story.[23] This is the gist:

In 1924, Ostman set out for several weeks to prospect for gold on the Toba Inlet, British Columbia. One night, a Sasquatch grabbed Ostman's sleeping bag and carried the startled prospector for several hours, eventually dumping him inside a valley. It was, he said, eight or ten acres in area; the only way in or out was a V-shaped break in one wall of the montane bowl, not far from where he had been emptied from his bag. Four Sasquatches blocked him from his escape—a family unit that Ostman came to study over the next six days. Ostman named the Sasquatch that grabbed him the Old Man. He was the patriarch, huge and wary. His mate, the Old Lady, had "very wide hips, and a goose-like walk. She was not built for beauty or speed. Some of those lovable brasseries [*sic*] or uplifts would have been a great improvement on her figure," Ostman opined. The Old Lady collected grass and twigs and nuts for the rest of the family to eat. A young male, the son, liked to sit, grab his feet, and scoot along on his rump while his parents relaxed. He was inquisitive, the first to befriend Ostman. The prospector offered the young male an empty snuff can, which he took to his sister, the two of them learning how to open and close it. The young female was excitable and, unlike her mother, flat chested—"no development like young ladies," Ostman wrote. The Sasquatches communicated with each other through grunts and what Ostman thought was a primitive language. They tried to make their intentions known to him, as well. He thought that he might have been captured as a mate for the young female.

In time, the Old Man became curious about Ostman's snuff but seemed to "think it useless to only put it inside my lip." One morning, just after Ostman had taken a dip, the Old Man grabbed the can from him "and emptied it into

**22.** Ibid., 13, 16–20 (quotation, 18).

**23.** Green, *Sasquatch*, 97–99.

his mouth. Swallowed it in one gulp. Then he licked the box inside with his tongue." A few minutes later, his eyes rolled back; he put his head between his knees and rolled forward—giving Ostman just the break he needed. The prospector ran. The Old Lady tried to stop him, but Ostman scared her with a shot from his rifle and made it out through the V-shaped opening.[24]

In the annals of Sasquatchiana, Ostman's story is among the most bizarre—absurd on its face and in the subtlest details. The Sasquatches could talk? Really? They lived as a standard-issue nuclear family? On a diet of twigs and grasses? But creatures that big could not survive on such scanty fare. Either they would need to constantly forage or eat lots of meat. Ostman had a rifle, but didn't use it for fear of angering the male—so instead he decided to live with these creatures for days? That's an odd trade-off. And the Old Man just happened to suffer from the traditional ape weakness, the one that had been the undoing of so many Yetis—an irresistible urge to mimic? (It's no coincidence that one meaning of the word *ape* is *imitate*. Monkey see, monkey do.) Ostman's tale reeks of the campfire. It's the kind of yarn that would be fun to hear on a long night in the dark woods.[25]

Green admitted that the story "defie[d] belief," and he and Dahinden ignored it when a newspaper published the account. But eventually they were convinced by another journalist to talk with the prospector. The journalist assured Green "if Ostman was lying, he couldn't be sure he had ever interviewed anyone who was telling the truth." How could Green deny such a testimony? He had a deep and steady confidence in journalistic skills. So Green and Dahinden interviewed Ostman, measured his tale with their own b.s. detectors, and were impressed. Ostman's was only a story, like Roe's tale or the account of Jacko. There was little that could be tested. "There was no one else involved with whom [the events] could be checked," Green said, "and no records from the time and place where they were supposed to have happened." But Ostman seemed honest. And a few years on, Green met someone who said that he had first heard of Sasquatch from a trapper who'd heard tale of a Swede in the mountains of British Columbia captured by one.[26]

Green and Dahinden also borrowed from the legal profession a set of technologies for creating truth. They had a magistrate cross-examine Ostman, and his story held together. As well, they had him sign an affidavit attesting

**24.** Green, *Bigfoot*, 21–34; Hunter and Dahinden, *Sasquatch*, 49–58.

**25.** John Napier, *Bigfoot: The Yeti and Sasquatch in Myth and Reality* (New York: E. P. Dutton, 1973), 76–80.

**26.** Green, *Bigfoot*, 16–34 (first quotation, 16); idem, *Sasquatch*, 99 (second quotation).

to the truthfulness of his account—by now a request Green almost always asked those he interviewed. A number of those involved with the Ruby Creek Incident had sworn out affidavits, as had Roe. Green thought it impressive that people were willing to stake their reputation on these stories—hardly the action of liars, pranksters, or those who were only recounting a legend.[27]

It is too much to say that Ostman convinced Green and Dahinden, even with Green's friend-of-a-friend tale, the results of the legal investigation, and Ostman's plain-faced honesty. The story was just so ludicrous. But they were intrigued. And, as the years passed, Green came to appreciate that Ostman's yarn—while only a story—had a power to substantiate itself that was revealed slowly, as the investigation grew.

## "Occam's Razor Cuts on the Side of the Sasquatch"

Over the course of their investigation, Green and Dahinden dug through musty tomes, read through the writings of pioneers, and scoured old newspapers for mention of Sasquatch; they gathered contemporaneous reports of the wildman; they interviewed people who claimed to have seen the beast or its tracks. Some of these, like the article about Jacko, were little more than stories. Teddy Roosevelt, for example, wrote in one of his books about a hunter who was attacked by some monstrous thing in the Idaho wilderness. Other stories were more substantial. In thirty years, Green collected over two thousand reports of people who said that they had either seen the wildman or its tracks. Of these, a few became classic cases, repeated again and again in the Bigfoot literature: the Ruby Creek Incident, the story of Jacko, Roe's account, Ostman's yarn.[28]

To Green and Dahinden, the overstuffed dossier that they had gathered—what Green called the Sasquatch File—weighed against the possibility of fraud or April Fool's prank. One of the chief reasons to doubt the existence of the Loch Ness monster is the dearth of historical accounts: the creature was first sighted in 1933, uncomfortably late in the day for a beast that was supposed to be a hangover from prehistoric times. Green and Dahinden's research exempted Sasquatch from this criticism. They collected reports that dated back to the first half of the nineteenth century. If even one of these re-

**27.** Green, *Bigfoot*, 13, 16–20; idem, "The Case for a Legal Inquiry into Sasquatch Evidence," *Cryptozoology* 8 (1989): 37–42.
**28.** Hunter and Dahinden, *Sasquatch*, 12–17, 23–25; Green, *Sasquatch*, 29, 34, 83, 283, 334; Loren Coleman, *Bigfoot! The True Story of Apes in America* (New York: Pocket Books, 2003), 46–51.

ports was not from a prevaricator, prankster, or hallucinator, if only one out of two thousand was correct, then Sasquatch existed. "I think that to try to explain the existing phenomena by the only other explanation, that this is all a human production, becomes far more involved—Occam's Razor cuts on the side of the Sasquatch," Green said. Sasquatch was not a legend (and not a race of giant, hairy Indians), but an undiscovered species of apes. They were "all animal," Green wrote. "Magnificent animals, completely self-sufficient on their physical endowments alone."[29]

It was in the course of creating this composite image of Sasquatch that Green discovered the power of Ostman's account. The prospector had provided a "wealth of anatomical detail"—the shape of Sasquatch nails, teeth, even the size of the Old Man's penis. Ostman's description, Green wrote, was "confirmed over and over" in the reports of others—and in the scientific literature: the Old Man's penis, according to Ostman, was small, unexpectedly so, given that the creature was a giant. As it happens, gorilla penises are also small. How could Ostman have guessed at such a detail, Green and Dahinden wondered. How could he have anticipated what they would find in the reports of pioneers from decades before and learn from people whom they interviewed years later, people who had never heard of Ostman? Ostman had been one of the first to report on Sasquatch. He had no pattern from which to work. So how could Ostman have given an account that matched others, unless, perhaps, he had seen a Sasquatch?[30]

Green wrote as though Ostman's tale compelled him toward conclusions that went against his better judgment, just as the Ruby Creek Incident had forced him to take Sasquatch seriously. In fact, there was a more complicated process underway. It seems clear that what Green and Dahinden discovered was a rich vein of North American folklore about wildmen—not unlike the wildman legends recounted around the globe, except that these had mostly been ignored by elites and never systematically collected before. These folkloric accounts were as diverse as the ones that Burns had collected. Ostman's Sasquatch, for instance, was more akin to the type that Burns had described—indeed, could have been patterned on one of Burns's stories—than to apes. But Green downplayed those elements of the story that made

**29.** Green, *Bigfoot*, 35; idem, *The Sasquatch File*, 48; Green, *Sasquatch*, 461 (final quotations); "Interview: Does the Sasquatch Exist and What Can Be Done about It?" *The ISC Newsletter* 8, no. 2 (1989): 2 (first quotation); David J. Daegling, *Bigfoot Exposed: An Anthropologist Examines America's Enduring Legend* (Walnut Creek, CA: Altamira Press, 2004), 26.

**30.** Hunter and Dahinden, *Sasquatch*, 58; Green, *Sasquatch*, 97 (second quotation), 111, 461 (first quotation); Coleman, *Bigfoot!* 191–92.

Sasquatch seem to be a race of giant Indians and focused instead on the mundane, anatomical details. Green wasn't being compelled by the evidence—it was the other way around. He forced the stories about Sasquatch—historical and contemporary, from Native Americans and whites—through as narrow an aperture as Burns had, but one with a different filter.

Throughout his investigation, Green was repeatedly confronted with the unreliability of witnesses and the ease with which a hoax could be perpetrated and then magnified by imagination into a story that sounded authentic, that could not be explained away. Late in 1970, for instance, he looked into the report of a giant apelike creature running through a forest at superhuman speeds. One man told Green that if he'd had a gun, he would have shot the thing. Better that he didn't, since the quicksilver ape was only a boy scout from a nearby camp who dressed in a dark blue turtleneck sweater, put a toque on his head, and stood by the side of the road. It wasn't "much of a costume," Green said, but conceded that if he hadn't found the scouts, "I guess we'd still be looking for tracks." The incident dramatically proved just how fallible Green was, how likely the evidence he had gathered might just be a hodgepodge of misidentifications, pranks, and legends.[31]

Green, however, learned to ignore the lesson, to squelch doubt. In 1989, he said, "For some years, when I would get away from this for a month or two, I would find that some mechanism in my mind had made me come to the conclusion that the whole thing had all been explained away, and I would really have to get down to reviewing the evidence again to realize that this conclusion was just an impression, just a sort of emotional feeling." From one perspective, Green was just restating the power of evidence to change his mind; from another, he was acknowledging that he chased away any suspicion that there were alternative explanations. He was convincing himself, learning to trust his own infallibility. Sasquatch was as fraught with as much linguistic and metaphysical confusion as the Yeti, but Green found a pattern among the diversity and transformed what had been an Indian legend into a prosaic object of biology.[32]

## Slick Eyes the Sasquatch

In June 1958, Tom Slick wrote his final article about the Yeti for the *Journal-American*. He was convinced the Abominable Snowman existed—actually, he

**31.** "Sasquatch Masquerade," *Bigfoot Bulletin*, 20 (August 31, 1970): 4.

**32.** Green, *Bigfoot*, 4; "Interview: Does the Sasquatch Exist and What Can Be Done about It?" 2 (quotation).

FIGURE 10. Oil magnate Tom Slick had a taste for science both prosaic and outré. In this picture from 1947, Slick showed off his Foundation for Applied Research. He also sponsored expeditions in search of the Yeti before turning his attention to North America's wildmen. (No. L-3438-F. *San Antonio Light* Collection, University of Texas at San Antonio's Institute of Texan Cultures. Courtesy of the Hearst Corporation.)

said that there were three species of them in the Himalayas—and was just as certain "there were many varieties of wildman, living in Indochina, Cambodia, Burma, and Indonesia." He argued that there must have been "thousands" of species of "missing links" and "the fact that all should be extinct is the amazing thing, not that some have survived." Slick seemed especially intrigued by the Sasquatch. "Some very plausible reports have come out of British Columbia," he told the newspaper's readers, most likely referring to the various stories that Green and Dahinden had found.[33]

In time, Slick would visit Canada to investigate for himself—but before that, word of another wildman spread through the mass media, this one in California. It was being called Big Foot, and its history became entangled with Sasquatch and the Yeti—the three wildmen legends braided into a single narrative.

**33.** Tom Slick, "Elusive Snowman Still in the Hills" *New York Journal-American*, June 22, 1958.

CHAPTER FOUR

# Big Foot 1958

On Monday, August 27, 1958, Jerry Crew left his home in the northern California hamlet of Salyer. Pictures of Crew taken six weeks later show a broad-chested, short-haired man with big glasses, a strong chin, and prominent ears. By all accounts, he was an earnest and sober individual. Crew drove west along California State Highway 299, the chief artery through this montane region, running some 150 miles between Eureka on the Pacific and Redding in the Central Valley. Crew was a catskinner for the Granite Logging Company and the Wallace Brothers Logging Company. The lumber industry employed about one out of every two workers in the county, generating more revenue than the rest of the economy combined.[1]

A few miles on, Highway 299 intersected with Highway 96 at Willow Creek, a gold rush town once known as China Flats and, by 1958, a regional hub that provided services for lumbermen that small towns such as Salyer could not, although by most standards Willow Creek was itself a small town. Like much of the area, Willow Creek was doing fairly well. Since 1949 lumber production in Humboldt County had almost doubled in response to the post–World War II housing boom. Per capita income in the county was on par with the rest of California, and above the national average.[2]

Crew turned north. State Highway 96 followed the Klamath River into the Shasta-Trinity National Forest, crossing into Del Norte County and continu-

**1.** Allen to Genzoli, n.d. (between September 26 and October 4, 1958), Bigfoot file 7, Andrew Genzoli papers, Humboldt State University, Arcata, CA; Marian T. Place, *On the Track of Bigfoot* (New York: Pocket Books, 1978), 1–5; Steven C. Hackett, "The Humboldt County Economy: Where Have We Been and Where Are We Going?" *Humboldt (CA) Times-Standard*, February 1999, http://www.humboldt.edu/~economic/humcoecon.html (accessed April 9, 2008).

**2.** Hackett, "The Humboldt County Economy."

ing on to Yreka. Along the highway, strung out between Willow Creek and Yreka like beads on a string, were a number of small towns, Weitchpec and Orleans and Happy Camp. Highway 96 was the main road servicing them, but it was not paved its entire length; Crew's ride was bumpy and slow. On one side of Highway 96 was a steep drop down to the river, on the other, a rocky cliff face. "Geological maps of the region," noted nature writer David Raines Wallace, "look . . . like the results of a jammed conveyor belt. . . . The ridges [are] not particularly high or craggy, rather a succession of steep, pyramidal shapes" that stretch "almost geometrically into blue distance." Thick stands of pine, spruce, and fir covered the mountains, ranging down to the water's edge.[3]

As he drove, Crew passed through the Hoopa Indian Reservation. The bucolic setting and current prosperity masked an ugly history of violence against Native Americans. In February 1860, a group of Eureka men, armed only with hatchets, clubs, and knives, slaughtered the native Wiyots while they were in the midst of a festival, killing women, children, infants, and the elderly. Unapologetic, the *Humboldt Times*, the local paper, defended the massacre. The U.S. Army gathered the remaining members of the tribe and moved them to the Hoopa reservation, and the region went about trying to forget the horrors of that night.[4]

Just beyond the Weitchpec Bridge, near the confluence of Bluff Creek and the Klamath, Crew turned onto Bluff Creek Road, a timber access route that the Wallace brothers were building on subcontract from the government. Crew had been on this job for two years. About thirty men worked here, whites from surrounding small towns and Hoopa Indians from the reservation. Some women and children were around, too. The commute from Salyer usually took two and a half hours. Many of the other men working on the road moved their families from Happy Camp and Salyer and the other small towns into the forests and lived in trailers during the construction season. Crew, however, returned home each weekend because he was so deeply involved in community and church affairs.[5]

**3.** Betty Allen, "Times Reporter Has a Look at Tracks—Says They're Real," *Humboldt (CA) Times*, October 9, 1958; Place, *On the Track of Bigfoot*, 1–5; David Rains Wallace, *The Klamath Knot: Explorations of Myth and Evolution* (Berkeley: University of California Press, 2003), 2, 23.

**4.** Lynwood Carranco, *Genocide and Vendetta: The Round Valley Wars in Northern California* (Norman: University of Oklahoma Press, 1981).

**5.** Allen, "Times Reporter Has a Look at Tracks—Says They're Real"; Place, *On the Track of Bigfoot*, 1–5.

Most of what happened next is recorded only in Marian Place's *On the Trail of Bigfoot*. Place was a children's author and a believer in Bigfoot—sometimes credulously so. She wrote her book almost twenty years after the events of August 27. But she was a diligent researcher and what she reported is as trustworthy as anything else written on Bigfoot—indeed, decidedly more trustworthy than much else. According to Place, Crew saw the foreman, Wilbur "Shorty" Wallace, at the construction site's main camp and honked his horn lightly. Wallace waved him on. Crew worked at the far end of the road, a quarter mile beyond the camp (about twenty miles from the highway), bulldozing brush and stumps left behind by the loggers who were clearing the path, and roughly grading the land.[6]

Crew parked near his bulldozer, traded his moccasins for work boots, and put on his aluminum hardhat. He noticed a few footprints in the leveled earth but thought nothing of them until he climbed onto his tractor and looked down upon them. The prints were big and manlike. They pressed deeply into the earth. Was someone pulling a prank? he wondered. Crew drove back to tell Shorty what he had seen.[7]

## The Folkloric Origins of Bigfoot

Some of the other men working on Bluff Creek Road gathered around and listened to Crew talk with Shorty. They had their own gossip about giant, humanlike tracks to pass on. One man mentioned that similar tracks had been found on another Wallace worksite along the Mad River. Twenty-five workers claimed to have seen those. More tracks had been found in Trinidad, up the coast. It's unknown whether anyone mentioned it—although it seems likely—but only a few months before the *Redding Record-Searchlight* had run a story about giant footprints found along a Pacific Gas and Electric Company right-of-way back in 1947.[8]

Shorty suggested that whatever had made the tracks around Crew's workstation also might be responsible for other . . . disturbances. The summer before, he said, on a lower section of the road, a 450-pound drum of diesel fuel had gone missing; only its impression and large footprints had been left in the dust. The drum had been found a little while later at the bottom of a gully—into which it must have been tossed, since the foliage on the hill-

**6.** Place, *On the Track of Bigfoot*, 6–8.

**7.** Ibid.

**8.** Garth Sanders, "Big Footprint Can't Be Tracked Down," *Redding (CA) Record-Searchlight*, March 4, 1958, 12:12; Place, *On the Track of Bigfoot*, 11–15.

side was unbroken. Not unlike the 700-pound spare tire for the road-grading machine that had somehow found its way into a ditch, Wallace reminded the workers. The men had rescued the tire, and were told that vandals had pushed it. But maybe not. Maybe the tire, like the drum, had been tossed by some *thing*. Some thing that left immense tracks. Something big and strong. But what?[9]

According to Place, the men debated the possible culprit for a time. There was no consensus about what had made the various tracks, no coherent legend of a mysterious trackmaker, no Sherpa to tell Crew and the rest what they had seen. Finally, Shorty "winked broadly" and interrupted the debate, telling the men "to be sure to let him know if they saw any apes skedaddling through the timber. Meantime, he'd sure appreciate it if they got to work."[10]

The men did return to work; they also continued to discuss those tracks and their maker. They called him (and no one doubted that the owner of those large feet was a *he*) Big Foot, two words. Journalist Betty Allen, who visited the camp in late September, found a bevy of stories about Big Foot. The men accused Big Foot of vandalism, and if something went missing he was the presumed thief. Some of the stories, Allen said, were "hair raisers." For example, some time in October four dogs were lost, and Big Foot was accused of killing them. Supposedly, a few of the workers and their families did take the tales seriously. Allen reported that some of the men kept "their guns handy at night" because a creature that could toss drums of diesel fuel was something to be feared. But the worriers seem to have been the exception. "A lot" of the tales, Allen said, were "quite fictitious." They had a "legendary flavor." When Jess Bemis, another Salyer resident, took a job clearing land on Bluff Creek around this time, he and his wife Coralie joined the fun and, in Coralie's words, "added fuel to the story by passing on bits of information," although at the time neither believed Big Foot was real.[11]

**9.** Place, *On the Track of Bigfoot*, 11–15.

**10.** Andrew Genzoli, "Those Tracks Again!" *Humboldt (CA) Times*, September 26, 1958; Place, *On the Track of Bigfoot*, 15 (quotation).

**11.** Rocky Bemis, "The Rocky Bemis Story With Old Articles," http://www.bigfootencounters.com/stories/rocky.htm (accessed April 9, 2008); "New Bluff Creek Mystery Puzzles Indian; 4 Dogs Found Ripped to Pieces," *Humboldt (CA) Times*, October 19, 1958; Andrew Genzoli, "Someone, Please Answer," *Humboldt (CA) Times*, September 21, 1958 (fifth quotation); Genzoli, "Those Tracks Again!" (first through third quotations); "'Bigfoot' Making Big Tracks in U.S. News," *Humboldt (CA) Times*, October 7, 1958 (fourth quotation); Betty Allen, "Bigfoot Still on March As 'Experts' Try Explanations," *Humboldt (CA) Times*, October 31, 1958.

Lumberjacks, hunters, trappers, and other working-class men had long told stories of such prodigies. For decades, seasoned veterans had funned greenhorns with tales of sidehill dodgers and mosquitoes so big that they sucked cows dry and by having them fetch the equally legendary left-handed wrench. Or they sent them to hunt snipes. Around the turn of the twentieth century, Eugene Shepard, a Wisconsin lumberjack, raconteur, and prankster, announced that he had caught a hodag, the rhino of America's north woods. Shepard photographed a group of friends killing the beast with picks and axes. The picture was made into a postcard; hundreds of thousands were sold; tourists flocked to Rhinelander, Wisconsin; reportedly, the Smithsonian even expressed interest. Seeing is believing. But the hodag was just a woodcarving. It was all a humbug. American history is rife with such practical jokes, stories of giant turtles and panthers, jackalopes and sea serpents, agropelters and snow wassetts—an entire bestiary of legendary animals. The tradition continued long after the frontier closed. In 1950, for example, the men's adventure magazine *Saga* introduced a feature called "Sowing the Wild Hoax" and encouraged the blue-collar men reading it to send in examples of "particularly fiendish" and "unusually funny" practical jokes.[12]

Over the years, fake footprints have been a favorite hoax and tales of giant wildmen common; this was the folklore—the tales and newspaper reports—that Green and Dahinden discovered and collected. Elgin Heimer, a resident of Myrtle Point, Oregon, probably thought that he was just making a joke, but he expressed an important truth when he suggested to the *Humboldt Times* that Crew's mysterious tracks had been left by "Paul Bunyan's two-year-old boy." Bigfoot was Paul Bunyan's heir.[13]

Such joshing, especially among working-class men, served to initiate novices and cement relations on the job. Teasing was a way of testing and proving one's masculinity—coming up with a joke showed cleverness, withstanding the ribbing (and responding in kind) displayed strength, which was necessary to fitting in. Tales about legendary creatures also helped those who worked far from civilization to manage their anxieties. Inchoate fears about an un-

**12.** "Sowing the Wild Hoax," *Saga*, September 1950, 21; Curtis D. MacDougall, *Hoaxes* (New York: Dover, 1958), 17–18; Richard M. Dorson, *American Folklore* (Chicago: University of Chicago Press, 1959), 271–72; Constance Rourke, *American Humor: A Study of the National Character* (New York: Harcourt, Brace, Jovanovich, 1959); Daniel Hoffman, *Paul Bunyan: Last of the Frontier Demigods* (Lincoln: University of Nebraska Press, 1983), 12–18, 33–37.

**13.** Elgin Heimer to the *Humboldt Standard*, October 29, 1958, Bigfoot file 7, Andrew Genzoli papers.

FIGURE 11. Around the turn of the twentieth century, Eugene Shepard claimed to have captured a hodag—a legendary beast of the Upper Northwest. This picture was meant as proof, although it was obviously staged. The hodag was part of a long tradition among lumberjacks recounting the exploits of mythical monsters. Tales told about Bigfoot by woodsmen in northern California during the 1950s continued the custom. (Image WHI-36382. Wisconsin Historical Society.)

knowable nature were congealed into slightly ridiculous forms—the will-am-alone, for instance, was a kind of squirrel that dropped pellets of rolled lichen onto sleeping lumberjacks, causing nightmares—and thus the fear was made to seem absurd, too. In a very real sense, the men and women working and living on Bluff Creek Road told stories about Big Foot to scare themselves silly.[14]

**14.** E. E. LeMasters, *Blue-Collar Aristocrats: Life-Styles at a Working-Class Tavern* (Madison: University of Wisconsin Press, 1975), 136–40; Hoffman, *Paul Bunyan*, 12–18, 33–37; David Halle, *America's Working Man: Work, Home, and Politics among Blue-Collar Property Owners* (Chicago: University of Chicago Press, 1984), 180–85; David L. Collinson, *Managing the Shop Floor: Subjectivity, Masculinity, and Workplace Culture* (Berlin: Walter de Gruyter, 1992), 111–14; Marina Warner, *No Go the Bogeyman: Scaring, Lulling & Making Mock* (New York: Farrar, Straus and Giroux, 1998); Peter Richardson, "When Work Was Fun," working Paper no. 9 (Atlanta, GA: The Emory Center for Myth and Ritual in American Life, 2001).

## Big Foot Makes the Papers

In the middle of September, a new line of tracks appeared along Bluff Creek road, the first since Crew had found the prints near his bulldozer. A few of the men inspected the tracks and declared that they were neither fake nor the mark of bears. If Crew had once thought that he was the victim of a practical joke, he no longer did. Neither did the Bemises or about a dozen other men. Big Foot, whatever he was, existed. Crew started to hunt it. He also traced one of the giant footprints onto paper and took the rendering to Bob Titmus, a taxidermist in Anderson, not far from Redding.[15]

As Titmus remembered the meeting decades later, he told Crew that the trace lacked too much detail and taught him how to make a plaster cast. Later, Crew called Titmus and said that he had made a cast. It was sixteen inches long. Titmus and a fellow taxidermist, Al Corbett, visiting from Seattle, drove to Salyer and inspected the cast. He was not impressed and suggested—as he said later—"the other workers there at the site had been playing pranks on one another." Crew insisted that the tracks were real: there were too many, their impressions too deep, their detail too fine. Crew gave Titmus and Corbett a map to his worksite and told them to see for themselves. For one reason or another, the three did not make it to Bluff Creek Road that day.[16]

Around this time, Coralie Bemis sent word of the new tracks to Andrew Genzoli at the *Humboldt Times*. Genzoli was a Herb Cain-esque columnist who had worked for the *Times* in the 1930s after graduating from high school, and then set out to see the world. He'd returned in 1948 and had been given the job of writing a column that would be of interest to rural readers. Genzoli named it "RFD." He was known as an amateur historian—he claimed to have read through most of the paper's morgue during his first stint with the *Times*—and peppered his columns with liberal doses of nostalgia for a lost and simpler Humboldt County. Bemis thought that he was the type of person who would be interested in a wildman and might look into the matter; but Genzoli was dismissive, at least at first. He thought that someone was "pulling [his] leg" and set the letter aside. But, when the column that he was writing for September 21 came up short, he decided to print the letter. "Maybe we have a relative of the Abominable Snowman of the Himalayas," he wrote in his column that day, "our own Wandering Willie of Weitchpec." It was a fate-

**15.** James O. Holley, "Big Tracks Sink Deep, Seem Real," *Houston Post*, January 12, 1960, Bigfoot file 3, Andrew Genzoli papers; Place, *On the Track of Bigfoot*, 5–7.

**16.** "Interview: Bob Titmus," *The ISC Newsletter* 12, no. 2 (1993–1996): 2.

ful moment: like the Yeti and Sasquatch, Big Foot was promoted by the press from local legend to international celebrity.[17]

Genzoli's column struck a chord. Around dinner tables, in barbershops, at the grocery, people talked about those mysterious tracks. The journalist found himself writing a couple more columns on Big Foot over the next few days, no longer reluctant to publish now that he'd seen there was a lot of enthusiasm for the subject. Big Foot, Genzoli had come to realize, was "good material for a good imaginative writer who is tired of space assignments." Betty Allen, a resident of Willow Creek, proud grandmother, and correspondent for the *Humboldt Times* was that writer. Amid the hubbub, she had Al Hodgson, proprietor of Willow Creek's general store, drive her to the Bluff Creek worksite so that she could investigate the tracks and talk to those who had seen them. She filed a number of articles with the paper about the county's most mysterious resident.[18]

On the first Saturday of October, Genzoli met Crew; the construction worker had come to Eureka looking for someone who would take his plaster cast track seriously, since Titmus had rebuffed him. Genzoli was impressed by Crew's demeanor. No longer reluctant to publish, he immediately arranged for Crew to have his picture taken with his trophy for a story, and Crew refused the photographer's request to smile—"If I did, then someone would accuse me of trickery," Crew reportedly said. The picture ran the next day, on the front page of the October 6 issue of the *Humboldt Times*, alongside an article that Genzoli penned (drawing on much of the reporting that Allen had been doing).[19]

"The men are often convinced that they are being watched," Genzoli wrote in the article. "However, they believe it is not an 'unfriendly watching.' . . . Nearly every new piece of work . . . finds tracks on it the next morning, as though the thing had a 'supervisory interest' in the project." Either Genzoli or Allen also interviewed Ray Wallace, Shorty's brother and one of the Wallaces running the logging company, who claimed to have measured the creature's stride: fifty inches while at a stately pace, nearly ten feet while

**17.** Genzoli, "Someone, Please Answer" (second quotation); Lynwood Carranco, "Andrew Genzoli—Journalist, Historian," *Humboldt Historian* 24 (1976): 24–25; Andrew Genzoli, "Bigfoot Story Captured the Nation 19 Years Ago," *Humboldt (CA) Times-Standard*, September 21, 1977 (first quotation); "Andrew Genzoli and Bigfoot," *Humboldt Historian* 32 (1984): 5.

**18.** Andrew Genzoli, "Now, About Those Foot Prints," *Humboldt (CA) Times*, September 25, 1958; idem, "Those Tracks Again!" (quotation).

**19.** Andrew Genzoli, "What Is It?" *Humboldt (CA) Times*, October 7, 1958 (quotation); Place, *On the Track of Bigfoot*, 19–22.

TIMES. EUREKA. CALIFORNIA. SUNDAY. OCTOBER 5, 1958

# Iuge Foot Prints Hold Mystery Of Friendly Bluff Creek Giant

erald Crew of Salyer who made the plaster-of-paris cast of the big-footed anderer in the vicinity of Bluff Creek, shows the size of the impression ith the use of a 15-inch ruler. The foot measured 16-inches from heel big-toe tip. Andrew Genzoli, regional editor and RFD columnist for e Humboldt Times, who has been featuring stories about the big feet since September, brushes dust from the mould to obtain a better This imprint was made either Wednesday night or early Thursday ing by "Big Foot." The impression was made by Crew Thursday ing, also.

## uestions Surround The rigin of Many Tracks ound In Wooded Country

FIGURE 12. Andrew Genzoli (left) and Jerry Crew examine the cast that Crew took of a Big Foot track. Crew's discovery had the same effect on northern California's wildman as Shipton's had on the Abominable Snowman: thrusting the monster into the limelight. This picture accompanied Genzoli's story in the *Humboldt Times*. (By permission of Humboldt State University—Special Collections and the *Eureka Times-Standard*.)

running. Someone had also contacted Titmus, who by now had been out to the worksite and revised his previous opinion: these tracks had not been faked, he said. "Who is making the huge 16-inch tracks in the vicinity of Bluff Creek?" Genzoli wondered. "Are the tracks a human hoax? Or, are they the actual marks of a huge but harmless wild-man, traveling through the wilderness? Can this be some legendary sized animal?" Genzoli called the mysterious trackmaker Bigfoot, one word, which he thought played better in newspapers.[20]

Years later, Genzoli said that he thought that the tale of Crew's giant plaster cast and rumors about the mountain wildman "made a good Sunday morning story." But it was more than that. It was a sensation—more so, much more so, than the publication of Bemis's letter had been. The article was sent over the newswires and, like the tracks that Shipton found on the head of the Menlung Glacier, Crews's cast caught the world's imagination. "On Monday, Tuesday, and for the rest of many days," Genzoli said, "we had reporters from all the wire services pounding on our doors. There were representatives from the *New York Times*, the *Los Angeles Times*, *Chronicle* and *Examiner*, San Francisco [*sic*], and many, many more." Less than two weeks after the article appeared, the television game show "Truth or Consequences" offered $1,000 to anyone who could explain how the tracks had been made. In the year after Bigfoot's big debut, Genzoli received more than 2,500 letters. Here was an Abominable Snowman—not inhabiting the frigid, faraway Himalayas, but in California! Here was Bigfoot.[21]

## The Confirmed and Converted Confront Bigfoot

Bigfoot was *the* topic in Humboldt County during the autumn of 1958. The *Humboldt Times* featured stories about Bigfoot in eighteen issues during the month of October, some issues carrying more than one story. "Almost every conversation one hears around here," Betty Allen wrote from Willow Creek, "either begins on the subject of Bigfoot—or soon swings around to him." There were a lot of "local experts," most of whom, she noted slyly, relied on that old authority, the invisible, ubiquitous, but always credible *they*. "'They say,'" she wrote, "is a favorite expression, and the 'they say' authorities are filled with theories" about what was making those tracks. A bear, some said. Pranksters, other insisted. Perhaps it was an ape or a wildman. There were

**20.** Genzoli, "Giant Footprints Puzzle Residents along Trinity River," *Humboldt (CA) Times*, October 5, 1958.

**21.** Genzoli, "Bigfoot Story Captured the Nation 19 Years Ago."

rumors that a race of giants inhabited Shasta—called Lemurians, they were supposed to be lost relatives of Atlanteans, possessors of secret knowledge. Maybe they were haunting the camp.[22]

Allen classified the various they-sayers into two categories, the confirmed and the converted. The confirmed were those certain in their opinion that the whole thing could be easily explained. Probably, many said, bears left the tracks. Perhaps, said Scott Bell, a timber feller from Eureka, "It's a big-footed Swede. Most Swedes are lumberjacks and I suspect a lumberjack of making the tracks." Many others thought that the Wallace brothers had made the tracks. They were well-known pranksters—indeed, fake footprints had been found at another of their worksites and, apparently, even at the time some in the Wallace clan admitted that the tracks Crew found had been faked. The Wallaces could have attached fake feet to the treads of tractors or the roller used to smooth the road, Willow Creek residents speculated to Allen. That would press the tracks deeply into the ground. Shasta Sam, the smart-alecky alter ego of the *Redding Record-Searchlight* opinion page editor, wrote in an open letter to the creature, "Some folks think you're an Abominable Snowman strayed from the Himalayas. I think you're an Abominable Snowjob." Eight-year-old Jack Rubyn of Eureka seconded Shasta Sam's assessment. On Halloween, he told the *Humboldt Times*, "I don't believe in witches, goblins, or Bigfoot, either. I'm too grown up for that." The Hoopa Valley Hospital, trying to raise $50,000, pleaded with the prankster to confess, collect the $1,000 from "Truth or Consequences," and donate the reward.[23]

Certainly events that fall made the tracks seem less than serious. For example, in the middle of October, George Smith claimed to have seen Bigfoot while he was driving along Bluff Creek Road. It was eight feet tall, he said, and broad shouldered, a monster that scared him so much that he wouldn't get out of his car. But he was certain that it was just a man in a gorilla costume, because the thing's fur sagged like a badly tailored suit. An eight-foot-tall man in masquerade! Running through the woods! The tracks weren't a

**22.** Allen, "Bigfoot Still on March As 'Experts' Try Explanations" (quotations); Sumathi Ramaswamy, *The Lost Land of Lemuria: Fabulous Geographies, Catastrophic Histories* (Berkeley: University of California Press, 2005), 71–89.

**23.** Genzoli, "Giant Footprints Puzzle Residents along Trinity River"; "The Inquiring Reporter," *Humboldt (CA) Times*, October 28, 1958 (first quotation); "To Bigfoot," *Redding (CA) Record-Searchlight*, October 11, 1958 (second quotation); "Has Bigfoot a Big Heart?" *Humboldt (CA) Times*, October 22, 1958; Allen, "Bigfoot Still on March As 'Experts' Try Explanations"; Marjorie Gray, "Abominable Snowjob," *True*, April 1960, 8; "The Inquiring Reporter," *Humboldt (CA) Times*, October 31, 1958 (third quotation).

mystery—just a bit of fun. Genzoli said that Bigfoot was a "light-hearted sort of thing": "at a time when the world is filled with struggles, aches and pains," Bigfoot offered "the chance to talk about something else . . . curious, maybe funny, and for some, thrilling."[24]

Allen numbered herself among the other class of Bigfoot experts, the converted, those who believed that the prints had been made by some mysterious thing. "If those tracks are the work of a prankster, he's an artist," she said. "I looked down on at least twenty of the tracks that had been made . . . and they were just as perfect as those made by anyone else walking around the road. You could see the toes and the rest of the foot very plainly. There was no exaggeration to be found, that I could see." Also included among the converted were Crew and Titmus, Crew's mother-in-law, the Bemises, and Olive Curtiss, a Eureka housewife who wanted "to go up there and hunt whatever it is," because "these types of things fascinate me."[25]

Judging by the news coverage, the most common explanation of the tracks among the converted—offered by Allen, at least in the autumn of 1958, Titmus's friend Al Corbett, and many others—was that they had been left by a gigantic, retarded Indian. Some said that Indian had run away from a Civilian Conservation Corps camp; others claimed that about twenty years before, the Humboldt County sheriff had discovered a couple who kept their fourteen-year-old son chained to a tree. In one telling, "The boy was entirely naked except for a wide leather belt to which the chain was attached. He was over six feet tall, with immense hands and feet and weighed about 250 pounds, then. He had rather long, reddish brown hair and also a lot of hair on his body. He looked like an ape, and could not talk. In fact, he was an imbecile." While the authorities were deciding what to do with the boy, he went missing. At the time, it was believed that the father had killed him. But perhaps the boy had broken free and still haunted the forest . . . .[26]

This tale may have been a version of an even older legend—and so not an explanation of the tracks that Crew found. Back in 1878, William T. Andrews,

**24.** Genzoli, "What Is It?" (quotations); "Bigfoot's a Man in Bearskin, Eurekan Claims," *Humboldt (CA) Times*, October 18, 1958.
**25.** "'Bigfoot' Making Big Tracks in U.S. News" (second quotation); Allen, "Bigfoot Still on March As 'Experts' Try Explanations" (first quotation); "The Inquiring Reporter," *Humboldt (CA) Times*, October 28, 1958 (third quotation).
**26.** Betty [Allen] to Andy [Genzoli], n.d. (1958), Bigfoot file 7, Andrew Genzoli papers; "Wild Indian Gets Credit for Gigantic Footprints," *Fresno (CA) Bee*, October 8, 1958; "Bigfoot Still Topic of Interest in Humboldt and Abroad," *Humboldt (CA) Times*, October 14, 1958 (quotation).

a carpenter, said that he had witnessed Bigfoot's death. Bigfoot in those days was a legendary outlaw who harried travelers as they moved over the Oregon Trail, a huge man, well over six feet tall and reportedly weighing three hundred pounds, with enormous feet that made large tracks in the frontier dust. "He was the scapegoat for unsolved crimes, the bogey man for small children," and the subject of doubtful stories "where the thirsty citizenry . . . bent its elbow along polished bars," according to historian Porter Ward. Even after Bigfoot's career ended, some Native Americans reportedly faked huge moccasin tracks to scare their adversaries into believing that the giant outlaw was still rampaging.[27]

Andrews contended that Bigfoot was not a legend but a real man. He said that he was in Silver City, Idaho, when John Wheeler, a part-Cherokee crackshot, gunned Bigfoot down, and he heard Bigfoot tell his life story as he lay dying. Bigfoot said that he had been born in Oklahoma of mixed parentage, part white, part black, and, like Wheeler, part Cherokee. In the 1850s he joined a wagon train heading West; he fell in love with a white woman but was rejected, and so killed her other suitor, then lit out and became leader of a gang.[28]

Andrews's story was published in the *Idaho Statesman*, but at the time he was a resident of Humboldt County. It's not hard to imagine that he told the stories to friends and neighbors, and that the tale of the outlaw Bigfoot became part of the local lore. Over time, the story may have been changed into the tale of an enormous, retarded Native American haunting California's north coast. Like the stories that the construction crew told about Bigfoot the mysterious trackmaker, these tales about Bigfoot the avenging Indian may have been a way of expressing anxieties, in this case, white worries about living closely with a still-unfamiliar people. Perhaps there was even a little guilt in them, Bigfoot taking justifiable revenge for the sorry history of white attempts to exterminate Native Americans.[29]

In time, Jerry Crew developed his own idea about what left those mysterious tracks near his tractor: Bigfoot was a prodigy created by God to confuse those who believed in evolution. Bigfoot proved the inadequacies of secular knowledge and the folly of turning away from God—just as it was stated in

**27.** Porter Morgan Ward, "Bigfoot: Man or Myth?" *Montana: The Magazine of Western History*, April 1957, 19–23 (quotation, 20); "Bigfoot" (Idaho State Historical Society, Reference Series #40, 1970); Bennett L. Williams, *The Legend of Big Foot: The Shoshone Indian Chief Nampuh* (Homedale, ID: Owyhee Publishing Company, 1992).
**28.** Ward, "Bigfoot: Man or Myth?" 19–23.
**29.** Ibid., 20.

Paul's epistle to the Romans, Crew noted: "When they knew God, they glorified him not as God, neither were thankful; but became vain in their imaginations, and their foolish heart was darkened. Professing themselves to be wise, they became fools, and changed the glory of the uncorruptible God into an image made like to corruptible man, and to birds, and fourfooted [*sic*] beasts, and creeping things. . . . And even as they did not like to retain God in their knowledge, God gave them over to a reprobate mind, to do those things which are not convenient." Crew explained his theory to Betty Allen in 1959, and it may be that he only developed it later; or, he may have reached his conclusion then, in September or October. Crew certainly acted like a man motivated by religious convictions, determined to prove Bigfoot's existence.[30]

Others of the converted thought Bigfoot neither born of man nor God, but nature—he was a wildman, just like the Yeti or Sasquatch. After Genzoli printed Coralie Bemis's letter, a historian from adjacent Del Norte County dug up a newspaper story from 1896 about a wildman living in the area. "The thing was of gigantic size—about seven feet high—with a bull-dog head, short ears and long hair," reported the newspaper. "It was also furnished with a beard, and was free from hair on such parts of its body as is common among men. Its voice was shrill, or soprano, and very human, like that of a woman in great fear." Was this wildman Bigfoot's ancestor? And what about those tracks from 1947 that the *Redding Record-Searchlight* had mentioned? The man who saw them thought that a monster had made them. Was that the same as a wildman? There was also a newspaper report that a California geologist estimated that eight hundred pounds of pressure was needed to make the tracks—too much for even the heaviest human.[31]

Allen found support for this contention when she interviewed some Hoopa Indians and discovered that they—like their First Nations neighbors to the north—had stories about giant wildmen. In some of the stories, the creatures were outsized, but otherwise humanlike. Creek devils, they were called. Creek devils lived in the forest, sometimes stealing women from the Hoopa. There was also the tale of a cave in a granite cliff that supposedly housed a lost tribe of large-footed Indians. In other stories, the creatures were huge, hairy, and stinking—monsters. Omah, these creatures were called.

**30.** Betty Allen, "Pacific Northwest Expedition," 1959, Bigfoot file 7, Andrew Genzoli papers.

**31.** Sanders, "Big Footprint Can't Be Tracked Down"; Genzoli, "Now, About Those Foot Prints"; "'Bigfoot' Weighs More Than 800 lbs., Decides Geologist," *Humboldt (CA) Times*, October 11, 1960; Lynwood Carranco, "Three Legends of Northwestern California," *Western Folklore* 22, no. 3 (1963): 185 (quotation).

One of Allen's sources told her that omah had been common in the area until prospectors arrived. Edward von Schillinger, a young stake-setter employed by the Wallace brothers, heard from "an old Indian" that the creatures did not take kindly to those who tried to track them: they stood "on high bluffs and roll[ed] enormous rocks upon them, or bludgeon[ed] them to death with telephone-pole sized sticks."[32]

The legends, of course, proved nothing. The Hoopa themselves, according to Marian Place, admitted that omah were spirits as well as real creatures. But, like the similarly ambiguous tales about Abominable Snowmen told by Sherpas, Nepalis, and Tibetans, the stories were tantalizing. They suggested that it was not impossible that a mysterious being, a giant, manlike beast unknown to whites, haunted the area. The Indian tales provided the heft of history to what was otherwise a seemingly weightless affair. Green noted, "There are several dozen reports from California before 1958, but all are recollections of one sort or another. Not a single one is from a newspaper story in the 1900s. That is indeed odd." The Indian tales made the gap less odd.[33]

Titmus was among those who believed that Bigfoot was a wildman—the tracks, he said, had not been made by humans or any known animal. Like Green and Dahinden, his opinion was informed by his skill: as a taxidermist and outdoorsman, he claimed to be an expert in animal tracks. On some level, it seemed impossible that Indian legends were right, that Titmus's claims were true, that the forest was haunted by a wildman. But the atmosphere in Humboldt County was charged. Titmus's wife admitted to the *Sacramento Union* that she was "a little apprehensive" about her husband's new hobby.[34]

## Humbug!

Genzoli belonged to neither category, confirmed or converted. Instead, he emphasized the ambiguity of the story, the possibilities. Bigfoot, Genzoli thought, marched in "the never-dying parade of the mysterious" alongside "Flying Saucers, the Little Men from Mars, the giants on Mount Shasta," that is to say, the Lemurians. It was a position that he maintained throughout

**32.** "Bigfoot Not Fussy, Likes Culverts for His Boudoir," *Humboldt (CA) Times*, October 8, 1958 (quotations); Betty Allen, "R F D," *Humboldt (CA) Times*, October 10, 1958; Place, *On the Track of Bigfoot*, 25–32.

**33.** Place, *On the Track of Bigfoot*, 30; Loren Coleman, *Bigfoot! The True Story of Apes in America* (New York: Pocket Books, 2003), 53 (quotation).

**34.** "Taxidermists Go Hunting for Mountain Monster," *Sacramento (CA) Union*, October 6, 1958 (quotation); Place, *On the Track of Bigfoot*, 30.

his life—with a firmness that belied just how lighthearted the subject was to him. Genzoli had a paternal interest in Bigfoot. He had named the monster with what he thought was journalistic concision, brought it to the world's attention, and resisted any attempt to remove the creature from that procession of the bizarre. When he heard word that some people associated with the Hoopa Valley Hospital planned to fake a confession and claim the prize from "Truth or Consequences," he admonished against it in print. (The game show responded by donating $1,000 to the hospital.)[35]

And when, a few days after he broke the story to an international audience, a woman said that it was all a joke, the tracks were a hoax, Genzoli came out (as he said), "in defense of Bigfoot." He wrote in his October 9, 1958 column:

> I am slightly amused by a story I heard about a local county servant who offered a most unlikely "solution" to the Bigfoot story . . . . From out of her *True Story* magazines, or whatever she reads, she had found, just like fifteen or 20 others have before her, a picture of a rubber foot . . . . Being able to read, she discovered that this false foot can be slipped over the foot and worn for a hoaxy, thrilling sort of experience . . . . Topping this, milady dashed to the sheriff's office, aroused a deputy . . . . The lady said that anyone with these feet could astound anyone . . . . The deputy offered an "explanation" (again it wasn't original) like several others I had heard. He said he felt relieved such a "simple explanation existed . . . ." I'll bet he was . . . . Whether the deputy believes in the Bigfoot story or not doesn't seem to matter. This chapter in Humboldt's story is out of the law enforcement realm anyway, and if anyone is injecting himself into it from that viewpoint, then he is way out of order.[36]

Not everyone was as sanguine about the mystery as Genzoli, however. The story called up anxieties—real and imagined—and some wanted those worries subdued, wanted the mystery solved, wanted the police involved. On the same day that Genzoli's column appeared, the *Blue Lake Advocate*, a weekly newspaper from a small community along Highway 299, printed a story headlined "'Big Foot' Tracks Reported a Hoax." The newspaper considered the county official reliable and the mystery worth solving. The tracks had been made as a practical joke, the article contended, a bit of humbuggery among the construction crew, but the jesting had snowballed out of control.

**35.** Andrew Genzoli, "Still a Mystery," *Humboldt (CA) Times*, November 4, 1958 (quotations); idem, "A 'Tip-Off,'" *Humboldt (CA) Times*, October 30, 1958; idem, "Bigfoot Story Captured the Nation 19 Years Ago."

**36.** Andrew Genzoli, "In Defense of Bigfoot," *Humboldt (CA) Times*, October 9, 1958 (most ellipses in original).

And now the tales were supposedly scaring elderly residents. The *Blue Lake Advocate* hoped that proving the tracks fake would "relieve any nervousness which has been felt by visitors and residents of the Bluff Creek area." A television station followed up with a report promising an exposé. The sheriff's department took the matter seriously enough to arrange for a meeting with the man who had been implicated, hoping that he would confess, and the foolishness would stop. Given all the talk of shooting Bigfoot, it's reasonable to assume that the authorities were probably anxious to end the mystery before someone was accidentally gunned down as he or she walked through the forest. Law enforcement officials aren't usually as comfortable with the unsolved, rampant speculation, and rumors of monsters as newspaper columnists. They prefer the stable and boring to the mysterious.[37]

Four days later, Monday, October 13, before the sheriff could talk with him, the *Humboldt Times* learned the identity of the accused, interviewed him, and discovered that he denied the allegations. He refused even to speak with law enforcement officials. "Sheriff's Office Ends Up With 'Bigfoot' in Mouth," blared the front page headline on Tuesday, the wording almost certainly influenced by Genzoli's animosity toward the investigation. On October 16, the *Advocate* backed off its claim of a hoax since the county official could offer no additional evidence.[38]

Really, though, Genzoli need not have worried about the promised exposé. The accused was Ray Wallace, Shorty's brother—and he wasn't about to force Bigfoot out of the parade of the mysterious. The naturalist Robert Michael Pyle, who interviewed Wallace in the 1990s, concluded that Wallace not only bulldozed mountains but also built "mountains of bullshit," an assessment that was widely shared by those who met or knew Wallace. He was cut from the same cloth as Eugene Shepard: a storyteller, a prankster, a working man who liked to have fun.[39]

In his first interview with the *Humboldt Times*, Wallace acted as though he were incensed at being fingered. He hadn't even been in the area when

**37.** "'Big Foot' Tracks Reported a Hoax," *Blue Lake(CA) Advocate*, October 9, 1958 (quotation); "Promised Hoax Exposé of Mysterious Footprints Fails to Materialize," *Humboldt (CA) Standard*, October 14, 1958; "Bigfoot Advice: Shoot Only in Self-Defense," *Humboldt (CA) Times*, October 17, 1958.

**38.** "Promised Hoax Exposé of Mysterious Footprints Fails to Materialize"; "Sheriff's Office Ends up with 'Bigfoot' in Mouth," *Humboldt (CA) Times*, October 14, 1958; "To Our Readers," *Blue Lake (CA) Advocate*, October 16, 1958.

**39.** Robert Michael Pyle, *Where Bigfoot Walks: Crossing the Dark Divide* (Boston: Houghton Mifflin, 1995), 191.

FIGURE 13. Ray Wallace was one of three Wallace brothers overseeing the construction where Jerry Crew found the Big Foot tracks. Cut from the same cloth as Eugene Shepard, self-proclaimed capturer of the hodag, Wallace first denied believing that Bigfoot existed, then—after being accused of faking the tracks that Crew found—asserted that he had known of Bigfoot for years. (Dave Rubert Photography, courtesy of the Wallace family.)

the tracks were found, he claimed, so he couldn't have pulled the hoax. "I'm going to sue them for slander," he said. "I won't fool around about it!" Wallace had worked in the area for four years, he told the newspaper, and he'd seen these kinds of tracks before. "I used to think they were bear tracks," he said. "And I still do." But while Wallace was not worried by the tracks, he said that his men were. The footprints and rumors, he claimed, had driven several men from the job. Work had stopped. He was losing money. "I wish they'd let it die down," he said, so that he could get back to work, never mind the bears. It seems pretty clear from his subsequent actions, however, Wallace was conning the newspaper: he didn't want Bigfoot to die. He wanted him to live.[40]

A few hours after the *Humboldt Times* arrived at porches and newsstands on Tuesday, October 14, its front-page headline mocking the sheriff's department, a pair of men held a press conference about Bigfoot. They said that they had seen Bigfoot on Sunday night—two days before. Ray Kerr and Leslie Breazeale told reporters that they were driving down a country road, Breazeale napping in the passenger seat, when suddenly, "He—or it—bounded across the road." Bigfoot "ran upright like a man, swingin' long, hairy arms. It happened so fast," Kerr said, "it's kinda hard to give a really close description, but it was covered with hair. It had no clothes. It looked eight or ten feet tall to me." Breazeale said that he woke up when Kerr hit the brakes.

**40.** "Sheriff's Office Ends up with 'Bigfoot' in Mouth."

"I saw it leap into the brush," he said. "I don't know what it was, but it wasn't no man—that's definite." Kerr and Breazeale were both from McKinleyville, both employed by the Wallace brothers, and, according to Marian Place, both had been instructed by Ray Wallace to hunt the creature.[41]

The *Humboldt Times* contacted two of the Wallace brothers for comment; Shorty insisted that the tracks were neither a hoax nor bear prints; Ray was still angry—or feigning ire—at being accused of faking the tracks ("I'll sue the . . . for slander," he repeated), but otherwise had changed his story. He agreed with his brother: these were not bear tracks nor a practical joke, but the mark of a creature unknown to science. Wallace regaled the reporter with stories of a wildman that (he said now) he'd known for years. "Why, I've seen [such tracks] all over this country before, man," he said. He had seen a deer that had been ripped to shreds, heard tales of the wildman from Native Americans; his employees had found a herd of cattle in the woods that they were convinced the wildman shepherded. Wallace said that another employee had seen the beast—just last week. And then the guy had quit, part of what Wallace now claimed was a general exodus. Fifteen men had left the job, he said. "I've got three tractors sitting up there without operators, man, and the brush-cutting crew has all quit." Wallace had plans to catch the creature and end the terror. He said that he was going to "build a big cage, man, and put it out there in the woods baited with a fresh-killed calf. We'll kill off all the human scent on the trap and around and maybe we'll catch it." But the plans were curiously without urgency for someone who—if Marian Place's numbers are correct—had lost half his work force. Wallace thought that he'd get around to hunting the beast in the winter or spring.[42]

It's hard to resist the interpretation that Wallace had Kerr and Breazeale fake their revelation. Why else would their press conference come days after the alleged sighting but be so perfectly timed with the accusations against Wallace? And why would Wallace have not bothered to tell about the supposed worker who had seen the beast more than a week before? Why insist that it was a bear if he knew differently? Why? Because Ray Wallace was improvising, making up the story as he went along, taking advantage of the

**41.** "Bigfoot Is 10 Feet Tall, Covered with Hair Says Pair," *Redding (CA) Record-Searchlight;* October 16, 1958 (second quotation); " 'Bigfoot' Sighted by Two Workmen Near Bluff Creek," *Humboldt (CA) Times*, October 16, 1958 (first and third quotations); Place, *On the Track of Bigfoot*, 20, 35–37; Loren Coleman, "Was the First 'Bigfoot' a Hoax? Cryptozoology's Original Sin," *The Anomalist* 2 (1995): 19.

**42.** "Is Bigfoot a Man, a Bear, or a Hoax?" *Humboldt (CA) Times*, October 15, 1958.

newspapers' eagerness to publish new and exciting revelations about Bigfoot. Most likely, he wasn't even losing employees—that was just another detail he had invented. Wallace was playing a game that P. T. Barnum had perfected a century before, mixing the real and the unreal, telling obviously contradictory tales—anything to keep the story going and himself in the limelight. His newfound earnestness was part of the gag. Hoaxing requires it: only sincere professions will convince the greenhorn to hunt the snipe.

## "Maybe Bigfoot Is Lost Relative of Old 'Sasquatch'"

By November interest in Bigfoot had died down. On the first day of the month, Jess Bemis reported fresh tracks had been found. After that, Genzoli ran one more column about Bigfoot, and then the newspaper printed nothing more about the mysterious tracks for the rest of the year. According to the press, some of the men working on Bluff Creek Road, tired of publicity, bulldozed other tracks before they could be reported. Bigfoot was "still a mystery" Genzoli wrote in his final piece. Wallace had suggested that the creature might be a wildman, but he didn't seem too serious or reliable.

It was John Green who finally offered a solution to Bigfoot's identity. In the Sasquatch file that Green was compiling, newsman that he was, the story of Bigfoot at Bluff Creek became another classic case.[43] When Green first saw reports of Bigfoot, only a little over a year after he started investigating Sasquatch seriously, he thought that someone in California was "attempting to plagiarize" the Canadian monster, having a little fun, pulling a practical joke using Sasquatch as the model. There was evidence to suggest that the stories in British Columbia were more than humbuggery, more than legends. There were credible witnesses, details that made the creature seem a real animal, but Green doubted that was the case in California. Still, his experiences with the Harrison Hot Springs hunt taught him not to dismiss out of hand what at first seemed the ravings of cranks.[44]

Early in October, Green loaded his wife and a friend into his car and left for California. Dahinden had not yet become a Canadian citizen so couldn't go

**43.** "Bigfoot Walks Again!" *Humboldt (CA) Times*, November 1, 1958; Genzoli, "Still a Mystery"; "Giant Tracks Seen Again at Bluff Creek," *Humboldt (CA) Times*, October 29, 1958; Place, *On the Track of Bigfoot*, 39.

**44.** "Maybe Bigfoot Is Lost Relative of Old 'Sasquatch,'" *Humboldt (CA) Times*, October 11, 1958 (quotation); John Green, *Sasquatch: The Apes Among Us* (Seattle, WA: Hancock House, 1978), 83.

along. It was a long and dirty trip. The maps proved unreliable, sending them wandering, lost. On one mountain road, they were nearly run down by a truck. "Almost a quarter inch of dust" settled "on everything in the car, including us," Green said. As they approached the area where Bigfoot had been seen, locals warned them off. It was all a hoax, they said, and recounted the story of the retarded Indian. Green figured that there was no "tradition in that part of California of the Sasquatch" and so the Americans had invented "silly explanations," such as the one about the "crazy man who had run away to the woods."[45]

Arriving at Bluff Creek Road, the Canadians were told by Crew and Bemis that the best tracks had recently been destroyed, but they might find some older tracks if they looked around. "Oh sure, just what we expected," Green thought. To his surprise, however, they quickly found some. Ian MacTaggart, the zoologist with the British Columbia Provincial Museum, had told Green that the Ruby Creek prints were made by a bear, its front and rear tracks overlapping. But now, seeing these tracks along Bluff Creek, he knew that MacTaggart had been wrong. These weren't bear tracks. They weren't pranks. They weren't somebody plagiarizing Sasquatch tracks. The whole thing couldn't be dismissed as Indian legend. "Looking at them was quite an experience," Green said. "I realized that in spite of having undertaken a 2,000-mile trip just to see them for myself, deep down I had never expected that there would be anything to see. Fortunately my wife was with me. She might have found it more difficult to be so understanding over the years had she not seen the tracks herself."[46]

What particularly impressed Green was the similarity between the tracks that he saw in California and the sketch that he had of one from the Ruby Creek Incident. "They did not diverge more than half an inch at any point," he wrote. How could any of this be a hoax then? How could Wallace fake tracks—how could anyone in California fake tracks—that just so happened to look like mysterious footprints found in Canada seventeen years earlier? The *Humboldt Times* covered Green's investigation—more grist for the mill—and suggested that "maybe Bigfoot" was the "lost relative of Old 'Sasquatch.'" Green thought differently. Bigfoot and Sasquatch were more than relatives. "There was no doubt that the same thing was involved in both areas," he wrote.[47]

**45.** John Green, *Bigfoot: On the Track of the Sasquatch* (New York: Ballantine, 1973), 40 (third quotation); idem, *Sasquatch*, 66–67 (first, second, and fourth quotations, 67).
**46.** "Canadian Satisfied Big Tracks Genuine," *Humboldt (CA) Times*, October 14, 1958; Green, *Bigfoot*, 39–45 (first quotation, 41); idem, *Sasquatch*, 67 (second quotation).
**47.** "Maybe Bigfoot Is Lost Relative of Old 'Sasquatch'" (second quotation); Green, *Bigfoot*, 41–42 (first quotation 41); idem, *Sasquatch*, 67 (third quotation).

Confirmation of a sort came the next month. Although the newspapers stopped covering Bigfoot, Titmus continued to hunt the beast, and in November, he found some more tracks. Titmus invited Green down for a look-see—the two had met on Green's earlier visit and gotten along well, and so Green made the long trip south again. These tracks were deeply impressed—Green had to jump from a log to make an indentation as deep—and the ball of the foot appeared to be split, or doubled. "The original track," Green wrote, "could conceivably have been that of an enormous human with a very wide foot and fallen arches," but whatever had made these tracks, with their split ball, was nothing like a flat-footed human. Years later, the anthropologist John Napier studied these tracks and found that although they were bizarre, and suggested that Bigfoot walked in a completely different way than humans, they were internally consistent. "Who other than God or natural selection is sufficiently conversant with the subtleties of the human foot and the human walking style to 'design' an artificial foot which is so perfectly harmonious in terms of structure and function?" he asked rhetorically. These tracks were also shorter than the original Bigfoot tracks, fifteen inches instead of sixteen. The difference in the length indicated that there was more than one sport walking the woods. Bigfoot was a member of a species, the same species as Sasquatch.[48]

On this trip, Green also spoke with Betty Allen and learned that the local Native Americans did indeed have tales about hairy wildmen. The legends just weren't widely known. "There had been no equivalent of Mr. Burns to tell the white community this story," he noted. Green came to fill Burns's role in regard to Bigfoot, just as he had replaced Burns as the interpreter of Sasquatch. He distanced the wildmen from legends. He brought the weight of history and numbers against the claim that the tracks were fakes. There was no Bigfoot in a zoo, no Sasquatch in captivity (at least not anymore), but their existence could be verified using the tools of law and journalism. It wasn't science, but these were other ways of making truth, and the truth was, Green argued, that Bigfoot was not an Indian myth, not a hoax. Bigfoot was a real creature, a Sasquatch, an ape.[49]

**48.** Green, *Bigfoot*, 41–42; John Napier, *Bigfoot: The Yeti and Sasquatch in Myth and Reality* (New York: E. P. Dutton, 1973), 120–23 (second quotation, 123); Green, *Sasquatch*, 67 (first quotation); Christopher L. Murphy, *Meet the Sasquatch* (Surrey, BC: Hancock House, 2004), 181.
**49.** Green, *Bigfoot*, 40 (second quotation), 42 (final quotation); idem, *Sasquatch*, 67 (first and third quotations).

CHAPTER FIVE

# ABSMery 1959–1961

During the summer of 1959, John Green visited the area around Bluff Creek again. He met with Titmus, for whom Green was developing a great deal of respect—the taxidermist had a vast fund of knowledge and was dedicated to the search. No doubt, he was easier to be around than the notoriously prickly Dahinden. Green and Titmus spent a lot of time in the woods, camping, looking for tracks. California was where the action was, where North America's great ape had been most active. It seemed to Green that it would be only a matter of months before the beast was caught. What a scoop![1]

On the trip, Green also met Ivan Sanderson. A forty-eight-year-old Scottish naturalist, Sanderson had come of age before the sun had set on the British Empire; he had attended Eton and Cambridge (where he studied geology and biology) then set out to explore the world, collecting specimens for august British institutions. During World War II, he served as a press analyst at British Ministry of Information's office in New York City. Maybe it was the seductions of Manhattan, maybe it was discovering the power that the modern media had to shape reality—it's not clear why, but Sanderson redirected his prodigious energies. He quit the life of the British explorer, moved to New York, gained citizenship, and became master of a new empire—the empire of American media. Between 1950 and 1957, he published a book each year (except 1954), starting, appropriately enough, with *How to Know American Mammals*; during the same period, he also wrote at least twenty-five popular articles for publications ranging from *American Druggist* to *True*. The magazine *Sports Afield* called him a "nature writer of real distinction, lucid,

**1.** John Green, *Sasquatch: The Apes Among Us* (Seattle, WA: Hancock House, 1978), 69; Loren Coleman, *Bigfoot! The True Story of Apes in America* (New York: Pocket Books, 2003), 219.

FIGURE 14. Like Tom Slick, the naturalist Ivan Sanderson (shown here) had an interest in both mainstream science and that which was decidedly beyond the pale. In the late 1950s, he took an interest in Bigfoot and helped to sustain national interest in the creature, and in wildmen more generally. (Courtesy of the American Philosophical Society.)

lyric, and humorous." Sanderson was also a fixture of radio and television, on which he discussed scientific topics. Charming, he looked the part of the British professor, tall and slender with a thin moustache and noticeable accent. He had a seemingly inexhaustible supply of stories.[2]

The meeting between Green and Sanderson was an important one for Bigfoot's history. Sanderson—like Green—wanted to make the study of Sasquatch into a proper science, what Sanderson called ABSMery, and the naturalist seemed to have the resources to do so. An acquaintance of Tom Slick, Sanderson helped to convince the millionaire to hire Green, Dahinden, and others to hunt Bigfoot—science in action. Neither the hunt nor Sanderson's ABSMery, however, lived up to their promise—Bigfoot remained uncaught, outside the borders of science. But that doesn't mean ABSMery was a failure. Sanderson helped to make the monster into an enduring icon, especially among white working-class men, a symbol of resistance to the indignities of modern life.

**2.** "Biography, Ivan Terence Sanderson," n.d., Bigfoot file 7, Andrew Genzoli papers, Humboldt State University, Arcata, CA; Hubert Saal, "Sanderson, Incorporated," *Sports Afield*, October 1957, 80–84, 133–34 (quotation, 81); Ivan T. Sanderson, *Green Silence: The Story of the Making of a Naturalist*, ed. Sabina W. Sanderson (New York: David McKay Company, Inc., 1974).

## The (Weird, Wacky) Wonderful World of Ivan T. Sanderson

Sanderson was not a run-of-the-mill naturalist. In addition to his affection for the natural world, he was fascinated with outré theories and odd speculations. Among other topics, he was intrigued by tektites, luminous snow, plant emotions, Norse runes in North America, sea monsters, and flying saucers. Sanderson considered himself a Fortean, that is to say, a devotee of the writer and scientific gadfly Charles Fort. Back in the early part of the twentieth century, Fort spent his days seated in the New York Public Library sifting through old newspapers and scientific journals looking for reports of bizarre, unexplained events—"damned" things, he called them, because they had been cast outside of science, but nonetheless seemed reported in good faith: teleportation, mysterious lights in the sky, rains of frogs, disappearing people. After his death, Fort became something of a cult figure, with Forteans continuing his task of compiling miscellanies of the bizarre. In the late 1950s and early 1960s, Sanderson committed almost all his attention to Forteana. He wrote for the science fiction magazine *Fantastic Universe*, which had Fortean leanings (Sanderson edited an issue on UFOs in 1957) and, after it folded, moved on to another Fortean magazine, *Fate*, where, between 1962 and 1966, he had published forty thousand words' worth of material (and had rejected another thirty-five thousand). He also wrote a number of books on matters Fortean. In 1965, he established the Society for the Investigation of the Unexplained, a Fortean organization that collected reports of, and investigated, unexplained phenomena. The society published the magazine *Pursuit*.[3]

There was a strong antiscience bent to much Fortean writing. Tiffany Thayer, for example, founder of the first Fortean Society, thought that science was a conspiracy foisted on a naive public. Sanderson did not truck with such theories. "Apart from the occult and all forms of mysticism, religion, and suchlike," he said once, "there is nothing I abominate more than pseudo-science." Rather, he thought that Fort invigorated science—made science more scientific. Science had become insular, he said, professional, as rote, as ossified as any bureaucratic desk job, all "bottle-washing and button-pushing." From their ivory towers, scientists could not see what Fort did:

**3.** Sanderson to Ollie, February 14, 1966, and Sanderson to Paul R. Reynolds, February 29, 1960, both in Agents #1 file, Ivan T. Sanderson papers, B Sa3, American Philosophical Society, Philadelphia, PA; Charles Fort, *The Books of Charles Fort* (New York: Henry Holt, 1941); Jim Steinmeyer, *Charles Fort: The Man Who Invented the Supernatural* (New York: Jeremy P. Tarcher/Penguin, 2008).

FIGURE 15. Charles Fort was an early twentieth-century writer who collected evidence of phenomena that science could not explain—what he called "damned facts." Fort developed something of a cult following, and among his devotees were Ivan Sanderson and a number of other Bigfoot aficionados. Here he is shown with Theodore Dreiser (left) in 1931. (Theodore Dreiser Papers, Rare Book & Manuscript Library, University of Pennsylvania.)

that the world is pregnant with the unexplained. They dismissed such *things*, as Sanderson frequently called these wonders. Sanderson thought of himself as writing against what Forteans termed the "wipe"—the disregard for the bizarre by officials and scientists, the neglect of things that did not fit into current theories. He was rescuing damned facts, burnishing them, and considering them as they ought to be considered. "Is not all science a search for new facts, for things that were not known or not understood before?" he asked.[4]

Among those damned facts that Forteans attempted to rescue were wildmen. Sanderson's interest in wildmen was first sparked when a classmate at Cambridge drew his attention to the subject. (It's likely that Sanderson's mentor on the subject was Gerald Russell, the future Yeti hunter.) From that point on, Sanderson said, he "collected anything I could find that was said or published about this matter," amassing a huge file on wildmen reports from

**4.** Sanderson to Martin Gardner, March 26, 1968 (first quotation), Gardner, Martin file, and "The Borderlands of Science," n.d. (final quotation), Borderlands of Science file, both in Ivan T. Sanderson papers; Ivan T. Sanderson, "More About the Abominable Snowman," *Fantastic Universe*, December 1959, 27–37 (second quotation, 35–36); Ivan T. Sanderson, *Investigating the Unexplained* (Englewood Cliffs, NJ: Prentice-Hall, 1972), 2–4; Doug Skinner, "Doubting Tiffany," *Fortean Times*, August 2005, 48–52.

around the world—like Green and Dahinden, he had discovered the largely uncataloged folklore about the legendary beasts. Sanderson came to believe that there were several undiscovered species of wildmen throughout the world, in Africa and Asia and Europe and South America.[5]

In the early 1950s, Sanderson met Elliott Rockmore, a science fiction fan and flying saucer enthusiast who had also been collecting stories about wildmen, and who shared with Sanderson some stories about Sasquatch. A few years later, Sanderson began corresponding with a cadre of Russians led by the historian Boris Porshnev. They had been studying reports of wildmen in their country, had sent out expeditions, issued reports, and established what seemed to be an institute at the Darwin Museum. Around the same time, Bernard Heuvelmans, by now a friend, shared with him the material that he had gathered for his book *On the Track of Unknown Animals*. Sanderson was also friendly with Ralph Izzard, the Yeti hunter, who, like Russell, had been a classmate at Eton.[6]

Of course, Sanderson heard of Bigfoot's appearance in northern California, word apparently reaching him from a European colleague who read of the wildman in the newspaper. Perhaps inspired by the attention lavished on Bigfoot, Sanderson began a series of articles on wildmen for *Fantastic Universe*. The editor of *Fantastic Universe* was a friend and allowed Sanderson to "say anything that [he] liked" without editorial interference; he often used the magazine to work out his thoughts and try out new ideas in front of a friendly but small audience. His three articles on the wildmen for *Fantastic Universe* were experiments, attempts to make sense of all the reports he had collected and figure out the relations between the various species of wildmen; the arguments were abstract, with few specific examples, and confused, not quite contradictory, but not quite coherent, either.[7]

At about the same time that he was writing these articles, Sanderson had a chance to go out and investigate the matter for himself. Random House com-

**5.** "Interminable Woodsmen: Legend Come to Life," August 1968 (quotation), Wandering Woodsmen file, Ivan T. Sanderson papers; Loren Coleman, *Tom Slick and the Search for Yeti* (Boston: Faber & Faber, 1989), 100.

**6.** "Interminable Woodsmen: Legend Come to Life," August 1968, Wandering Woodsmen file, Ivan T. Sanderson papers.

**7.** Sanderson to Paul R. Reynolds, February 29, 1960 (quotation), Agents #1 file, Ivan T. Sanderson papers; Ivan T. Sanderson, "The Abominable Snowman," *Fantastic Universe*, October 1959, 58–64; idem, "More About the Abominable Snowman," 27–37; idem, "There Is an Abominable Snowman," *Fantastic Universe*, February 1960, 16–26; idem, *Abominable Snowmen: Legend Come to Life* (Philadelphia, PA: Chilton, 1961), 20, 129.

missioned him to write *The Continent We Live on*, a natural history of North America so he, the ornithologist Robert Christie, and the geologist Wendell Skousen spent almost a year driving across the United States and Canada in a station wagon, seeing what they could see. Sanderson brought with him what he called a "brief" on Fortean items put together by a pair of his friends and sorted according to state, so that along with North American natural history he could explore reports of the bizarre as well. In the brief, "under 'California' (and again Oregon, Idaho, Washington, and British Columbia) were page after page of references" to wildmen.[8]

The three men arrived in Humboldt County during August 1959, Sanderson confident in his ability to suss out the truth about Bigfoot in an hour or so—the amount of time, he claimed, that it took him "to spot a genuine phoney." As a Fortean, Sanderson thought himself free of the dogmas and theoretical commitments that blinded most scientists to the world's wonders, but neither was he a yokel. He had academic credentials, had spent extensive time exploring, and was—although he didn't use the phrase—an adept at Zadig's method. "I was trained in scientific methodology and born a complete pragmatist," he bragged, "and after many years of police and intelligence training I can spot a phoney as well as most." He didn't expect to *see* Bigfoot, nor did he need to. "Half a dozen questions," he told Betty Allen, "are enough to raise sufficient doubt in our minds to decide if the matter is no longer worth pursuing."[9]

Certainly, it didn't take Sanderson long to conclude that Ray Wallace was a "menace." Sanderson didn't explain why he distrusted Wallace, although it seems fair to surmise that his assessment had something to do with Wallace's changing story, first denying the existence of a wildman, then going on about how he had known of such a creature for a long time; and, probably, his appraisal had something to do with the way Wallace's yarns became increasingly fabulous—Bigfoot not just an ape, but a shepherd, say. Additionally, according to Sanderson, Wallace spread scurrilous rumors about him, although, again, he left the details vague. Sanderson also distrusted Bob Titmus. He heard that the taxidermist was selling plaster casts of Bigfoot tracks

**8.** "Interminable Woodsmen: Legend Come to Life," August 1968 (quotation), Wandering Woodsmen file, Ivan T. Sanderson papers; Ivan T. Sanderson, *The Continent We Live on* (New York: Random House, 1961); Marian T. Place, *On the Track of Bigfoot* (New York: Pocket Books, 1978), 53–54.

**9.** Sanderson to Allen, August 1959 (first and final quotations), Bigfoot 7 file, Andrew Genzoli papers; Sanderson to Martin Gardner, March 26, 1968 (second quotation), Gardner, Martin file, Ivan T. Sanderson papers.

to tourists for $3.50 each. That was more than tacky—it raised questions about Titmus's objectivity.[10]

Sanderson, however, didn't leave Humboldt County after an hour. Instead, he became convinced that the area was populated by "some living examples of a race of humanoid creatures." Part of what decided him was the landscape—the vast stretches of montane forests impressed him as being able to support and hide wildmen. Other people he met on his trip also convinced him. Allen and Crew and Crew's family seemed genuine; the Native Americans with whom he spoke came across as "consummate naturalists and complete realists." In Sanderson's opinion, laymen, unburdened by the biases of scientists, were often better observers of reality than elites. "As a working reporter, having now been privileged to travel extensively," he said, "I would state that I find the so-called 'native' in some respects on the whole more reliable than the foreigner, and the white foreigner in particular." Finally, Sanderson became convinced of Bigfoot's existence because—*mirabile dictu*—as he was investigating, word came that Titmus had found more tracks. He rushed out to the site and deemed the prints fully authentic.[11]

## ABSMery

While in California, Sanderson penned an article about Bigfoot—the first in what would be a rush of writing, three more articles and a massive book, *Abominable Snowman: Legend Come to Life*, all published by the fall of 1961. He also extended his travels so that he could meet Green and Roe and Dahinden and the Chapmans and Ostman in Canada. Sanderson's articles and book built on ideas he had played with in *Fantastic Universe*—indeed, his last essay for the pulp was very similar to a draft of the article that he wrote in California—but the airiness was replaced with specific anecdotes and his claims cohered.[12]

**10.** Sanderson to Allen, August 1959, (quotation), Bigfoot 7 file, Andrew Genzoli papers; Sanderson to Slick, May 18, 1962, Slick, Tom file, Ivan T. Sanderson papers; Loren Coleman, "Was the First 'Bigfoot' a Hoax? Cryptozoology's Original Sin," *The Anomalist* 2 (1995): 15–17.

**11.** Sanderson to Allen, August 1959, Bigfoot 7 file (quotations), Andrew Genzoli papers; John Green, *The Sasquatch File* (Agassiz, BC: Cheam, 1973), 22 (these tracks are mistakenly labeled as coming from 1958); Place, *On the Track of Bigfoot*, 56–59.

**12.** Sanderson, "Our Missing Relatives," n.d, Our Missing Relatives file, Ivan T. Sanderson papers; idem, "The Strange Story of America's Abominable Snowman," *True*, December 1959, 40–43, 122–26; idem, "A New Look at America's Mystery Giant," *True*, March 1960,

The main thrust of his argument was that humanity's ancestry was complex—the ascent from ape to human was not linear, but a branching net, with many species—and the world was vast, so it was possible that some of humankind's distant relatives had survived extinction and escaped detection by Western science. Some of these unknown wildmen were gorilla-like apes, while others were primitive humans, engaging in trade with modern humans, perhaps even interbreeding. There was a precedent for such cases, he pointed out, examples of peoples or creatures thought extinct suddenly being found. The okapi, a relative of the giraffe, was unknown to anyone but locals in what is now the Democratic Republic of the Congo until 1901. In 1911, Ishi, the last of the Yahi tribe of Native Americans, emerged from the mountains of northern California and into the world of whites for the first time. In 1938, scientists found the coelacanth, a fish that was thought to have been extinct for millions of years. Why not, then, a hidden wildman?

Sanderson called the study of wildmen "ABSMery," a neologism based on the abbreviation of Abominable Snowman: ABSM. He hated the name Abominable Snowman—like the Holy Roman Empire that was neither Holy nor Roman nor an empire, Sanderson insisted that the beast was neither abominable nor a creature of the snow, nor a man—but, ever the salesman, he understood its power. The Abominable Snowman was known worldwide and had in its favor the word of mountaineers and scientists. Sanderson wanted to expand the aura of possibility that cloaked the snowman to include other wildmen. If one existed in the Himalayas, why not the wilds of Canada? Indeed, the Asian and North American wildmen might be the same creature. Sanderson argued that the large *dzu-teh* and North America's wildman were *Gigantopithecus*, the New World version having crossed the Bering Strait when there was a land bridge. The suggestion added scientific gravity to Genzoli's offhand remark that California had found its own Abominable Snowman and it provided a title for Sanderson's first article, "The Strange Story of America's Abominable Snowman."

This article, and the ones that followed it, did not appear in *Fantastic Universe* or *Fate* or any Fortean publication; it appeared in *True*, as did two others. (Another ran in *Sports Afield*.) *True*, certainly, and *Sports Afield*, to an extent, belonged to a class of magazines that existed only from the 1950s and 1970s, adventure magazines for men. These were successors to the science fiction and detective pulps of the 1920s, '30s, and '40s—often owned by the

........................

44–45, 101–2, 115; idem, "The Ultimate Hunt," *Sports Afield*, April 1961, 66–69, 113–18; idem, "Abominable Snowmen Are Here!" *True*, November 1961, 40–41, 86–92.

same companies and employing the same talent—but had a more upscale look and, unlike the pulps, were frantic in their assertion that what they published was true: titles included, *True Action*, *True Adventures*, *True Danger*, *True Men*, *True Strange*, *True War*, *True Weird*, *Real Action*, *Real Adventures*, *Real Combat Stories*, *Real Life Adventures*, *Real Men*, *Real War*, and *Real*. All told, well over a hundred such magazines were published each month during the late 1950s. They were quite popular, with *True* selling about two million copies each month, catering mostly to an audience of white working-class men.[13]

In the 1950s and 1960s, American working-class white men were doing quite well financially, but, broadly speaking, they were leery of the changing culture. Mass media made the world seem fake and the postwar emphasis on family and "togetherness" (a word coined in 1954) could feel stultifying. In addition, the economy was changing from an industrial one to one based on services and consumption—and that changed what it meant to be successful, even what it meant to be a man. Working-class men valued themselves for their skills, their pragmatism, their ability to do things with their hands—"More than sexuality," wrote historian Joshua Freeman, "craft, strength, and the ability to endure made a man a man"—but the economy increasingly did not prize those qualities. Instead, identity was more and more tied to the womanly act of buying things. Lillian Rubin, a counselor as well as sociologist (and so a keen interpreter of working-class life), noted that white working-class men thought, "Without notice, the rules of the game have been changed; what worked for their fathers no longer works for them. They only know that there are a whole new set of expectations—in the kitchen, in the parlor, in the bedroom—that leave them feeling bewildered and threatened."[14]

Men's adventure magazines like *True* tapped into these anxieties. Some advertising preyed on masculine insecurities about the new world, in which looks and personality were more important than competence. Mostly, though, the magazines fanned embers of resentment burning in their readers' breasts and flattered them for rejecting frivolous things—such as shopping—and embracing the masculine, the true (although at the same time the maga-

**13.** Adam Parfrey, ed., *It's a Man's World: Men's Adventure Magazines, the Postwar Pulps* (Los Angeles: Feral House, 2003).

**14.** Lillian Breslow Rubin, *Worlds of Pain: Life in the Working Class Family* (New York: Basic Books, 1976), 120 (quotation); Barbara Ehrenreich, *The Hearts of Men: American Dreams and the Flight from Commitment* (New York: Anchor Books, 1983); Loren Baritz, *The Good Life: The Meaning of Success for the American Middle Class* (New York: Perennial Library, 1990); Joshua B. Freeman, "Hardhats: Construction Workers, Manliness, and the 1970 Pro-War Demonstrations," *Journal of Social History* 26, no. 4 (1993): 725–45 (quotation, 725).

FIGURE 16. The men's adventure magazines for which Ivan Sanderson wrote were aimed at a working-class audience that was relatively resistant to consumer enticements. The magazines played to that resentment while simultaneously trying to lure their readers into consumer culture. This advertisement for Slim Gard showed readers that their looks were important—and that problems with their appearance could be solved through shopping. (From *Saga*, July 1969.)

zines were themselves consumer products, filled with advertisements that encouraged men to buy). Articles denigrated the plastic and corrupt world in which men lived and celebrated another one, where competence still mattered. Some did so by showing that office jobs could be as rewarding as jobs that tested the mettle of a man: the world, although it had changed, still needed John Wayne. Others—the bulk—recounted adventurous tales of everyday men confronting—and defeating—nasty villains, from Nazis to Commies, from rampaging Indians to homosexuals, from sharks to tigers. Lurid covers showed endangered, half-dressed women watching as their brawny saviors battled these terrors, and headlines screamed "My God! My Guts Are Coming Out," "I Ride with the Desert Raiders," "I Shot Mussolini," "Weasels Ripped My Flesh," or "I Battled a Giant Otter," all while the magazine's banner claimed the stories were *True*, were *Real*.[15]

Sanderson's articles about wildmen fit neatly into *True*'s rubric, championing the role of skill against the bookish knowledge of scientists. For example, Sanderson told readers that he asked Jeannie Chapman "leading questions," using his superior knowledge to see if she was lying or not, and determining that she was honest. Sanderson also claimed the ability to make connections that no one else could. While talking with the Chapmans, Sanderson had George repeat the Sasquatch's call. It was "exactly the same strange gurgling whistle that men in California" had made when they mimicked Bigfoot's call, Sanderson wrote in *True*. "If all this is a hoax or a publicity stunt, or mass hallucination," he asked, "how does it happen that this noise—which defies description—always sounds the same no matter who has tried to reproduce it for me?" His expertise led him to where science feared tread, he suggested: to truth. Elites didn't know as much as they thought, and the working class could know a great deal more if they applied their skill—and read his articles. The world was a large place, with many hidden mysteries yet to explore.[16]

## The Pacific Northwest Expedition

When Sanderson visited Green in Canada, conversation swung around to sponsoring a Sasquatch hunt. Sanderson was an acquaintance of Tom Slick—he had served as one of his advisors during the Yeti expeditions and the two

**15.** Tom Pendergast, *Creating the Modern Man: American Magazines and Consumer Culture, 1900–1950* (Columbia: University of Missouri Press, 2000), 208–42; Parfrey, *It's a Man's World*.

**16.** Sanderson, "A New Look at America's Mystery Giant," 44–45, 101–2, 115 (quotations); Sanderson, *Abominable Snowmen*, 51–52, 436.

FIGURE 17. The journalist Betty Allen conducted seminal investigations on Bigfoot for the *Humboldt Times* and also helped to coordinate meetings between Bigfoot hunters such as Tom Slick and Bob Titmus. (Humboldt State University—Special Collections.)

had been negotiating a joint venture to exploit South American minerals and medicinal plants. In the months after Sanderson left Canada, he and Slick must have contacted each other, because two days before Halloween 1959 Sanderson called Betty Allen and asked her to reserve a room at Wyatt's Motel for Slick and arrange for Crew and Titmus to meet with him.[17]

The oil magnate reached the hotel at 10:30 Halloween morning, a Saturday, along with his personal secretary Jeri Walsh. Allen was there to greet him—a hostess, but still also a reporter: she was keeping notes. Not long after Slick arrived, Crew drove up in his Volkswagen and the four motored up Highway 96 to Bluff Creek. At the construction site, Slick, Walsh, and Crew inspected the frontier edge of the road while Allen spoke with some of the workers. Leslie Wallace, one of the Wallace brothers, told her that there had been no tracks that summer or fall. Ernie Killinger, the construction superintendent, was, according to Allen, "very skeptical" of the whole phenomenon. Afterwards, Crew drove them to where he had seen those famous tracks in August of 1958; there was no sign of them now. They then headed to a spot where the creek was accessible. Slick, Walsh, and Crew hiked down to the water.[18]

**17.** Alma to Solomon M. Chester, July 27, 1959, and memo, n.d, both in Slick, Tom B. file, Ivan T. Sanderson papers; Betty Allen, "Pacific Northwest Scientific Expedition," November 1959, Bigfoot file 7, Andrew Genzoli papers; Green, *Sasquatch*, 69.

**18.** Betty Allen, "Pacific Northwest Scientific Expedition," November 1959, Bigfoot file 7, Andrew Genzoli papers.

A few moments later, Titmus rolled up. He was with a friend, Ed Patrick. They checked on some trip-wired cameras that Titmus had set up a month before—nothing—and then met with the others by the creek. They stayed down there for two hours. Apparently, Allen remained on the roadside or in the Volkswagen, maybe because she was older and couldn't manage the walk, maybe because "the men," as she called them—although Walsh was a woman—were talking business and didn't want a reporter around. When the meeting ended, Crew drove Allen and Patrick back to Willow Creek; Slick and Walsh went with Titmus. At the hotel, Titmus showed off the casts that he had made in August and the smaller ones that he had made the previous November—the ones that had convinced Green that Bigfoot was a species of animal. Titmus also brought out a jar of what he claimed was Bigfoot feces. It was an iconic moment in Bigfoot history: a taxidermist, a creationist, a reporter, a millionaire, and his secretary in a motel room staring at mysterious shit.[19]

Slick did some more exploring the next day, nosing around another logging operation before flying out that night; apparently, Titmus also left, going to Oregon to inspect some tracks up there. And so both missed out on fresh evidence: Slick's visit, just like Sanderson's, seemed to have roused Bigfoot. Ernie Killinger and Jack Skidmore, another catskinner, found new tracks Monday morning. Supposedly, they "made a believer" of Killinger. Word of the discovery passed to the *Humboldt Times* and from there to Allen. She doubted the authenticity of the tracks. The timing was suspicious; it would have been too easy for Killinger to fake them, especially if he thought that he had an audience. Crew, however, took the tracks seriously. He, the pastor of the Community Church of Willow Creek, and another man spent Monday night investigating. According to Allen, "It was pitch black in the canyon and at one time they heard the sound of heavy walking down the hill from the road." Spooked, they ran. The next afternoon, Crew returned to the hunt. This time, he told Allen, he was sure that he smelled the wildman.[20]

Titmus also believed that the tracks had been left by a real wildman. He did not have a chance to cast them until Thursday, after several days of hard rain, but was convinced anyway. The tracks, he determined, had been left by the smaller Bigfoot, the one with the fifteen-inch foot. No doubt excited that a hunt might be sponsored, Titmus invited Green to California. This time, Dahinden came, too—he was, by now, a Canadian citizen, and so could cross

**19.** Ibid.

**20.** Ibid. (quotations); "'Bigfoot' Leaves Fresh Tracks at Bluff Creek," *Humboldt (CA) Times*, November 3, 1959.

into the United States. The two Canadians met with Titmus, inspected what remained of the tracks from early November, and also saw what looked to be the impression of a knuckle.[21]

Slick's biographer, Loren Coleman, argued that Titmus's cast of the smaller track encouraged Slick to move his operations from the Himalayas to California, but that seems wrong. It appears that Slick's mind was already set, perhaps convinced by Sanderson, because he wrote George Agogino on November 4—a day before Titmus cast the track—saying how excited he was about hunting Bigfoot. If the tracks played any role in Slick's decision, it was as confirmation. During the ensuing week, there were apparently more discussions among Sanderson, Slick, and Green. They decided another meeting was necessary, and this time Green asked Allen to book a room for Slick at Wyatt's Motel.[22]

At 2:30 in the afternoon that Sunday, November 15, Allen, Slick, Green, Dahinden, the Titmuses, Ed Patrick, and a handful of others convened in the motel lobby to discuss what Slick called "The Pacific Northwest Expedition." Hunting Bigfoot, Slick stressed, was important patriotically, financially, and scientifically. Already a number of countries were hunting the Abominable Snowman of the Himalayas. It would be a national embarrassment if some other country caught the Asian creature while America ignored the wildman in its own backyard. Financial backers, he said, had put up $5,000 to begin the hunt. And he had scientists lined up to study the evidence—many of the same ones who had consulted on his Yeti hunt. Expedition members would have to swear themselves to secrecy, Slick said, so that he could sell the story and recuperate the money that he had sunk into this hunt and the ones in the Himalayas. In return, they would get a chance to reap a financial windfall, be involved with a scientific breakthrough, and experience adventure—the ultimate hunt. This was dangerous stuff. The beasts "in the Himalayas [were] exceedingly dangerous and intelligent too," Slick said. Bigfoot most likely was as well. The hunters, he warned darkly, would have to decide if they were prepared to gun down such a humanlike creature.[23]

**21.** Betty Allen, "Pacific Northwest Scientific Expedition," November 1959, Bigfoot file 7, Andrew Genzoli papers; Green, *Sasquatch*, 69; John Green, *On the Track of the Sasquatch* (Surrey, BC: Hancock House, 1980), 39–40; Coleman, *Tom Slick and the Search for Yeti*, 104.

**22.** Betty Allen, "Pacific Northwest Scientific Expedition," November 1959, Bigfoot file 7, Andrew Genzoli papers; Green, *The Sasquatch File*, 22; Coleman, *Tom Slick and the Search for Yeti*, 104–5.

**23.** Betty Allen, "Pacific Northwest Scientific Expedition," November 1959 (quotations), Bigfoot file 7, Andrew Genzoli papers.

The meeting was fractious and disorganized. According to Green, an elderly woman knitting in a corner of the room frequently interjected and one of the most vocal participants was not a prospective hunter at all but a man who had come with Slick to talk business. Dahinden was not yet ready to credit the beast's existence, while Titmus was certain that the tracks could not have been faked. Nobody other than Slick really thought Bigfoot should be shot. At 6:00 that evening, Crew burst into the room saying that he had found tracks. He'd been checking the cameras with some friends when they happened across the prints. Everyone headed out to inspect Crew's discovery. And so the meeting ended with, as Allen noted, "no definite policy or group" having "been definitely set up."[24]

Eventually, an expedition *was* formed. Titmus was named the leader, about ten men were hired—the exact number and exact costs, like so much else, a secret—and a helicopter was commissioned to drop food at the hunting camps. But manpower, machines, even a leader were not enough to bring order. Chaos seems to have been integral to Slick's leadership style. The hunt in the Himalayas had been disorderly, and the same was true in California. Bigfoot hunters, Green noted, tended to be "pig-headed": "People who will go hunting for an animal that is rejected by the world of science and almost everybody else are bound to be people who don't pay much attention to any opinion but their own, and who expect not only to have an opinion but to act on it." And so, as there had been on Slick's expeditions in the Himalayas, there was a lot of fighting, antagonisms that simmered and boiled. Dahinden called the Pacific Northwest Expedition a "comedy of errors." It was "a real mess," he said, "one insane thing after another." A few years later, a local newspaper editor joked, "I really think" the hunters "should have tried our good spring water instead of the type of refreshment they brought with them."[25]

One problem was that Slick's millions attracted grifters and con men. Some time near the beginning of the hunt, two men appeared with what Green said was a "cock and bull story" about seeing a seven-hundred-pound apish beast on a fire trail. They were John La Pe, who ran a Spanish restaurant, and George Gatto, who worked at an Italian restaurant, both in Eureka.

**24.** Ibid. (quotation); Green, *Sasquatch*, 69.

**25.** James O. Holley, "Expedition Hunts U.S. 'Snowman,'" *Houston Post*, January 10, 1960; idem, "Big Tracks Sink Deep, Seem Real," *Houston Post*, January 12, 1960; "Bigfoot Reports," *Blue Lake (CA) Advocate*, June 24, 1965 (final quotation); Don Hunter and René Dahinden, *Sasquatch* (Toronto: McClelland & Stewart, 1973), 87, 90, 92 (remainder of quotations); Green, *Sasquatch*, 69–71 (first quotation); Coleman, *Tom Slick and the Search for Yeti*, 105–6.

Slick signed them up, so that in case they found something he would own the story and the beast. But La Pe and Gatto weren't interested in the solemnity of the hunt or obeying the secrecy clause. They peppered the local press with reports of tracks, catching wisps of Bigfoot's musky odor, and hearing thunderous crashes in the woods. They said that they went into the forest armed with submachine guns. (A few months later, La Pe was arrested for passing bad checks.) "That was the first real intimation we had of Tom's strange failing as a judge of people," Green later wrote. "It must have been different in the business world, but when it came to monster hunting he was chronically doubtful about the people who really wanted to find the thing, but an easy mark for any con artist." The first intimation, perhaps—but not the last. Around the same time, Slick also hired Ivan Marx, a bear hunter, raconteur, and prankster on par with Wallace.[26]

Meanwhile, Titmus was not inspiring confidence as a leader. He set up more trip-wired Brownie cameras and tried to attract Bigfoot with bait laid out on trays or stashed in trees—nothing too odd about that. But, according to Dahinden, Titmus also "assumed the bizarre chore of raiding the women's rest rooms at service stations" for used sanitary napkins, which he then nailed to trees. (Apparently the theory that Bigfoot was a libidinous monster on the prowl for human females was not restricted to Native Americans.) Titmus also spent some time looking for another of Slick's mythical beasties—a nine-foot salamander said to inhabit the region. It was a quest that even La Pe ridiculed.[27]

Dahinden was disgusted with the proceedings. Hired at a salary of $350 per month, he lit out after two months, returned, and then left again, furious at the incompetence. At one point, a helicopter dropped him and another man at a camp abandoned by La Pe, Gatto, and Kirk Johnson Jr. (who was cosponsoring this expedition as well). Snow had fallen and blocked access to the campsite. The helicopter took the most expensive equipment; Dahinden and the other man were supposed to collect the rest and hike out to meet a truck. The truck, of course, was a day late, and so the men had to set up camp in the snow. It was so bad that the other man started talking about eating the two donkeys that they had rounded up. "I told him that if he suggested that

**26.** James O. Holley, "It Lurked on the Edge of Dim Forest," *Houston Post*, January 11, 1960; Garth Sanders, "Taxidermist Wants Information on Bigfoot," *Redding (CA) Record-Searchlight*, February 27, 1960; Hunter and Dahinden, *Sasquatch*, 88; Green, *Sasquatch*, 71 (quotations); Coleman, *Tom Slick and the Search for Yeti*, 106.

**27.** Hunter and Dahinden, *Sasquatch*, 88–90 (quotation, 89); Coleman, *Tom Slick and the Search for Yeti*, 106, 120–25.

again I would have to shoot him," Dahinden recounted. "I had the only gun." They warmed themselves with whiskey and a fire built "round a tree with full pitch. The flames must have been up thirty feet or more."[28]

Sanderson was also distressed by the expedition. This was not what he imagined ABSMery should be! He didn't think that dogs should be used—they would scare the beast from the area. He chafed at the news blackout—he was a writer, after all. He worried that Slick was cutting him out—the reports to him were thin and spotty. And he thought that Slick's decision to put the hunters on salary was a bad one—it was an incentive to *not* catch the beast, to stretch out the hunt, keep the money coming. Dahinden made the same point: "There was money involved, and . . . there were some among the hunters who were determined to find enough continuing 'evidence' of their quarry to ensure that the flow of financing would not be interrupted." Dahinden thought that Titmus was among those inventing evidence. That mysterious shit, he said, the droppings that had been displayed at Wyatt's Motel, that was just a pile of horse apples.[29]

Slick, however, seemed unconcerned by the chaos that swirled about—that engulfed—his expedition. He fired Gatto and La Pe, but otherwise treated the hunt as a bit of adventure. "Tom got a tremendous kick out of monster-hunting," Green said. He brought his kids to the campsites. He postponed meetings to stay out in the brush. One time, Green remembered, he and Slick were walking through the woods just after setting up camp. (Titmus was making supper.) "There wasn't much chance of any animal hanging around" amid the noise and action, Green said, "but Tom clutched his rifle at the ready and said . . . 'Boy! We're hunting the biggest game in the world.'"[30]

## Enter Peter Byrne

The expedition went on hiatus, or at least slowed—the reports, unsurprisingly, conflict—at the end of December in deference to the mountain snow. Around the same time, Slick ordered Bryan and Peter Byrne to shut down operations in Asia and encouraged them to join him in North America. It seems likely that Slick hoped that the Byrnes could bring some semblance

**28.** Hunter and Dahinden, *Sasquatch*, 88, 91–92 (quotations, 92).
**29.** Sanderson to Jeri Walsh, May 5, 1961; Jeri Walsh to Slick, May 16, 1961; Sanderson, memo, n.d., all in Slick, Tom B. file, Ivan T. Sanderson papers; Sanderson, "The Ultimate Hunt," 118; Hunter and Dahinden, *Sasquatch*, 87 (quotation), 89.
**30.** Green, *Sasquatch*, 72 (quotations); Coleman, *Tom Slick and the Search for Yeti*, 105.

of order to the expedition. Certainly, Peter Byrne thought that he was being offered leadership of the expedition. The brothers reached California some time during the first half of 1960—Peter's book on Bigfoot says early in the year, but an article he published says June.[31]

If Slick did indeed hope that the Byrnes would organize the expedition, then he was sadly mistaken. The Byrnes' arrival irritated Titmus; the taxidermist was still handling the finances, but suddenly this newcomer was giving orders. (It probably didn't help that Titmus was going though a divorce at the time). Eventually, Titmus joined Dahinden and Green in Canada. Peter Byrne fired everyone else, all the other trackers and hunters and hangers-on, so that the expedition was only him, his brother Bryan, Steve Matthes—a hunter—and Slick, when he dropped by. Members of this reduced expedition found a dozen sets of prints but never saw the beast.[32]

They also never rid the expedition of tomfoolery. Shortly after the Byrnes arrived, Ray Wallace called to say that he had *caught* a young Bigfoot. For $1,000,000, he would hand over the critter. Byrne contacted Slick, who said that he would pay $5,000 for a look at the animal. No dice, Wallace said, and the negotiations continued for weeks. "Then came the urgent calls," Byrne said. Wallace "was getting into difficulties. The only thing that the young Bigfoot would eat was Kellogg's Frosted Flakes and it ate them by the hundred-pound bag. It was, in fact, running them dry financially." Eventually, Wallace decided releasing the beast was the better part of valor.[33]

In Canada, Green, Titmus, and Dahinden regrouped. They were still interested in hunting the Sasquatch. All the stories they had collected, all the sightings, all the tracks—it was only a matter of time, a few weeks, a few months before a Sasquatch was captured. They couldn't give up when they were so close. So, they negotiated with Slick to sponsor an expedition in British Columbia. Green, at least, was wary of Slick's leadership. He wanted Sanderson named head of the expedition. That never happened, but Green did win one concession: his team would have veto power over Slick's hirings,

**31.** Sanders, "Taxidermist Wants Information on Bigfoot"; Peter Byrne, "Being Some Notes, in Brief, on the General Findings in Connection with the California Bigfoot," *Genus* 18 (1962): 55–59; Hunter and Dahinden, *Sasquatch*, 90–91; Peter Byrne, *The Search for Big Foot: Monster, Myth or Man?* (New York: Pocket Books, 1976), 103–4.

**32.** "Interview: Bob Titmus," 1–6; *Sasquatch Odyssey: The Hunt for Big Foot*, DVD, directed by Peter von Puttkamer (1999; West Vancouver, BC: Big Hairy Deal Films, 2004).

**33.** "Ex-Willow Creek Man Says Bigfoot Captured," July 1961, Bigfoot file 7, Andrew Genzoli papers; Byrne, *The Search for Big Foot*, 104–6; Coleman, *Bigfoot!* 72.

a reaction to the La Pe and Gatto fiasco and possibly to the Byrnes' usurpation of the Pacific Northwest Expedition.[34]

The so-called British Columbia Expedition was in operation from the summer of 1961 until the fall of 1962 with various members joining and resigning. Slick stopped in occasionally, as he had in California, to join the hunt. "Between arguments we had a lot of fun," Green wrote, "and nobody got hurt. We solemnly shipped off quantities of suspected hair and droppings for analysis, some of which remained unidentified, but all that was found of Bigfoot and his friend was tracks, and not very many of those." Green, Titmus, and another man found some tracks in July of 1961, and Titmus found some more in October.[35]

## The Wipe: Or *True*'s Trouble with Truth, and Ivan Sanderson's

Sanderson insisted that ABSMery was a science, rooted in anthropological and zoological knowledge. The claim was not without some basis: the Italian journal *Genus* published his articles and both Carleton Coon and George Agogino watched Sanderson's work with more than passing interest. For the most part, however, ABSMery was not a science but something else altogether—an imaginative reaction to the conditions of a consumer-oriented society. George Agogino, who acted as a consultant for Slick's North American expeditions as he had for the Himalayan ones, suspected that some of the hunters were pulling a con, even beyond Wallace's Frosted Flakes–eating Bigfoot. Early on, Agogino heard rumors that the feces Titmus had found showed the presence of unknown parasites. But Sanderson later found that the droppings submitted by Titmus—whether the same droppings or not is unclear—actually came from a moose. (W. C. Osman Hill, who analyzed the scat as well, only said that it was best not to comment on it at all.) Alleged Bigfoot fur also came from moose. Sanderson wondered if Titmus was faking data. There are no moose in California, so the fur couldn't have been picked up mistakenly, but, as a taxidermist, Titmus had ready access to it. Years later, Agogino also said that some of the cameras

**34.** Jeri Walsh to Slick, May 16, 1961, and Sanderson to Slick, June 24, 1961, both in Slick, Tom B. file, Ivan T. Sanderson papers; "Interview: A Candid Conversation with a Prominent Sasquatch Field Worker Who Says Exactly What He Thinks," *The ISC Newsletter* 4, no. 2 (1985): 4; "Interview: Does the Sasquatch Exist and What Can Be Done about It?" *The ISC Newsletter* 8, no. 2 (1989): 7.

**35.** Sanderson to Slick, June 24, 1961, Slick, Tom B. file, Ivan T. Sanderson papers; Green, *Sasquatch*, 72 (quotations); Coleman, *Tom Slick and the Search for Yeti*, 108.

had been tripped and the film removed, as if to destroy an incriminating photo.[36]

Could Titmus have been faking evidence? It seems plausible. Titmus had the means and plenty of opportunities. And he had motive: Titmus was profiting from selling Bigfoot prints and being paid to hunt; he had gotten his name in the newspapers and may have even felt a jolt of perverse satisfaction at having fooled so many people. These benefits would only continue as long as enough of the right people believed that Bigfoot existed, as long as evidence continued to accumulate. He may or may not have faked the tracks found by Killinger and Skidmore—there were too many suspects for that crime—but he had ample opportunity to shape them so that they seemed to be just like the ones he had found the previous November. Titmus had a paternal interest in those smaller tracks and made certain that they would not be mistaken for any other prints. On the cast that he gave to Slick he wrote, "This is the 15" track—Not to be confused with the 16" track." Allen said that Titmus had made much of these smaller tracks at his Halloween meeting with Slick. It could have just been pride in finding something so unusual, but given that the moose fur and droppings indicated there had been deliberate fraud, there's reason to suspect other, less noble motives. Titmus also made questionable claims later, including saying that he had actually seen a Bigfoot himself back in the 1940s but had been so traumatized that he suppressed the memory for decades.[37]

Sanderson, however, didn't seem bothered by all the shenanigans. He presented himself as a hardheaded skeptic out to reform science, but that was something of a cover—what he was really out to do was sell stories of the bizarre to an American mass media that had an insatiable appetite for the novel and outrageous. He wrote about all kinds of silly, obviously untrue *things*. Back in 1948, for example, he investigated three-toed tracks found on a Florida beach that he claimed had been left by a giant penguin that had lost its way. (Later, the tracks were revealed to be the work of pranksters.) And in the 1960s Sanderson proposed that flying saucers did not come from

**36.** Agogino to Coon, January 5, 1960, General Correspondence A–D 1960 file, Carleton S. Coon papers, National Anthropological Archives, Smithsonian Institution, Suitland, MD; William C. Osman Hill, "Abominable Snowmen: The Present Position," *Oryx* 6 (1961): 95; Coleman, *Tom Slick and the Search for Yeti*, 110; Coleman, "Was the First 'Bigfoot' a Hoax? Cryptozoology's Original Sin," 15–17.

**37.** Betty Allen, "Pacific Northwest Scientific Expedition," November 1959, Bigfoot file 7, Andrew Genzoli papers; Coleman, *Tom Slick and the Search for Yeti*, 104 (quotation); "Interview: Bob Titmus," 2.

outer space but were the technology of an intelligent civilization living . . . at the bottom of the ocean! Sanderson offered these accounts as scientific conclusions, and maybe he really did think them true, but he also knew that arguing for the factuality of these stories was what put butter on his bread. "This is the day of monsters and other mysteries and . . . all youngsters from age nine to ninety just gobble them up and are crying for more," he told his agent. Fortean stories let him tap that market: according to his own reckoning, he had seven thousand fans who waited anxiously for his reports of the bizarre. His *Abominable Snowman*—despite a hefty $7.50 price tag (by comparison, an 18-ounce jar of peanut butter cost 55¢)—sold fifteen thousand copies in five years.[38]

Nor was *True*, despite its name, bothered by the untruths. The pressure to produce at such magazines was intense: a single editor might be responsible for as many as five magazines—that was fifty stories a month. There were only so many true stories in the world, though, and it took so long to track them all down and research them and fact check, all the while the clock was ticking, ticking, ticking, and the magazines needed to be *out*, at the printers, on the newsstands, while completely new stories had to be written for the next issue and the next. And so the magazines routinely invented stories out of thin air, making up new World War II battles and heroic adventures. For all the clamorous claims that the stories in these magazines were *true*, they were often just dressed up Westerns, spy stories, or science fiction tales.[39]

Readers didn't mind that their *True* (or *Real*) was full of lies. Truth in these magazines was not about facts or correspondence with reality but resisting changing values and valorizing an older tradition, when men were men and honored for their skills. Bigfoot was the perfect embodiment of this notion of true—probably just an invention but still seemingly authentic, as gritty as a cowboy, self-reliant, living on its own terms, far from the corrupting influences of a feminized and weak society. One man, for instance,

**38.** Sanderson to Ollie, February 14, 1966 (quotation), Agents #1 file; Sanderson to Sterling, October 8, 1966, Chilton file; Sanderson to Oliver Swan, October 10, 1966, Agents file; Richard Heller to Oliver G. Swan, June 13, 1969, Pyramid publications file, all in Ivan T. Sanderson papers; Ivan T. Sanderson, *Invisible Residents: A Disquisition upon Certain Matters Maritime, and the Possibility of Intelligent Life Under the Waters of This Earth* (New York: Ty Crowell Company, 1970); "Florida 'Giant Penguin' Hoax Revealed," *The ISC Newsletter* 7, no. 4 (1988): 1–3; Scott Derks, *Working Americans, 1880–1999*, vol. 1: *The Working Class* (Lakeville, CT: Grey House Publishing, 2000), 385.

**39.** Bruce Jay Friedman, "Even the Rhinos Were Nymphos," in Parfrey, *It's a Man's World*, 13–19.

reported to *True* that a Bigfoot-like creature appeared at his snowed-in logging camp. "He gave the 'dozer a tremendous kick sending it end-over-end down the mountainside. He then flipped our log bunkhouse over on its roof and grabbed my fifth of Four-Star Hennessy, a calendar picture of Marilyn Monroe, and MY COPY OF TRUE. He stowed these treasures in his despicable pouch and disappeared." It was a joke, of course, but a revealing one. Bigfoot did the things that white working-class men wished that they could do: live a life free of obligation, accompanied only by a woman who certainly would not nag or keep him from going fishing. Sanderson may not have been very scientific, but he knew his audience, and he gave them what they wanted.[40]

So, to make his claims stronger, Sanderson simply erased all of the problems with the evidence in his articles and, later, in his book. Titmus warranted not a single mention in Sanderson's writing—he was simply gone from the story, never mind that Green and many in Willow Creek put a lot of faith in what he said. Sanderson also hid evidence of Titmus's possible duplicity by lying. In *Abominable Snowman*, he argued that the mysterious droppings Titmus found "present[ed] one of the most positive bits of evidence for the existence of an ABSM, whatever it may be. Just about the only thing that can *not* be manufactured—at least to fool a medical man or veterinarian—is fæces." Of course, the droppings fooled no one. But Sanderson was fooling his readers, transforming a known fraud into a piece of irrefutable evidence.[41]

In Sanderson's hands, Ray Wallace also underwent an amazing transformation. Not a bullshit artist or menace, he became, in both the first *True* article and Sanderson's book, "a hard-boiled and pragmatic man." Wallace's contradictory statements were smoothed into a conventional conversion story. When Crew's story first went national, Sanderson wrote, "Wallace was convinced that somebody was trying to disrupt his work, and this made him furious." According to Sanderson, Wallace maintained his skepticism even against his brother Shorty's stories of moving drums of diesel fuel, disappearing culverts, and tossed tires. He only came to believe that Bigfoot was a real creature "when he stopped for a drink at a spring" and "stepped right into a mass of Mr. Bigfoot's tracks in the soft mud."[42]

**40.** Hugh Magone, "Abominable Snowmannerisms," *True*, March 1960, 4 (original emphasis).

**41.** Sanderson, *Abominable Snowmen*, 338 (quotation; original emphasis); Coleman, "Was the First 'Bigfoot' a Hoax?" 16, 20–22.

**42.** Sanderson, "The Strange Story of America's Abominable Snowman," 122–23 (final quotation); idem, *Abominable Snowmen*, 131 (first quotation).

Just before Sanderson's first article appeared in *True*, he wrote Andrew Genzoli saying that he was "utterly disgusted" by editorial "distortion and unauthorized insertions." The protest, though, seems to have been insincere: Sanderson's drafts hardly varied from what *True* published, and he went on to repeat some of the most obvious falsehoods—such as the quality of Wallace's testimony—and add even more outrageous tales.[43]

In the end, Sanderson's greatest accomplishment was not reporting the truth or making ABSMery into a respectable science—it was never that, at least for a time. Rather, he helped to free wildmen from the confining niche of Forteana, *True*'s circulation was about twenty times that of *Fate*, which was by far and away the largest of the Fortean publications. After Sanderson's first *True* article appeared, Andrew Genzoli's office was flooded with correspondence again—more than a thousand letters. Sanderson made the beast into a consumer object—by making it stand against consumerism. Sanderson's Bigfoot appealed to those who were looking for the real reality behind the plastic one, who saw in the past an era when men could test their mettle against the world, who felt that the new order deprived them of dignity.[44]

**43.** Lynwood Carranco, "Three Legends of Northwestern California," *Western Folklore* 22, no. 3 (1963): 179–85 (quotation, 183).

**44.** Place, *On the Track of Bigfoot*, 106; Carl Llewellyn Weschke, "The Face of Fate," *Fate*, June 1989, 46–49.

CHAPTER SIX

# Melting the Snowman 1961–1967

On July 29, 1959, shortly before Green and Sanderson first met, Sir Edmund Hillary was at the Savoy Hilton in New York accepting the "Giants of Adventure" award from the men's adventure magazine *Argosy*. Hillary told the gathering that he wanted to return to the Himalayas, this time not to reach Everest's summit but to spend a winter at between 16,000 and 20,000 feet studying human physiology in a low-oxygen environment. "A very topical problem in this rocket age," he said another time. While his team was high in the mountains, he also wanted to investigate the legend of the Yeti. "Something exists," he told those at the ceremony, although he wasn't sure what, exactly. He had never seen tracks himself, but he trusted Shipton and some of the others who had reported seeing Yeti prints. "I believe there is sufficient evidence to warrant a closer search for the maker of these tracks," he wrote in the *New York Times* several months later.[1]

Not long after the award ceremony, Field Enterprises Educational Corporation, publishers of *World Book Encyclopedia*, invited Hillary to Chicago to talk more of his plans and have dinner with John Dienhart, Field's director of public relations. The encyclopedia was experimenting with publishing more topical books—books that addressed not just the past but what Field's CEO Bailey Howard called "living history." The corporation had recently started producing its *Year Book*, an annual record of news, and would soon put out *Science Year*, with articles on space travel, genetics, and medical technology. Dienhart was impressed by Hillary's plans and Field became the expedition's

**1.** Walter Sullivan, "Space Tests Seen in Everest Climb," *New York Times*, July 30, 1959 (second quotation); Edmund Hillary, "Abominable, and Improbable?" *New York Times Magazine*, January 24, 1960, 66 (first and third quotations).

sponsor. A way of "making history and not just recording it," the mountaineer said.[2]

Whatever other merits the expedition may or may not have had, it certainly changed the history of modern wildmen. Hillary returned from the mountains convinced that the Yeti did not exist, and that conclusion nearly put an end to interest in ABSMery. Bigfoot was driven from mass culture and found refuge only in the embrace of Forteans and the popular culture of small-town America.

## Melting the Snowman

With backing secured, Hillary put together a team of twenty-two scientists and mountaineers, most of them involved with the physiological research. He chose Marlin Perkins to lead the Yeti investigation. Perkins was head of Chicago's Lincoln Park Zoo and a well-known celebrity, having hosted NBC's "Zoo Parade." Larry Swan—who had argued with William Strauss over the possibility of the Yeti's existence in the pages of *Science*—also joined the Abominable Snowman hunt. Swan was familiar with the Himalayas, having been raised in Nepal by his Methodist missionary parents; like Perkins, he was also a television personality, hosting a science show for kids on San Francisco's public broadcast station. Rounding out the hunting team were six Sherpas, one hundred and fifty porters, Bhanu Bannerjee (an Indian translator), journalist Desmond Doig (creator of the *Bing, The Abominable Snowbaby* comic), and John Dienhart. Hillary asked Shipton and Agogino to join him, as well. Shipton thought it "tempting" to climb with Hillary again, especially "on a trip which is so well endowed with cash," but opted to explore Patagonia instead. Agogino was busy and declined but heralded the expedition as "the best qualified group to enter the area that I know of."[3]

**2.** Roy Gibbons, "Chicago Firm Will Finance Himalaya Trip," *Chicago Tribune*, December 22, 1959; William Murray, *Adventures in the People Business: The Story of World Book* (Chicago: Field Enterprises Educational Corporation, 1966), 232–33 (first quotation, 232); Edmund Hillary, *Nothing Venture, Nothing Win* (New York: Coronet, 1977), 282–84 (second quotation, 283).

**3.** Tom Benet, "Bay Man Ready for Snowman," *San Francisco Chronicle*, Lawrence Swan clipping file, San Francisco State University, San Francisco, CA; Agogino to Coon, January 5, 1960, General Correspondence A–D 1960 file, Carleton S. Coon papers, National Anthropological Archives, Smithsonian Institution, Suitland, MD; Edmund Hillary and Desmond Doig, *High in the Thin Cold Air* (Garden City, NJ: Doubleday, 1962), 11–14; Marlin Perkins, *My Wild Kingdom: An Autobiography* (New York: E. P. Dutton, 1982), 136–37; Peter Steele, *Eric Shipton: Everest & Beyond* (Seattle, WA: Mountaineers Books, 1999), 223 (quotation).

The team left Katmandu on September 13, 1960, after the monsoon season ended. Most of the expedition headed to Mount Makalu while the Yeti hunters traveled to the Rolwaling Valley and Solu Khumbu area, not far from where Shipton had found his mysterious tracks in 1951. The 120-mile hike to the first base camp was especially hard on those whom Hillary called the "arty types": the Yeti hunters unfamiliar with life in the high mountains. (Even Swan admitted that he was not a "hot shot" mountaineer but of the "scared, conservative type.") "Trekking as one treks on an expedition is a selfish affair," Doig wrote. "It's your feet that matter, your weary shoulders, your ability to keep up with the rest—and damn the others." The surrounding area was desperately poor, many of the families existing solely on *chang*, the local alcoholic drink. Tom Nevison, the expedition doctor, tried to help children he saw in the Sherpa villages, "their bodies puffed up like water-filled balloons, their faces pinched and solemn," but could not do much to ameliorate the country's suffering.[4]

The trip proved too much for Dienhart. "Crew-cut and glib," in Doig's estimation, used to the "gay social whirl" of Chicago, he had prepared himself for the ardors of the expedition by shopping in New York and skiing in Argentina. It wasn't enough. His feet hurt; he wrenched a knee. A few days into the trek, Dienhart returned to Katmandu to fetch new tape recorders. He never returned, staying behind to dislodge scientific apparatuses stuck in customs and oversee the production of Field Enterprises's press releases. "Hillary-o-Grams," these were called, the title printed in a faux-Oriental font. They seemed a bit insubstantial compared to the horrendous, seemingly irremediable deprivation confronted by the expedition, and Dienhart's staff seemed to have a bit too much fun preparing them. In one, apparently for internal use only, they compiled a series of jokey anecdotes—for example, about a Yeti of "beautiful proportions" named Smokhee being kept as a pet in a village where the favorite song was "On Top of Old Smokey," and another one about a fabulous pure white Yeti living in a cave of white chalk that said to Dienhart, "Did you know you're a nigger?" (possibly a reference to Dienhart's interest in jazz). But the "Hilary-o-Grams" were indicative of more serious matters. When the British first set out to conquer Everest, the mountaineers were leery of publicity, concerned only with their own endeavors and the nationalism inherent in the climb; now, they embraced it: Hillary's expedition, to simplify matters somewhat, combined elements of the old British imperialism—in which mountaineering was a form of geopolitics—with elements

**4.** Benet, "Bay Man Ready for Snowman" (second quotation), Lawrence Swan clipping file; Hillary and Doig, *High in the Thin Cold Air*, 11 (first quotation), 18–32 (third quotation, 22; fourth quotation, 23).

of the American empire of mass media—in which conquering was done with the pen, not the sword—or the ice ax.[5]

Thirteen days after the expedition left Katmandu, it reached the village of Beding, small and poor, "some twenty stone houses, crudely built and strung together by a complex of stone walls that delineate, rather than protect, the village potato fields." Here, according to later accounts, the expedition found the first evidence that made them skeptical of the Yeti's existence. For several days, Doig "scoured the village for information," Hillary wrote, "like Sherlock Holmes." Early in October, Doig purchased what locals assured him was a Yeti skin. The fur, however, turned out to be the pelt of a blue bear, a rare native of the area. The misidentification led Doig to believe that the Yeti was not a real animal but Sherpa fantasy. Of course, as anthropologist John Napier noted, even if the Abominable Snowman was only legend, "the Sherpas' reiterative and tedious insistence that all tracks seen and all skins and scalps discovered are those of a Yeti" might not have been a symptom of ignorance or enthrallment to primitive religion but the mark of "extreme sophistication"—a way not only to accommodate Westerners who desired to see traces of the Yeti but also to attract Western funds without the dishonor of resorting to charity.[6]

On October 4, the team moved six miles to the summer village of Na, higher up in the mountains. A week after that the Yeti hunters moved onto the Ripimu Glacier, where, Doig said, "even yaks are afraid to tread." Seeking the Yeti was of secondary concern for Hillary, and so he moved the team more frequently than the arty types would have liked. Perkins and Swan, however, fit some investigating into the interstices of climbing. On the Ripimu, Perkins set up trip-wire cameras while Swan put together a makeshift laboratory. Five days later, however, the camp had to be broken; Hillary wanted the team farther up the glacier. The hike was tough going, the altitude—they were approaching 18,000 feet—making Doig "feel as limp as a chewed string." Bannerjee blacked out and when he was revived was temporarily blind.[7]

**5.** "Hillary-O-Gram," n.d., *World Book Encyclopedia* scientific expedition to the Himalaya papers, World Book, Inc., Chicago; Hillary and Doig, *High in the Thin Cold Air*, 11–12 (first and second quotations, 11), 18 (third quotation), 20–21, 26–27, 33–36; Walt Unsworth, *Everest: The Mountaineering History* (Seattle, WA: Mountaineers Books, 2000), 33–35.

**6.** "Storms Slow Hillary," *New York Times*, October 9, 1960; Hillary and Doig, *High in the Thin Cold Air*, 28–30 (first quotation, 29), 36–41, 72–73, 78–79, 117–24; John Napier, *Bigfoot: The Yeti and Sasquatch in Myth and Reality* (New York: E. P. Dutton, 1973), 59 (final quotations); Hillary, *Nothing Venture, Nothing Win*, 285 (second quotation).

**7.** Hillary and Doig, *High in the Thin Cold Air*, 4, 12, 41–47 (first quotation, 43; second quotation, 47); Perkins, *My Wild Kingdom*, 143–45; Lawrence W. Swan, *Tales of the Himalaya: Adventures of a Naturalist* (La Crescenta, CA: Mountain Air Books, 2000), 37–41.

While moving the camp up the Ripimu, the Sherpas came across tracks of the Abominable Snowman. "Leaving their loads," Doig wrote, "they descended the three difficult miles to Base Camp in an avalanche of excitement. It took only minutes to ignite equal enthusiasm in camp." The tracks looked human to Doig and Hillary both. "But that's not saying they're Yeti," Hillary said. "I would like a lot more convincing proof."[8]

Three days later, Swan, woozy from the altitude but inspired by the mountaineers hiking at higher elevations, climbed a nearby 19,000-foot peak. He planted a bamboo stick into the summit, decorated it with a strip of blue toilet paper, and then plopped to the ground. After a while, he noticed a line of tracks just below him. What happened next is unclear. Swan says that he followed the tracks; Doig says that nightfall was approaching so Swan returned to camp, leaving Doig and Perkins and Bannerjee to follow them the next day. Whatever the chronology of events, the results are not in dispute: the tracks were the second bit of evidence and, in the expedition's judgment, the most important, disproving the Yeti's existence.[9]

The Sherpas assured the hunters that, yes, these were the tracks of the Abominable Snowman, pointing out where the creature had dropped to all fours, where it had reared onto two feet, and where, on brittle snow, it had crab walked. As the trail continued, however, the footprints resolved into a rosette of pugmarks. Not a Yeti, then; a small quadruped—a dog, perhaps a fox or a small wolf—had made the tracks. How had those small marks turned into large Yeti footprints? The answer, Swan figured, was altitude. Above 18,000 feet, snow did not always melt under the sun but sublimated—transformed directly from solid to gas. When a depression was made in the snow (by a fox, a wolf, a raven, a rock), part of the mark was exposed to the sun, part kept in shadow. The snow exposed to the sun sublimated, lengthening and widening, but unevenly, causing the tracks to develop what looked like toes. Because there was no melting, the outlines of the track remained sharp.[10]

On the glacier, Swan followed some of the tracks as they made a circle. Since the "toes" always faced the same direction, it looked as though a bipedal creature had walked first forward, then sideways—and then backwards! Sherpas had said that the Yeti could turn its feet backwards to confuse trackers. "Perhaps here was that fine sword," Swan wrote, "the rapier that cut to the heart of the tale of the Yeti. The explanation had that quality of art and

**8.** Hillary and Doig, *High in the Thin Cold Air*, 48.

**9.** Ibid., 49; Swan, *Tales of the Himalaya*, 38–39.

**10.** Hillary and Doig, *High in the Thin Cold Air*, 50; Perkins, *My Wild Kingdom*, 145–46; Swan, *Tales of the Himalaya*, 39–40.

refinement, enough to fit the elegant quality of the legend itself." It was an explanation as ingenious as that proffered by Zadig, a solution as clever as any by Sherlock Holmes. Combining all the subtle clues, using all of their expertise, the team had reconstructed the unknown animal from its traces. "Whenever we found [tracks],"Doig wrote, "it was the same story—tracks the Sherpas swore to be authentic Snowman were quite obviously those of some small unsuspecting quadrupeds promoted by sun and local imagination into the realm of Himalayan fantasy."[11]

Meanwhile, Hillary was considering pushing on further, to the Menlung Glacier where Shipton had found his famed prints. But the conditions that way were bad, the wind keen, the snow deep. So the team retreated to its first camp on the Ripimu and then—crossing the 19,000-foot Tashi Lapcha Pass—to the comparatively low-altitude village of Khumjung, near Thyangboche and Pangboche. Here, in Khumjung, the expedition uncovered the final clues that convinced them that there was no such thing as an Abominable Snowman.[12]

Some of the team examined the purported Yeti hand at the monastery in Pangboche, concluding, as Hillary said, "This is essentially a human hand, strung together with wire, with the possible inclusion of several animal ones." In Khumjung, they saw the village's Yeti scalp, allegedly harvested long ago. "It looked like a hairy, pointed, brimless helmet," Perkins said. The next day, Ang Temba—one of the Sherpas on the expedition—brought for sale a number of interesting trinkets: a horse's horn, lightning excreta, a petrified lama penis, a human tail, and two pelts. The Yeti pelt they recognized to be the fur of a blue bear. The other was the hide of a serow, a goat-antelope native to the area. It had a dark mane and coarse black and red bristles that were reminiscent of the fur on the Yeti scalp; under a magnifying glass, the hair from the scalp and the fur from the serow hide "compared exactly." The team bought the bear pelt and serow hide (but not the other curiosities) and sent a few bristles from both the serow hide and supposed Yeti scalp to Osman Hill in London, who was still studying the Abominable Snowman. Perkins, with the help of local craftsmen, molded another piece of the serow hide into the shape of an Abominable Snowman scalp to show that the goat-antelope might be the real source of the relics. "The results were excellent," Doig wrote. To Swan's amazement, once the scalp was finished, one of the Sherpas who had helped to mold the serow skin "approached

**11.** Hillary and Doig, *High in the Thin Cold Air*, 51; Swan, *Tales of the Himalaya*, 39.

**12.** Hillary and Doig, *High in the Thin Cold Air*, 74–78; Perkins, *My Wild Kingdom*, 147–50.

it in awe with palms appressed as if this obvious fake was a true and holy object."[13]

Convinced that the riddle of the Yeti had been solved, Doig negotiated a six-week loan of the actual scalp from Khumjung for scientific study. In return, the expedition agreed to pay for renovations to the village monastery and to attempt to raise funds for a school. Hillary was away with the physiology team but hurried back when he heard news of the agreement: the deal was a good way, he thought, "to repay, in some measure, at least, the cheerful, courageous, and faithful service that mountaineering expeditions have had for so long from the Sherpa people." On November 25—less than two months after the expedition arrived in Beding—the hunt was over. Hillary, Perkins, Doig, and Khunjo Chumbi, a Khumjung elder and the scalp's chaperone, left on a whirlwind, worldwide tour, drawing attention to the plight of the Sherpas. Khunjo Chumbi was celebrated in the United States, in London, in Paris. The publicity brought in enough money to build a school in Khumjung, the village's first, improving the Sherpas' lives and, Hillary said, strengthening their resistance to Communist Tibet, where schools were supposedly being built, tempting the seminomadic Sherpas across the border. "World Bookers" everywhere—as one public relations executive with Field Enterprises said—could feel as though they were "part of the events shaping our daily lives."[14]

In Chicago, zoologists at the Field Museum confirmed that the Yeti skins were in fact bear pelts. In Paris, a cadre of anthropologists, zoologists, and criminologists—including Bernard Heuvelmans—studied the scalp. It was generally and provisionally agreed that the relic had not come from a Yeti. In London, Osman Hill met the travelers at the British Museum (Natural History). He had received the bristles that Doig sent, both those from the scalp and those from the hide—although Doig had not told him which was which—and concluded that they had all come from the same genus of animals,

**13.** Hillary and Doig, *High in the Thin Cold Air*, 74–78, 81–87 (third and fourth quotations, 82), 131 (first quotation); Marlin Perkins, "The Search for the Abominable Snowman," in *The World Book Year Book 1962* (Chicago: Field Enterprises Educational Corporation, 1962), 108–11 (second quotation, 109);Swan, *Tales of the Himalaya*, 42 (final quotation).
**14.** Edmund Hillary, "Fact Sheet," Sir Edmund Hillary lecture tour, 1962, and "An Idea Materializes," February 1963, both in *World Book Encyclopedia* scientific expedition to the Himalaya papers; idem, "Hillary Reports: The Official Account of The World Book Encyclopedia Scientific Expedition to the Himalaya," in *The World Book Year Book 1962* (Chicago: Field Enterprises Educational Corporation, 1962), 101 (first quotation); Hillary and Doig, *High in the Thin Cold Air*, 85–105; Murray, *Adventures in the People Business*, 233 (second quotation).

FIGURE 18. In 1960, Sir Edmund Hillary led an expedition that, in part, hunted the Yeti. The team returned convinced that the beast did not exist—and that relics such as this supposed Yeti scalp, which Hillary (right) is showing to Bernard Heuvelmans—were fabricated from other animals. Hillary's renown was such that interest in the Abominable Snowman died out for a decade. (Image 06675. © Musée de Zoologie—Lausanne/Agence Martienne.)

probably an ungulate. The findings weren't official yet—full analysis would take weeks—but Hillary concluded that the Abominable Snowman was only a myth. "We do not think the Yeti exists," he said in Amsterdam, on his way to return the scalp. "It's been a fascinating story," Swan said, "and I hate to be the one to destroy it." But, he added later, "Human insistence cannot transform Nature and its laws."[15]

## Sanderson's Failed Debunking of the Debunking

Not everyone was so certain that the matter was settled. W. C. Osman Hill thought that the evidence against the creature was thin and the judgment "hasty." There were some differences between the serow pelt and scalp, he said. Although they might be trivial or explicable, the fact remained that

**15.** "Snowman Melted," *New York Times*, December 30, 1960 (first quotation); "Abominable," *New York Times*, January 1, 1961 (second quotation); Hillary and Doig, *High in the Thin Cold Air*, 100–3; Swan, *Tales of the Himalaya*, 41 (final quotation).

they had not been explained. Additionally, while the pelt harbored well-known ruminant parasites, the scalp had unusual mites. The vermin also might indicate nothing—or they might be a clue to the creature's identity. Whichever, he said, they should not have been so easily dismissed.[16]

Carleton Coon agreed that Hillary's conclusions were questionable and the debunking did not shake Bernard Heuvelmans's belief that the Abominable Snowman existed. Peter Byrne knew that he had desecrated the Pangboche hand, and so also knew Hillary's observation that some of the bones appeared human might only reflect his meddling. Byrne also claimed that the Khumjung scalp was a known fake, created by monks jealous of Pangboche, and so studies of it were beside the point.[17]

George Agogino lost all respect for the "bee keeper," as he disparagingly referred to Hillary, and the expedition. It wasn't a scientific investigation at all, but a "deliberate attempt to pull fast publicity, nothing more," he said. "If they were scientifically interested in this problem, they would have awaited results of the hair analysis from at least one expert . . . but the Hillary 'mob' did not wait for results . . . before shooting off their big mouths." The evidence, he said, did "not change the overall picture of the possibility of Abominable Snowman."[18]

Such grumblings were restricted to private musings and obscure scientific journals—with one exception: Ivan Sanderson. He took on Hillary in the pages of *Sports Afield* and in his book. The expedition was doomed to failure, he said. It was too short—"the group was hardly gone before it was back"—too large—the army of porters almost certainly scaring away an animal as skittish as the Yeti—and focused on the wrong areas—high snowfields rather than the lower montane forests where the Abominable Snowmen actually lived. The scalp proved nothing, he claimed, agreeing with Byrne that it was likely a fake: such relics were made in *imitation* of Yetis, but not *from* Yetis. As

**16.** William C. Osman Hill, "Abominable Snowmen: The Present Position," *Oryx* 6 (1961): 94.
**17.** Betty Allen, "Bigfoot Explorers," 1959, in Bigfoot file 7, Andrew Genzoli papers, Humboldt State University, Arcata, CA; Carleton Coon, "Exercises in Unnatural History," *Natural History* 61, no. 1 (1962): 4; Peter Byrne, *The Search for Big Foot: Monster, Myth or Man?* (New York: Pocket Books, 1976), 101–2; Loren Coleman, *Tom Slick: True Life Encounters in Cryptozoology* (Fresno, CA: Craven Street Books, 2002), 107–8, 127–35.
**18.** Agogino to Coon, January 5, 1961, General Correspondence A–F 1961 file; Agogino to Coon, January 13, 1960 [actually, 1961] (second quotation), General Correspondence A–D 1960 file; and Agogino to Coon, January 17, 1961 (remainder of quotations), General Correspondence A–F 1961 file, all in Carleton S. Coon papers.

for Swan's theories about the footprints, Sanderson (wrongly) claimed that foxes didn't even live in the area. There was no reason to doubt the existence of Yetis, he said, and certainly no reason to doubt that there were other wildmen throughout Asia, in Africa, Europe, and the Americas. Sanderson even slyly suggested that the whole expedition was just a cover to spy on the Chinese![19]

But Hillary's celebrity was too much for Sanderson. Hillary was a knight, a mountaineer, the man who had conquered Everest, and that was proof of sober judgment and a good character unpersuaded by the temptations of filthy lucre or petty politics. "He is a prime example of earth-bound man in his last phase of development," Field Enterprise's publicity material intoned. "His values are the old, stanch values. His ways are the tried and trusted ways. He has made himself famous by the flex of his brawn and the imperturbability of his guts." Denying the Yeti's existence—whatever Hillary's critics said—seemed to go against the demands of publicity and commerce: the name Abominable Snowman was forever linked with newspaper sensationalism, so to dispute the creature's existence was to stand on the side of science, opposed to media inventions.[20]

Sanderson, by contrast, couldn't even convince his most ardent scientific supporters that he knew what he was talking about. Coon thought Sanderson's *Abominable Snowman* so outrageous that the author, "for reasons of his own, [was] deliberately trying to be disbelieved." His forays into anthropology were "sub-professional." Agogino, who wrote the foreword to the book without first reading it, sighed when he finally did wade through the bulky tome, "I am glad that my Introduction was such that I didnt [*sic*] stick my neck out in any way." He told Coon, "I am afraid it takes a sick mind" to accept Sanderson's contention that wildmen live in North America. "Rather soon Sanderson will claim that *Ishi* was really a Yeti." Even the rather odd conclusion that Yeti tracks had been left by foxes—could Hillary have found a less likely culprit?—seemed sensible compared to what Sanderson offered. "No story seems too wild or improbable for Mr. Sanderson," wrote a reviewer for the *San Francisco Chronicle*. "Why, oh why, Mr. Sanderson, not use just a little restraint? Why not admit that there just might not be an Abominable Snowman?" Why not accept, the reviewer asked, that Larry Swan had found

**19.** Ivan T. Sanderson, *Abominable Snowmen: Legend Come to Life* (Philadelphia, PA: Chilton, 1961), 329–50, 483–89; idem, "The Ultimate Hunt," *Sports Afield*, April 1961, 113–18.
**20.** "Sir Edmund Hillary Background Information," 1960, *World Book Encyclopedia* scientific expedition to the Himalaya papers.

"a reasonable explanation of the footprints usually associated with the Yeti"? Why not? Because, the reviewer suggested, Sanderson was not a scientist, and the Abominable Snowman was not an object of science—just as Hillary had said.[21]

Immune to criticism, Hillary's views became the conventional wisdom. "Snowman Melted," the *New York Times* declared. *Life* magazine, which once offered Slick money for photos of the Yeti, published Hillary's "Epitaph to the Elusive Abominable Snowman." At the end of January 1961, the Chicago scientists confirmed their initial impressions: the Yeti artifacts came not from an unknown primate but from known animals. Dienhart's publicity staff sent out a news release titled the "Death of an Abominable Snowman." Even *Fate* magazine was reluctant to contradict Hillary. "Believers may have to face the fact that the Abominable Snowman is also a fake," wrote the editor in his monthly column.[22]

In 1962, Marlin Perkins began hosting a new television show, "Wild Kingdom," a program that promised a realistic look at the animal world. There is "nothing Disneyish about the goings-on," said *Time* magazine. "No frogs dancing the frug, no kissing coots." The premiere debunked belief that the Yeti was a real animal, as did a later episode. The *World Book Encyclopedia* described the expedition in its 1962 *Year Book* and rewrote its article on the Abominable Snowman to make clear that the Yeti was only a myth. There was no more debate over the beast in *Science* or *Nature*. In a testament to Hillary's influence, American newspapers did not report on the Abominable Snowman for more than a decade.[23]

**21.** Agogino to Coon, November 6, 1961 (second quotation; original emphasis), General Correspondence A–F 1961 file, Carleton S. Coon papers; Arthur C. Smith, "The Abominable Snowmen—No Story Seems Too Improbable," *San Francisco Chronicle*, December 10, 1961 (final quotation), in Reviews file, Ivan T. Sanderson papers, B Sa3, American Philosophical Society, Philadelphia, PA; Carleton Coon, "Exercises in Unnatural History," 4–5 (first quotations).

**22.** "Death of an Abominable Snowman," January 25, 1961, *World Book Encyclopedia* scientific expedition to the Himalaya papers; "Snowman Melted"; Edmund Hillary, "Epitaph to the Elusive Abominable Snowman," *Life*, January 13, 1961, 72–74; "Now Let's Face It," *Fate*, June 1961, 11–12 (quotation, 11); Marca Burns, "Report on a Sample of Skin and Hair from the Khumjung Yeti Scalp," *Genus* 18 (1962): 80–88.

**23.** "The Fang and Fin Hour," *Time*, December 4, 1964; Curtis D. MacDougall, *Superstition and the Press* (New York: Prometheus, 1983), 255; Gregg Mitman, *Reel Nature: America's Romance With Wildlife on Film* (Cambridge, MA: Harvard University Press, 1999), 131–56; Coleman, *Tom Slick*, 135.

## The Quiet Years

Saturday, October 6, 1962, the private plane carrying Tom Slick crashed near Dell, Montana, killing the oil magnate and his pilot. The plane crash was nearly fatal for Bigfoot, as well. Hillary had just made belief in wildmen more preposterous than ever—if the Himalayan ABSM did not exist, what chance was there that one inhabited North America? Now, Bigfoot's most powerful champion was gone, and with him went the money to hunt Sasquatch. The expeditions in California and Canada disbanded: Peter Byrne returned to Asia, John Green focused again on his newspaper and on politics, Bob Titmus, struggling under the best of circumstances to balance monster-hunting and earning a living, gave up the search and became a taxi driver. ("That damn, stupid business," he said later.)[24]

Others involved with Bigfoot also turned their attention elsewhere. Jerry Crew moved to San Francisco. The Wallaces left Willow Creek. Genzoli stopped writing about the wildman, as did much of the national media. Sanderson had planned to convert a number of essays about ABSMs left unpublished when *Fantastic Universe* unexpectedly folded into articles for *Sports Afield*, but the series never came to be. Nor did he place any more articles about wildmen in *True*. There was some talk of publishing a sequel to his *Abominable Snowmen*, but that also never came to pass.[25]

Green called this time after Slick's death "the Quiet Years." But, although these years were quiet, they were not mute, and although interest in Bigfoot waned, it did not disappear. Deprived of a benefactor and made ridiculous by Hillary, Bigfoot still had its supporters and its refuges, where it could hide until the mass media again had reason to turn its spotlight upon the beast.[26]

Betty Allen was not dependent on Slick's millions, but answered only to her own curiosity, and so continued to follow the story where it took her. In the course of her research, she uncovered the tale of Fred Beck, a miner who said that in 1924 his camp had been attacked by rock-throwing apes; the incident had been covered by the *Portland Oregonian* at the time and had since

**24.** Byrne, *The Search for Big Foot*, 107; John Green, *On the Track of the Sasquatch* (Surrey, BC: Hancock House, 1980), 39; "Interview: Bob Titmus," *The ISC Newsletter* 12, no. 2 (1993–1996): 4 (original emphasis); "Building Beachfront," *Vancouver Sun*, November 2, 2000; Coleman, *Tom Slick*, 171.

**25.** Sanderson to E. McD., September 17, 1963, and Sanderson to Editorial Board, August 28, 1964, both in Wandering Woodsmen file; Sanderson to Paul R. Reynolds, February 29, 1960, Agents #1 file, all in Ivan T. Sanderson papers; Christopher L. Murphy, *Meet the Sasquatch* (Surrey, BC: Hancock House, 2004), 37.

**26.** Green, *On the Track of the Sasquatch*, 39.

taken on the quality of a legend, the area where the attack occurred coming to be called Ape Canyon. For Bigfoot enthusiasts, Beck's tale was evidence that the California wildman was not a new invention, a recent hoax, but a member of a species that had inhabited the region for a long, long time. Allen also continued to investigate mysterious footprints nearer to home. In 1963 and 1964, she looked into a rash of reported tracks found on a sandy bar along Bluff Creek, not far from the Notice Creek Bridge. Floods washed out the bar in 1964, but not before Allen heard of several fifteen-inch tracks, and one case in which fifteen-inch and ten-inch tracks were seen together, as though a mother and her cub had wandered along a favored path. Allen collected these reports into a book called *Bigfoot Diary*, which she privately printed and distributed.[27]

Dahinden was not so much answerable to his curiosity as he was a slave to what he called an "obsession." He found that he could pay his bills and still have time to hunt the beast by collecting and recycling lead shot from a British Columbia shooting range. His wife was not thrilled by the arrangement—an immigrant herself, having followed Dahinden to Canada after meeting him during his *wanderjahrs*, she had to withstand the knowing winks and small-town gossip that attended her husband's hobby. Tired of the heckling and desiring more security, she finally demanded that Dahinden curtail his activities, but the demand had a perverse effect: Dahinden took to spending more time in the woods. "I was scared stiff of losing my guts, or whatever you want to call it," he admitted. "I was afraid of getting submerged in all the small things people get submerged in." His obsession was a cruel master and he countered his wife's demand with his own: she needed to accept his hunting as part of who he was. According to Dahinden, she responded by insisting that he make a choice: his family or his quest.[28]

The Dahindens separated and eventually divorced. "Of course I still miss her and our two sons," René said later, "but I couldn't knuckle under to an ultimatum like that. If you're not doing what you like to do, you're a prostitute, not a man, and your kids will realize it." Dahinden spent his increased free time hunting the Sasquatch.[29]

**27.** John Green, *Bigfoot: On the Track of the Sasquatch* (New York: Ballantine, 1973), 70–71; idem, *The Sasquatch File* (Agassiz, BC: Cheam, 1973), 30–31; Christopher L. Murphy, *The Bigfoot Film Controversy* (Surrey, BC: Hancock House, 2005), 73–96.

**28.** Don Hunter and René Dahinden, *Sasquatch* (Toronto: McClelland & Stewart, 1973), 95–110 (quotations, 96).

**29.** James Halpin, "Sasquatch," *Seattle*, August 1970, 31–34, 58–59 (quotation, 58); Hunter and Dahinden, *Sasquatch*, 110.

Bigfoot also remained an object of fascination for Forteans, occultists, and mystery mongers. Around this time, flying saucer enthusiasts began to speculate that wildmen might not be from the earth at all, but visitors from other planets. "Where there are colossal, smelly, hairy, two-legged creatures, there are UFOs!" wrote paranormal investigators B. Anne Slate and Peter Guttilla. "The big question now is which came first? Is Bigfoot the desperate survivor of a bygone era whose only mistake is being in the wrong place at the wrong time? Have the elusive monsters become the unwitting slaves of ultra-advanced extraterrestrials? Are the giant anthropoids themselves from some distant or dying planet?" Fred Beck—still alive and now drawn back to Bigfoot matters—added support to these theories, self-publishing a book about his Ape Canyon experiences that embellished the story with occult folderol. His group had been led to the mine by spiritual beings, Beck said; he himself had had psychic experiences since he was a child. He saw UFOs. The giant apes that attacked the miners were "not entirely of this world," but were a mixture of psychic energy and matter, sometimes existing in this dimension, sometimes in other dimensions.[30]

In the mid-1960s, paranormalist Lee Trippett set out to contact Bigfoot via telepathy. Trippett was an electrical engineer from Eugene, Oregon, and member of the Western Research Foundation, which seemed to be a highfalutin name for the office in his father's ranch-style home. He theorized that Bigfoot and humans had evolved from a common ancestor thousands of years before and that the gulf between the two species widened when humans invented tools. Technology, Trippett said, allowed humans to develop their intellect, but in exchange they had lost contact with another part of themselves: the unconscious. The giants, on the contrary, developed that part of their mind—they were the id made flesh—but not their intellect and so came to recoil in terror from humans and their tools, which symbolized those parts of themselves that they had lost. By going camping in Oregon's Three Sisters Wilderness *sans* artifacts ("like a child of nature") and beaming thoughts of pure love into the ether, Trippett thought that he could entice the monster from the shadows.[31]

**30.** Fred Beck and R. A. Beck, *I Fought the Apeman of Mt. St. Helens* (n.p.: privately printed, 1967); Brad Steiger and Joan Whritenour, "Abominable Spaceman," *Saga*, February 1968, 34–35, 58–64; Peter Guttilla and B. Anne Slate, "The Strange Link between UFOs and Bigfoot," *Saga*, August 1974, 16–18, 54–60 (quotation,60); Loren Coleman, *Bigfoot! The True Story of Apes in America* (New York: Pocket Books, 2003), 46–51 (Beck quotations, 49–50).

**31.** Lee Trippett, "A Preliminary Report on U.S.A.'s Western Giants," October 24, 1964, Bigfoot file 2, Andrew Genzoli papers; George Draper, "Psyching the Giants," *San Francisco Chronicle*, December 8, 1965 (quotation).

It was here, among the Forteans and occultists and paranormalists and flying saucer enthusiasts, that Sanderson's arguments found their warmest reception. *Fate* continued to publish his Abominable Snowman articles—even after Hillary's debunking—until Sanderson finally turned his attention to other *pursuits*, as it were. The embrace of Sanderson by the odd and the outré further made his arguments seem silly and Hillary's pronouncements undeniable. When the national media did break the silence of the Quiet Years, stories about wildmen were reported with tongues planted firmly in cheek, as when the *Santa Rosa Press-Democrat* joked about the Fort Bragg monster or the *Orlando Sentinel* made hay with a Florida wildman dubbed the Abominable Sandman. It was another failure for Sanderson—not only had the Pacific Northwest Expedition collapsed, not only had ABSMery failed to take off, but even his attempt to get wildmen accepted beyond the Fortean community had come up short. Bigfoot returned to its original home.[32] And yet, the enthusiasm of Forteans, occultists, and other fringe groups ensured that the beast survived Slick's passing. Long after newspapers and magazines stopped writing about the beast, *Fate* still did, as did Fortean Frank Edwards. Radio legend Long John Nebel—the Art Bell of his day—continued to discuss the beast on his late-night program. Bigfoot had a home, had its admirers. Despite everything, Bigfoot still lived.[33]

## Big Foot Daze

There was another place where Bigfoot weathered Hillary's judgment and the passing of Tom Slick: Willow Creek. In 1960, the town started an annual festival over the long Labor Day weekend called "Big Foot Daze," which featured, among other events, a logging show, a parade, a barbecue, a ham shoot, dances, and a contest to become "Big Foot's gal." According to the *Willow Creek Advocate*, "crowds thronged in the gay spirit of the occasion, much to the satisfaction of the committee in charge." Big Foot Daze continued throughout the Quiet Years and beyond.[34]

The Willow Creek that started Big Foot Daze, however, was subtly different

**32.** "Skeptics and Enthusiasts Clash on Bigfoot Search," *Santa Rosa (CA) Press-Democrat*, February 18, 1962; Mark Pino, "'Horror' Story No Fool's Prank from '66 Vault," *Orlando (FL) Sentinel*, April 1, 2005.

**33.** Long John Nebel, *The Way Out World* (New York: Lancer, 1962), 130–144; Frank Edwards, *Strange World* (New York: Lyle Stuart, 1964), 11, 46–55.

**34.** "'Big Foot Daze' at Willow Creek Prove Huge Success," *Willow Creek (CA) Advocate*, September 8, 1960.

than the Willow Creek that had existed when Jerry Crew found those tracks—and that difference would grow through the years. The lumber industry's peak year had been 1959. The industry declined over the rest of the century, until by the late 1990s it employed less than 8 percent of the county's population. Replacing the lumber jobs were service and tourism-related ones. "Having been around this country a good long time," wrote the editor of the *Willow Creek Advocate* in 1965, "it is surprising when I realize just how much we have grown. . . . There was a time when we had to go to Eureka or Redding for just about everything and some of those things we could purchase locally were sky hight [*sic*]. This is not so today. Practically every type of merchandise and labor that you could ask for is now in the area and, thanks to competition, it is at, or very near, 'outside' prices." The transition from a timber to service economy brought stability to the area: the service industry, especially as it diversified, was better buffered against recession than the lumber industry had been.[35]

Bigfoot and Big Foot Daze were part of this new economy. They were advertising icons, commercials for local services and the area. Maps proclaimed the region "Bigfoot Country." In time, came Bigfoot burgers, Bigfoot Golf & Country Club, Bigfoot Lumber and Hardware, Sasquatch Second Hand, and the Bigfoot Curio Shop. The Union 76 station in town painted a Bigfoot mural on one of its walls. Jim McClarin, a Humboldt State College student, spent months carving a statue of the beast for Willow Creek's town center. Jim Wyatt built a Bigfoot cage for public display; later, his Wyatt's Motel became Bigfoot Motel. The editorial column of the *Willow Creek Advocate* was titled "Bigfoot Reports," and was written in the voice of Bigfoot. B.F. (the creature's nom de plume) shilled for local business, extolling the virtues of Frank Frost's Chevron station, Dick Brown's hardware store, Gary Roberts's auto parts store, and Al Hodgson's general store. "The boom is on," Genzoli wrote just after Sanderson's first article for *True* appeared on newsstands, "and one part of the north coast area is due for a population explosion. It's not a gold rush although it appears to be taking on aspects of such. . . . The reason for the influx is Bigfoot."[36]

The scent of humbuggery that swirled about Bigfoot was not a taint but a

**35.** "Bigfoot Reports," *Willow Creek (CA) Advocate*, July 8, 1965 (quotation); Steven C. Hackett, "The Humboldt County Economy: Where Have We Been and Where Are We Going?" *Humboldt (CA) Times-Standard*, February 1999, http://www.humboldt.edu/~indexhum/projects/humcoecon.htm (accessed April 9, 2008).

**36.** John Hart, *Hiking the Bigfoot Country* (San Francisco: Sierra Club, 1975); Marian T. Place, *On the Track of Bigfoot* (New York: Pocket Books, 1978), 106 (Genzoli quotation); Joe Cempa,

powerful attractant. "If the people of the United States don't know about this [area] by this time, I am crazy," a Willow Creek businessman said. "You can't buy this kind of publicity. Hoax or fact—I like it." In 1965, Betty Allen ceded rights for her *Bigfoot Diary* to the town, and for a long time the book was sold at the annual festival. In 1967, Big Foot Daze made $2,000 over the Labor Day weekend. Fifteen years later, seven thousand people swarmed into the town during the festival, tripling the town's population. Willow Creek made $12,000. At the 2005 Big Foot Daze—by then sometimes called Bigfoot Days—the Bigfoot in the parade, according to the *Eureka Times-Standard*, "didn't miss the chance to show off its corporate wares. He wore a Re/Max Humboldt Realty shirt."[37]

But Bigfoot was more than an advertisement for the community, and Big Foot Daze was more than a tourist attraction. The creature and its festival also helped to ease the strains of Willow Creek's transition from lumber town to service economy. Although they were more stable, the new jobs did not pay as well as the old, so the area declined relative to the rest of California, and parts of Humboldt County remained too sparsely populated to support important services. Willow Creek, for example, had trouble attracting a doctor to the town in the mid-1960s. There was, as well, a pervading sense that something vital had gone missing, an up-by-the-bootstraps frontier feeling that may never have existed but that was central to the region's self-identity. In one column, B.F. wondered what the Founding Fathers would think "of a government that builds dams, sells electricity, insures medical care and an income for the aged, pays farmers not to plant, buys and stores what they do plant, decides labor disputes, pays out welfare claims, is the largest land owner in the nation, has commissions to regulate and control aviation, television, radio, banks, interstate commerce, public utilities and dozens of other enterprises, and then taxes the heck out of its citizens to pay for all this and even with the taxes it keeps going in the hole every year." There was a sense that the world could not be controlled any more, that it was too

"California Town Basks in Glory of Bigfoot," *Toronto Star*, August 27, 1988; Hugh Dellios, "California Town Promoted as Bigfoot's Vacation Spot," *Houston Chronicle*, July 5, 1993.

**37.** "Bigfoot Still Topic of Interest in Humboldt and Abroad," *Humboldt (CA) Times*, October 14, 1958 (first quotation); "Big Foot Daze Clears over $2,000 for W. C. Recreational Project," *Blue Lake (CA) Advocate*, September 28, 1967; "Death Takes Betty Allen," *Klamity Kourier* (Willow Creek, CA), October 28, 1970; Brendan Riley, "Sasquatch Celebration Brings Money—and a Bigfooted Trucker," Associated Press, August 31, 1982; John Driscoll, "Bigfoot Leaves His Footprint on Willow Creek," *Eureka (CA) Times-Standard*, September 4, 2005 (final quotation).

complicated, too frightening. "Probably," B.F. concluded, "those fellas that dumped that tea in Boston harbor" would "burn the whole District of Columbia to the ground."[38]

Fearsome as Bigfoot looked, the wildman actually may have been comforting in these times, dispelling fright, making difficult changes easier. The creature retained the earthy, backwoods connotations that were being lost as the town grew, and so by celebrating Bigfoot, Willow Creek also celebrated that part of itself. In 2000, the Willow Creek–China Flats Museum opened. It consisted of artifacts dating back to the days of the forty-niners and a wing devoted to Bigfoot, mostly displaying items that Titmus willed to the museum after his death. Opening a museum was supposed to attract tourists and money, certainly, but there was more at stake than just cashing in. The museum was a way of instantiating and preserving Willow Creek's identity, of asserting that there was an essential part of the region that persisted despite the superficial changes, an incorruptible and immortal soul.[39]

If that's a valid interpretation of one of the museum's functions, then it's possible that all the references to the monster around town did something similar—hid the town's businesses beneath images of a beast that could not be chased out, a reminder of a thing that resisted civilization, resisted time and history. Through Bigfoot, Humboldt County residents gained access to that essential, inviolable part of themselves that seemed lost but, like the beast, was only invisible. Speaking in Bigfoot's voice allowed the *Willow Creek Advocate* editor to channel the past, to tap into the eternal wisdom of a wildman. Big Foot Daze was a masquerade, letting participants shuck their outer selves and become, for a moment, Bigfoot: embody the wildman's spirit.

Banished by Hillary from mass culture, Bigfoot found refuge in the popular culture that still survived in America's hinterlands. The festival was small, social, and reflective of the community's everyday experiences. As Big Foot Daze evolved over the years, becoming ever more commercial, the festival's activities changed, but they remained interactive: sometimes there were group breakfasts, sometimes there were beard-growing contests, and sometimes a person would don a Bigfoot costume, chase girls, and scare "children out of seven year's growth," as one wag put it. These were far from the concerns of *Science* or the *London Times*, more like the stories that the road

**38.** "Bigfoot Reports" (quotations); Hackett, "The Humboldt County Economy."

**39.** Donna J. Haraway, *Primate Visions: Gender, Race, and Nature in the World of Modern Science* (New York: Routledge, 1989), 26–58; Michelle Locke, "Bigfoot: Town Pins Hopes on Exhibit about Elusive Ape," *Toronto Star*, August 19, 2000; Blair Anthony Robertson, "They Take Sasquatch Seriously," *Sacramento (CA) Bee*, September 12, 2003.

FIGURE 19. Whatever controversies existed about its existence, Bigfoot became an icon of the Pacific Northwest. In 1989, Washington State chose Harrison Bigfoot as the mascot for its centennial celebration. (Centennial Commission Photographs, Washington State Archives, 1989.)

builders first told about Big Foot, like the Harrison-Hot Springs Sasquatch hunt, Burn's Sasquatch Days, or Barnum's What-Is-It.

Over the years, other parts of the Pacific Northwest adopted Bigfoot and put the creature in the service of similar ideals. Hoopa, Weaverville, and Happy Camp joined the festivities, hosting their own celebrations over the Labor Day weekend. For years, Seattle's professional basketball team, the Supersonics, used Sasquatch as its mascot. More statues followed McClarin's: Richard Beyer sculpted a Sasquatch for Seattle's Pike Place Public Market, another for Seattle University's playground. Native American artist Smoker Marchand created a Sasquatch statue for the Coulee Corridor. In 1989, when Washington celebrated its centennial, the state chose "Harrison Bigfoot" as the official state animal for the year. "Some people say that living in the Northwest is boring and dull," wrote one amateur historian. "They say that we have nothing but a bunch of trees and a lot of fish—plus Tonya Harding.

That may be true but what I say to them is this, 'You can plant a tree anywhere, everywhere there is water there are fish, and Tonya Harding may just hop on a plane and move to your neighborhood, but Bigfoot will always be unique to the great Pacific Northwest!' "[40]

If nothing else had happened, then this is likely where Bigfoot would have stayed, in Willow Creek, in the pages of Fortean magazines—a relic from another time, a creature that existed on society's fringes. But something did happen, something that brought Bigfoot back into the fold of the mass media and relaunched the monster's career. Bigfoot was captured.[41]

**40.** "Correspondence and Memos on the Centennial Mascot, Harrison Bigfoot," 1981–1990, box 13, AR154_1_15_10, Washington State Archives, Olympia; Ollie Welch, "Bigfoot: A Unique Northwest Mystery," *Clark County History* XXXVII (1996): 88–92 (quotation, 92); Margaret W. Beyer, *The Art People Love: Stories of Richard S. Beyer's Life and His Sculpture* (Pullman: Washington State University Press, 1999), 22–23, 48, 63; Jack McNeel, "Sasquatch Sculpture Erected," *Indian Country Today*, November 30, 2005.
**41.** Richard M. Dorson, *Man and Beast in American Comic Legend* (Bloomington: Indiana University Press, 1982), 75.

CHAPTER SEVEN

# The Return of Bigfoot 1967–1980

Late in the 1960s, around the time that Dahinden's marriage was crumbling, John Green eased himself back into the hunt for Sasquatch. During the early months of 1967, Green and Dahinden traveled up and down the West Coast, talking to those who claimed to have seen Bigfoot or its spoor, or those just interested in the subject—among them Lee Trippett, the man who tried to contact Bigfoot telepathically. As had been the case a decade earlier, the two men were impressed with the credibility of witnesses. One was the head of the audiovisual department at the University of Oregon; another became a deputy sheriff. Even seemingly incredible stories—such as the one about a gang of youth that regularly cruised rural roads in eastern Washington shooting at a "white demon"—seemed more convincing after they talked to those involved.[1]

At the end of August, Green visited northern California to investigate some Sasquatch footprints, this time without Dahinden, but with a tracking dog and her handler. The tracks were too old to interest the dog, a German shepherd named White Lady, so Green spent some time reacquainting himself with the locals then returned home. The next day, Monday August 28, he received a call from Bud Ryerson, the contractor building the road where the tracks had been seen, telling him that there were fresh prints. Green was unsuccessful in convincing any local scientists to come down with him, but the *Vancouver Sun* did pony up $500 for a charter flight, so he, Dahinden, White

**1.** Don Hunter and René Dahinden, *Sasquatch* (Toronto: McClelland & Stewart, 1973), 110; John Green, *Sasquatch: The Apes Among Us* (Seattle, WA: Hancock House, 1978), 73, 114; idem, *On the Track of the Sasquatch* (Surrey, BC: Hancock House, 1980), 41–49 (quotation, 49).

Lady, and her handler flew into the airport at Orleans. Al Hodgson sent someone to meet them, along with groceries and $100 cash.[2]

That night, White Lady was raring to go, but by the time Green and the rest had reached the construction site, night had fallen and, in his words, "none of us wanted to follow those tracks into the bush in the dark." Instead, Green called Dr. Clifford Carl, head of the Provincial Museum of Natural History and Anthropology, and asked that someone be sent down to see the tracks. The next day, White Lady refused to run the tracks. (The dog handler, perhaps defending White Lady's poor performance, later claimed that she had been called off again.) But the trip was not all a loss: word came that the museum would send Don Abbott, a cultural anthropologist. It was "the first time," Green noted, "that a representative of any scientific institution was ever sent to study the big tracks anywhere in North America."[3]

Weather delayed Abbott so that he didn't arrive until late on Wednesday, August 30. "I was laughing at the whole idea all the way down," he said, certain that the tracks were a hoax. Staying at Wyatt's Motel, he couldn't help but notice the giant Bigfoot cage in the courtyard, couldn't ignore the beginnings of McClarin's statue out on Highway 299, couldn't miss that Big Foot Daze was being held *that very* weekend, with Wyatt as Grand Marshall. And yet, there was a discordant note. All the people whom he met "seemed genuinely embarrassed at . . . the coincidence" of tracks appearing during Big Foot Daze. Maybe it wasn't just a hoax.[4]

Abbott went out to see the tracks on Thursday, while Dahinden, White Lady, and the dog trainer returned to Canada. There were two sets of tracks, as there had been earlier in the month, one fifteen or sixteen inches long—accounts vary—the other thirteen inches, both with a four-foot stride. The tracks emerged from a wooded area, followed the road for a while, turned right at a leftward jog in the road toward some equipment that seemed to have been tossed about, and then headed along the hill again before disappearing on hard-packed ground. The footprints were flat, Abbott noted, and featureless—which made him suspect that "they were man-made"—but

**2.** "Finders of New Big Foot Tracks Asked to Prevent Contamination of Scent or Track Destruction," *Willow Creek (CA) Advocate*, August 31, 1967; Green, *Sasquatch*, 73–75; idem, *On the Track of the Sasquatch*, 54–55.

**3.** Green, *On the Track of the Sasquatch*, 56.

**4.** Donald N. Abbott, "Report to the Minister on Recent Investigations into the Supposed 'Sasquatch,'" November 9, 1967, M-3 Archaeology file, Royal British Columbia Museum Archives.

their depth, their placement, and "above all," he said, "the fact that there were distinct differences between successive prints indicating movement of the toes" made him think otherwise. They had clearly not been made by a bear or any other known animal.[5]

Nonplussed, Abbott convinced four biologists and a physicist from Humboldt State College to come out Saturday, but they did nothing to solve the mystery. The physicist reluctantly admitted that it would have taken three or four hundred pounds to leave impressions as deep as the ones they saw; the biologists were convinced it was all a ruse, but they couldn't explain how it had been perpetrated. Locals to whom Green introduced Abbott had no such trouble. It was a giant ape, they said, and, to Abbott's surprise, they seemed believable. When he left Saturday, the first day of the Bigfoot celebration, it was not only Willow Creek that was in a daze—he was, too.[6]

In his official report, Abbott recommended that the museum investigate Sasquatch, perhaps hiring Green and Dahinden for the project. The suggestion carried some weight, and when he met with the Sasquatch hunters, Clifford Carl, the museum's director, said that the institution would now collect reports of sightings—but would not actively support Bigfoot hunters. This small change was not enough for Dahinden. Frustrated, he left for San Francisco early in the fall to see if he could drum up interest there. Before he could get any results, though, events overtook him. Word came that Bigfoot had been filmed. It was the end of the Quiet Years and the beginning of a new phase in ABSMery. This time, Bigfoot did not fade away after a few years, but—thanks to Sanderson and the development of working-class entertainments—became a cultural icon.[7]

## Bigfoot Filmed!

Late in the evening of Friday, October 20, 1967, Roger Patterson and Bob Gimlin roused Al Hodgson. They had been in the forest, they said, near Louse Camp, where Bluff Creek met Notice Creek. They had seen a Sasquatch, a female, big and hairy and stinking. They had cast her footprints. And they had filmed her! As they spoke, they said, the film was on its way to Washington, where Patterson's brother-in-law, Al DeAtley, was going to have it developed. Patterson was agitated, excited, uncertain, which made Hodgson believe

**5.** Ibid. **6.** Ibid.
**7.** Ibid.; Green, *On the Track of Sasquatch*, foreword; Hunter and Dahinden, *Sasquatch*, 114–15; Green, *On the Track of the Sasquatch*, 56.

that he was telling the truth: the two men had actually seen a Sasquatch, had captured it—on film, at least.[8]

Patterson had been making a film about Bigfoot back in his hometown of Yakima, Washington; Gimlin, who was of Apache descent, had been playing the "Indian tracker" in Patterson's movie, although, he later said, he doubted that Sasquatch existed. Patterson had met Green and Dahinden on their tour and had heard tell of the tracks that Abbott had investigated. Presumably, he and Gimlin came to California to film those tracks—but arrived to find that rains had destroyed them. It was discouraging but not fatally so for the trip. "I had two weeks off in between jobs," Gimlin said, and Patterson, as usual, wasn't working. "So, we said let's stay down here a couple of weeks and see what we can come up with." During the day, they rode their horses into the canyons, away from the bulldozers and logging trucks that were working in the area. At night, they slowly drove Gimlin's pick-up truck along the empty roads looking for tracks.[9]

According to Gimlin, that fateful Friday morning he awoke first and rode out to explore while Patterson slept. Some time later he returned to tack a loosened shoe on his horse. Patterson, who had been gone, returned and led Gimlin to an area that they'd been to before, a few canyons down, where the Bluff Creek ran. As Gimlin described it, early in the afternoon—the exact time is in dispute—the two men rounded "this bend in the creek bed. There was a fallen tree and as we came around it there was this creature standing by the creek. That's when everything started happening. The horses started jumping around, raising the devil and spooking from this creature. Roger, well his horse was rearing up and jumping around . . . he slid off him." They were, he said, sixty to eighty feet from the creature.[10]

Recounting the story that night, Patterson remembered the events differently. He said that he and Gimlin saw the beast at about the same time, Patterson yelling, "Bob, lookit!" He later told a different version of their

**8.** Abbott, "Report to the Minister on Recent Investigations into the Supposed 'Sasquatch,'" November 9, 1967, M-3 Archaeology file; John Green, "Interview with Bob Gimlin, 1992," September 30, 1997, http://www.Bigfootencounters.com/interviews/john.htm (accessed April 9, 2008); Christopher L. Murphy, *The Bigfoot Film Controversy* (Surrey, BC: Hancock House, 2005), 189–90.

**9.** Green, "Interview with Bob Gimlin, 1992" (first quotation); Kenneth Wylie, *Bigfoot: A Personal Inquiry into a Phenomenon* (New York: Viking, 1980), 179–82 (second quotation, 182); Greg Long, *The Making of Bigfoot: The Inside Story* (Amherst, NY: Prometheus, 2004), 157–61; Murphy, *The Bigfoot Film Controversy*, 183–94.

**10.** Green, "Interview with Bob Gimlin, 1992."

first-contact story: his horse was skittish as they entered the canyon. Coming around the bend, an overturned stump obscured his vision. His horse stopped—and out of the corner of his eye he saw something. But before he could get a good look his horse reared and then fell to its side. (Gimlin insisted later, "No, no, [Patterson's] horse never fell down. No.") Grabbing the horse's reins to steady it, Patterson saw what had scared his ride: "This creature was on my left, about 125 feet across the creek. . . . Its head was very human, though considerably more slanted, and with a large forehead and wide, broad nostrils. Its arms hung almost to its knees when it walked. Its hair was two to four inches long, brown underneath, lighter at the top, and covering the entire body except for the face around the nose, mouth and cheek. And it was female; it had big pendulous breasts."[11]

Patterson was quite short but strong, restless, and physical—a halfback in high school, a boxer in the U.S. Army, and a sometime-rodeo rider. He grabbed the camera and started filming as he ran toward the beast. Gimlin forded the creek, pulled his rifle from its scabbard, stepped down from his horse, and watched. "I never raised the rifle like I would shoot or anything like that, just held it in my hand and with the other hand held my horse to keep him from getting away from me," Gimlin said. He and Patterson had spoken "many times" about what they would do if they ever saw a Sasquatch and, as Gimlin said, they decided "unless it was necessary, we would never shoot. In other words, unless it was violent or attempted to attack us or something in that sense of the word." Patterson was convinced that Sasquatches were not apes or *Gigantopithecus* but primitive humans.[12]

The creature started walking away, turning to look back at one point. Patterson told the *Los Angeles Times* that he thought the Sasquatch seemed "curious, like it wondered what was making that noise," but, otherwise, "it didn't seem real startled, like it had seen people before, like we weren't anything special." He told Green, on the contrary, that the creature was wary: as he approached, it shot him a look. "You know how it is when the umpire tells you, 'One more word and you're out of the game!' That's the way it felt."[13]

And then they ran out of film. Gimlin said that they weren't really expecting to see Sasquatch and so had wasted a lot of footage filming autumn

**11.** Ibid.; James B. Shuman, "This Is Bigfoot," *Los Angeles Times West Magazine*, December 15, 1968, 25–26, 29 (remainder of quotations, 26, 29); Murphy, *The Bigfoot Film Controversy*, 194 (first quotation).

**12.** Green, "Interview with Bob Gimlin, 1992" (quotation); Shuman, "This Is Bigfoot," 25–26, 29; Green, *Sasquatch*, 115.

**13.** Shuman, "This Is Bigfoot," 29 (first quotation); Green, *Sasquatch*, 116 (second quotation).

leaves. Patterson hunched under a poncho to change film; Gimlin started to follow the creature into a copse of trees. Patterson called him off: "It kind of spooked me," he told the *Los Angeles Times*. "I didn't want to be out there alone without a weapon." Gimlin later added that there was some worry that other Sasquatches might be in the area, the mate or offspring of this one.[14]

Gimlin gathered the horses while Patterson finished changing the film. "It took quite a while," Gimlin said. Ready again, they followed the Sasquatch, but, in Gimlin's words, "didn't have much luck doing it." Afterwards, they returned to the film site, taking plaster casts of the tracks. The footprints were fourteen and a half inches long and so deep that Gimlin could only make similar impressions by jumping from a fallen tree. With the casting done, they drove to town and talked with Hodgson; Patterson then called a reporter from the paper in Eureka (now combined with its sister publication and called the *Humboldt Times-Standard*) and recounted the day's miracle. Late that night, Patterson and Gimlin returned to Louse Camp; it started to rain fiercely. Gimlin tried to protect the tracks with cardboard boxes, but the rain melted them and he resorted to covering them with tree bark. "The little creek that was six or seven feet across was now ten or twelve feet across and four feet deep!" Gimlin said. Early in the morning, they packed up and left, soaking wet but happy.[15]

In the meantime, at Patterson's request, Hodgson called Abbott and asked him to come down with some tracking dogs. Abbott decided against making the trip, but he was intrigued and contacted Al DeAtley that night. Abbot told Patterson's brother-in-law that while he understood that the film would be sold to the highest bidder he would like to have it shown to some of the staff at the museum and scientists from the University of British Columbia. DeAtley said he'd talk it over with Patterson. Report of the filming also reached Green. He left a message for Dahinden in San Francisco and, by his own account, spent $100 calling scientists trying to get them to visit Bluff Creek, all to no avail.[16]

Dahinden rushed to Willow Creek, only to find that the filmmakers had already departed. So he and Jim McClarin, the Humboldt State College stu-

**14.** Green, "Interview with Bob Gimlin, 1992"; Shuman, "This Is Bigfoot," 29 (quotation).

**15.** Donald N. Abbott, "Report to the Minister on Recent Investigations into the Supposed 'Sasquatch,'" November 9, 1967, M-3 Archaeology file; Green, "Interview with Bob Gimlin, 1992" (quotations); idem, *Sasquatch*, 118.

**16.** Donald N. Abbott, "Report to the Minister on Recent Investigations into the Supposed 'Sasquatch,'" November 9, 1967, M-3 Archaeology file; Green, *Sasquatch*, 118; Daniel Perez, *Bigfoot at Bluff Creek* (Norwalk, CA: Center for Bigfoot Studies, 2003), 10.

dent who was carving Willow Creek's Bigfoot statue, followed Patterson and Gimlin to Yakima. The headline of a front-page article in that day's *Humboldt Times-Standard* blared, "Mrs. Bigfoot is Filmed!" But not everyone in the newspaper's office was happy. In an unsigned article printed a few months later, Andrew Genzoli—the writing style attests to the authorship—dismissed the movie as a fake. "The Abominable Snowman has already become part of our local folklore and belongs to the strange realm of fantasy and fiction." Patterson was encroaching on Genzoli's paternal role, turning his creation into something that Genzoli thought it should never be.[17]

That sour note, however, had no effect on the proceedings, and by Sunday all of the principals convened in Yakima: Dahinden, Green, Patterson, Gimlin, DeAtley, and Jim McClarin. The *Times-Standard* article had been carried on the wires and now there was a buzz, reporters calling on Abbott, on DeAtley, on Patterson. DeAtley had developed the film and showed it at his house—to everyone except Gimlin, who, curiously, begged out of the premiere to sleep. Patterson had shot twenty-four feet of film on that creek bed, less than one minute. The film started jerky and out of focus but then stabilized, showing the monster looking just as Patterson had described it. She walked away from the camera, glancing over her shoulder. "I knew what I was going to see," Dahinden later said. "I'd had the thing described often enough, but it still gave me a hell of a shock when I saw it." The Sasquatch hunters were ecstatic. "We had the film," Green later said. "We thought it was all over."[18]

## Making Sense of the Movie

If Jerry Crew was the quintessential solid citizen, sober and civic minded, then Roger Patterson was his opposite. By all accounts—with one notable exception—he was a confidence man. Charming, constantly in debt, a man of grand vision but little patience for the mundanities of life, Patterson seduced people, convinced them to invest in his get-rich-quick-schemes, then was off again, following another dream, friends and marks left to clean up the messes he left, to pay the bills. Inspired by Ivan Sanderson's first Bigfoot article in *True*, he gathered stories for a 1966 book *Do Abominable Snowmen of*

**17.** "An Abominable Snow Job?" *Humboldt (CA) Times-Standard*, January 22, 1968 (quotation); Hunter and Dahinden, *Sasquatch*, 118; Green, *Sasquatch*, 118–19; Perez, *Bigfoot at Bluff Creek*, 10; Murphy, *The Bigfoot Film Controversy*, 191–96.

**18.** Hunter and Dahinden, *Sasquatch*, 119 (first quotation); Perez, *Bigfoot at Bluff Creek*, 17 (second quotation).

FIGURE 20. Before Roger Patterson and Bob Gimlin supposedly filmed Bigfoot, there was only one other picture of the beast—and that was barely more than a smudge, published in the *San Francisco Chronicle*. Otherwise, images of Bigfoot came from the minds of illustrators—such as this depiction of Roe's encounter with Sasquatch by Mort Künstler for one of Ivan Sanderson's articles published in *True*. (From the original painting by Mort Künstler, *Spotting Big Foot*. © ca. 1960 by Mort Künstler, Inc. http://www.mkunstler.com.)

*America Really Exist?*, adorned it with pictures that he drew, conceived the idea for an Abominable Snowman of Club of North America—and then left a friend to finish his incomplete manuscript, pay the publishing fees, market the book, and run the club while his attention turned to making a movie about Bigfoot, collecting investments, casting the locals, and then going to California.[19]

As Genzoli suggested, the film that Patterson brought with him out of the northern California woods was, almost without a doubt, one of his scams. "Roger certainly had the artistic talents, if he wanted to fake something," DeAtley said. "If he wanted to build a suit, he was probably artistic and talented enough to do that. He was a deep thinker, so if he was going to do that, he would definitely make it grainy. He'd jump it around." Patterson even had

**19.** John Green, *Bigfoot: On the Track of the Sasquatch* (New York: Ballantine, 1973), 70–71; idem, *Sasquatch*, 115; Perez, *Bigfoot at Bluff Creek*, 5–8; Long, *The Making of Bigfoot*, 43–144, 195–207.

FIGURE 21. Roger Patterson and Bob Gimlin's movie bore an uncanny resemblance to Künstler's drawing for Sanderson's article—so much so that Bernard Heuvelmans thought that Patterson had probably modeled his film on the illustration. In Patterson's case, however, as shown in this still, it was the filmmaker looking over a fallen log at the beast through the lens of a camera rather than a hunter gazing down the sighting of a gun. (Courtesy of the Fortean Picture Library.)

a model for his film: the movie almost exactly re-created an illustration of Roe's encounter with a female Sasquatch from Sanderson's second *True* article. (That picture impressed Patterson so much that he also redrew it for his book.)[20]

The movie was a ticket out of rural, working-class Yakima to middle-class respectability. Patterson planned to take the film to Hollywood, to New York, sell it, become rich and famous. "He'd hit his homerun," his brother-in-law said. All the years of petty scheming and now he could finally collect. DeAtley, too, wanted to turn the movie into cash. He had supported his ne'er-do-well brother through the years—mostly, it seems, to keep peace in the family—and now what had looked like a lousy investment was about to pay off, and pay off big. "I was money-motivated," DeAtley said. "Whether [Bigfoot] existed or didn't exist, I couldn't care less."[21]

**20.** Long, *The Making of Bigfoot*, 255–59 (quotation, 256).
**21.** Ibid., 255–59.

At the time, Green and Dahinden were not concerned with Patterson's character—they were convinced the film was genuine—but they were worried about his plan to sell the movie. "Go to New York," Dahinden remembered saying, "and they'll laugh you out of town. You'll be considered only a freak with a monster movie." The two Canadians were striving to make Bigfoot—and themselves—respectable in a different, contradictory way: by cultivating scientific support. If scientists knew that Patterson was just out to make a buck they would dismiss the movie out of hand, even if it wasn't a hoax. All of Green's and Dahinden's work, the ridicule that they had endured, the motes of encouragement that they had gathered, their success in finally getting a few scientists to take the subject seriously, all that would be in vain if Patterson tried to cash in too quickly.[22]

Their entreaties proved momentarily persuasive, and on Thursday, October 26, Patterson showed the film twice at the University of British Columbia, once to a small selection of scientists, once to a broader audience. (Later in the evening, he also showed it to journalists, but at the Georgia Hotel because the university refused to allow a press viewing in its halls.) To Green and Dahinden's irritation, the showing proved anticlimactic: it made no converts, inspired no scientific expeditions. It was decidedly not all over. Abbott remained as perplexed as he had been before. "It is about as hard to believe the film is faked as it is to admit such a creature really lives," he said. Others conceded that the film was cleverly made but were not so twisted by conflicting emotions. The anthropologist Robin Ridington suggested that maybe Patterson "made the film as a kind of dramatic re-enactment of a reality in which he firmly believed but had been unable to demonstrate to the world of science."[23]

Frank Beebe, a naturalist and the Provincial museum's illustrator, admitted that there was nothing in the film *per se* that could be used to disprove it, but that the evidence still suggested the movie was a hoax. The creature, Beebe noted, although presumably a female, walked with a male gait; and while the beast on the film had a sagittal crest—in essence, a large, pointed head—it lacked a protuberant belly: a "suspect structural contradiction," he said. Sagittal crests, like Cuvier's ruminant hooves or Sherlock Holmes's fingerprints, were clues. The large crest was where powerful jaw muscles con-

**22.** Hunter and Dahinden, *Sasquatch*, 119 (quotation); Green, *Sasquatch*, 119.

**23.** "Canada Story Reveals Interesting Detail of Big Foot Film Viewing," *Blue Lake (CA) Advocate*, November 2, 1967 (first quotation); Robin Ridington, "Literalism and Symbolism in Anthropological Understanding: The Sasquatch Image," *Zetetic Scholar*, no. 5 (1979): 66–72 (second quotation, 69).

nected to the skull, the muscles necessary to chew lots of low-calorie, fibrous leaves and vegetable matter. A protuberant belly was needed to hold the long intestines used to digest that kind of food. The wildman on the screen was impressive unless you understood biology, in which case it started to look like an impossibility.[24]

Unsatisfied with the scientific response, Bigfooters set out to prove the film's authenticity themselves. Bob Titmus visited the film site nine days after Patterson and Gimlin made their big announcement. "I knew if there was any hokus-pokus [*sic*] about the Patterson film the tracks would tell," he said later. McClarin also inspected the site and in the spring he and Green attempted to remake the film with McClarin playing the role of the Sasquatch. Patterson's film was uncalibrated; McClarin was six feet five; by matching a film of him walking with the film of Patterson's creature, Green could get a sense for how big the Sasquatch was. It was a "fussy business," Green later said, lining up the two films, and they were handicapped because most of the tracks were gone, leaving McClarin to re-create the beast's path from the memory of a visit to the area the previous November. Eventually, they synchronized the shots.[25]

These studies, however, raised more questions about the credibility and gullibility of the investigators than they answered about the film. Titmus said that he could reconstruct the Sasquatch's path from its tracks and even determine where Patterson stood. But this made no sense in light of Gimlin's claim that the torrential downpour following the filming was strong enough to destroy cardboard. Worse, as Bigfoot hunter Danny Perez noted, Titmus's description of the Sasquatch's path is completely different than Patterson and Gimlin's. Most tellingly, although Titmus visited the area specifically to study the film site, he didn't bring a tape measure! All he could do was estimate distances. Green came away from his recreation of the film convinced "beyond a doubt" that "the Sasquatch was not much under seven feet." But, a few years later, after additional analysis, he was forced to revise his estimate and admit that the Sasquatch may have been shorter than McClarin. It was another case where Green was to quick to believe in his own infallibility.[26]

**24.** Frank L. Beebe, November 9, 1967, "An Analytical Report on a Film Sequence Purporting to Show a New Species of Large Bipedal Primate in Northern California," M-3 Archaeology file.
**25.** Shuman, "This Is Bigfoot," 25–26, 29 (first quotation); Green, *Sasquatch*, 123–25 (second quotation, 124); idem, *On the Track of the Sasquatch*, 59–60, 73–74.
**26.** Green, *Bigfoot*, 73–74 (quotation, 74); idem, *Sasquatch*, 124; Wylie, *Bigfoot*, 186–87; Perez, *Bigfoot at Bluff Creek*, 12; Daegling, *Bigfoot Exposed*, 111–12, 121–49.

## The Return of Bigfoot

After trying Green and Dahinden's way and failing, Patterson, Gimlin, and DeAtley traveled to Hollywood to try their own. On November 1, with the help of noted entertainment lawyer Walter Hurst, they incorporated Bigfoot Enterprises and started talking with producers about making a feature-length film, while Patterson also took to the talk-show circuit to drum up interest. But the film failed in Hollywood just as it had in Vancouver. It wasn't looking good for Patterson. Fortunately for him, there was a man who had built a career out of bringing notice to such bizarre things, a man who wouldn't dither over whether the film was a hoax or genuine.[27]

Ivan Sanderson had heard about the filming from Jim McClarin on that monumental day in October, and then heard nothing until late in November, when Patterson, DeAtley, and Gimlin arrived in New York and telephoned. *Life* magazine, apparently unsoured on the prospect of monster pictures, had flown them out to discuss the possibility of buying publishing rights, but lost interest when scientists at the American Museum of Natural History deemed the movie a fake. Sanderson, as he later said, saw his opportunity and "pounced." He had cultivated a relationship with *True* magazine's competitor *Argosy* and convinced the magazine to buy a copy of the film and the right to publish pictures from it. Adopting his most pious tone, he said that *Argosy* had agreed to the deal "so that Bob, Roger and Al could get home for a couple of days for Thanksgiving." But there was more than magnanimity in the act, and Sanderson was not just being solicitous. He dubbed Patterson and Gimlin "the boys," a not-so-subtle declaration of who was in power.[28]

Sanderson next went about rounding up opinions from his own chosen experts—Bernard Heuvelmans, the old Yeti investigator W. C. Osman Hill, and a cadre of scientists in Washington, D.C., among them John Napier, who was currently setting up the Smithsonian's primate department. For a brief moment, interest was so intense that even DeAtley occasionally realized that he was convinced Bigfoot was real "I started to believe," he remembered later. "It definitely walked like a man, and not like a woman, even though it had

**27.** Superior Court of Washington for Yakima County, Case no. 58594, Gimlin *v.* DeAtley and Patterson, Finding of Facts and Conclusions of Law, February 6, 1976; Long, *The Making of Bigfoot*, 260–61.

**28.** Ivan T. Sanderson, "First Photos of 'Bigfoot,' California's Legendary 'Abominable Snowman,'" *Argosy*, February 1968, 23–31, 127–28 (first quotation, 128); idem, "The Patterson Affair," *Pursuit*, June 1968, 9 (second quotation); Long, *The Making of Bigfoot*, 261.

breasts supposedly. I had trouble defining it myself. I had difficulties deciding whether they were or they weren't."[29]

Many of the responses that Sanderson received were more positive than those offered by scientists at the American Museum of Natural History, although, at best, they were equivocal. Surprisingly, Bernard Heuvelmans rejected the film as an obvious fake. He wasn't even certain that Bigfoot was a wildman—he thought it possible the reports referred to a giant sloth—and the movie did nothing to persuade him. The fur looked wrong he said, as did the walk. Buttocks were a human characteristic, but Bigfoot had them. Napier watched the film "at least half-a-dozen times" on December 2, 1967, and like both Beebe and Heuvelmans, thought that the creature on the film didn't look right. It failed Zadig's tests. Napier's trained eye picked out many of the problems that Beebe's had; he also found two other structural contradictions. First, the footprint size indicated that the beast was about eight feet tall, but its stride length pegged it as much taller. Second, the upper half of the body was apish, broad shouldered and thick necked, while the lower half was human—those buttocks again; a layman might imagine this combination to be evidence that the beast was a hybrid, but to an anatomist, with an intimate knowledge of how bones and ligaments and muscles worked together, the beast looked like a fake. Nevertheless, Napier felt that he could not just dismiss the movie. He "could not see the zipper," he said. Like Abbott, Napier was confounded by Bigfoot, thinking it a hoax, but unable to prove it.[30]

Early in 1968, Patterson called Sanderson again. The boys had made it to Yakima in time for Thanksgiving—where Patterson had been arrested for stealing, since he never returned the camera he had rented to make his movie—and then left for Hollywood when the legal issues were resolved. But, again, they won no firm commitment from movie producers. "Roger's pot of gold was quickly melting down," DeAtley said. Patterson asked Sanderson if he could help. Sanderson was ready. "I dozed the 'boys' into appointing me their agent," he bragged to another writer. "It had to happen sooner or later." Sanderson took over the foreign film rights for the movie and arranged for

**29.** Sanderson, "First Photos of 'Bigfoot,'" 23–31, 127–28; Long, *The Making of Bigfoot*, 256 (quotation).

**30.** Bernard Heuvelmans, "The Patterson Document," n.d. [1968?]; Heuvelmans to Krantz, March 1, 1985; Heuvelmans to Krantz, August 13, 1985, all in Heuvelmans file, Grover Krantz papers, National Anthropological Archives, Smithsonian Institute, Suitland, MD; John Napier, *Bigfoot: The Yeti and Sasquatch in Myth and Reality* (New York: E. P. Dutton, 1973), 90–92 (first quotation, 90), 95 (second quotation).

the BBC to make a documentary. John Napier agreed to be in it, lending the movie an air of respectability.[31]

In February, *Argosy* published Sanderson's article, "First Photos of 'Bigfoot,' California's Legendary 'Abominable Snowman,'" illustrated with stills from the film. Even by Sanderson's low standards of truth and consistency, it was slipshod. He got the time of the filming wrong, revised the taxonomy—reclassifying Sasquatch as a "primitive, full-furred human." He wiped away all signs of equivocation or dissent, never mentioning Heuvelmans—although he held the French scientist in the highest regard. And—just as he had done for Ray Wallace—Sanderson improved Roger Paterson's character, transforming him from a petty con artist to a taciturn and honorable cowboy. But the flights of fancy and egregious errors did not slow Bigfoot's return to American mass media. That issue of *Argosy* sold out in a week—somewhere around a million copies gone in seven days. According to his own calculations, Sanderson received six thousand letters in response to the article, independent of what the magazine received. Sanderson even heard tell that in public lectures the Smithsonian scientists recommended that attendees should read his article in *Argosy*.[32]

More exposure followed. In April, *Argosy* published another article by Sanderson on Bigfoot; that same month, *National Wildlife* magazine printed a story on Sasquatch. Sanderson also published a revised version of his ABSM book; according to his wife, that book sold about eight hundred copies per year well into the 1970s. In December, *Los Angeles Times West Magazine* reported on the movie; *Reader's Digest* republished that article the following month. At the time, *Reader's Digest* had a circulation of about eighteen million. Inspired by Patterson, Green authored his own Sasquatch book; Clifford Carl penned the foreword. In addition, Green and Dahinden bought the Canadian rights to the film from Patterson, and during the summer of 1969, Green toured the country, showing the movie and selling about three thousand copies of his *On the Track of Sasquatch* in five weeks.[33]

**31.** Sanderson to Tom Allen, March 29, 1968 (second quotation), Allen, Tom file, Ivan T. Sanderson papers, B Sa3, American Philosophical Society, Philadelphia, PA; Long, *The Making of Bigfoot*, 167, 261 (first quotation).

**32.** Sanderson to Tom Allen, March 29, 1968, Allen, Tom file, Ivan T. Sanderson papers; Adam Parfrey, ed., *It's a Man's World: Men's Adventure Magazines, the Postwar Pulps* (Los Angeles: Feral House, 2003), 286.

**33.** Editorial comment, *Pursuit* 4, no. 3 (1971): 73; "Sasquatch Book Taken East," *Bigfoot Bulletin* 12 (December 31, 1969): 5; Dick Kirkpatrick, "Search for Bigfoot," *National Wildlife*, April 1968, 42–47; Ivan T. Sanderson, "More Evidence That Bigfoot Exists," *Argosy*, April 1968,

During the 1970s, Bigfoot became a star. Hillary's opinion no longer held as much sway; *Atlantic Monthly*, for example, did a cover story on the Yeti in 1975—the Quiet Years were decidedly over. Children's culture embraced Bigfoot especially warmly. Marian Place wrote four books on Sasquatch for children. At least a baker's dozen more juvenile Bigfoot books were published by 1983. Libraries in Oregon and Ohio organized summer reading programs around the study of the beast, encouraging children to check out the explosion of literature. Throughout the decade, the monster appeared in cartoons; there were Bigfoot lunch boxes, Bigfoot board games, and Bigfoot action figures.[34]

At the same time, men's adventure magazines, tabloids, and cheap paperbacks started churning out Sasquatch stories for an adult audience. The beast was popular among these working-class entertainments for the same reason that Sanderson's Bigfoot articles had been accepted so readily by men's magazines for years. The monster fit well with the demands of rapid publishing, was always topical, and could not sue when stories were invented about it. Thus, *Stag* magazine photographed footprints in a snow-covered Central Park and used them to illustrate a story titled "Bigfoot Captured on Film!" And editors of the tabloid *National News Extra* fabricated a tale about a woman who was raped by a Bigfoot and gave birth to a little Sasquatch. The editors figured that they could chronicle the monster's life over the course of several months, decreasing the need to come up with a new headline-worthy story each week.[35]

Exactly how often Bigfoot appeared in men's magazines, tabloids, and paperback novels is difficult to determine exactly. Most of these publications—like so many mass-produced goods—were meant to be disposable. They were literally trash. But a rough estimate is possible. According to Danny Perez's

72–73; James B. Shuman, "Is There an American Abominable Snowman?" *Reader's Digest*, January 1969, 179–86; David Abrahamson, *Magazine-Made America: The Cultural Transformation of the Postwar Periodical* (Cresskill, NJ: Hampton Press, 1996), 27.

**34.** Box 11, folder 7, Marian T. Place papers, Arizona State University, Tempe; Edward W. Cronin Jr., "The Yeti," *Atlantic Monthly*, November 1975, 47–53; Danny Perez, *Big Footnotes: A Comprehensive Bibliography Concerning Bigfoot, the Abominable Snowmen, and Related Beings* (Norwalk, CA: privately printed, 1988); Peter Hartlaub, "Sasquatch: Kitsch of Death," *San Francisco Examiner*, August 7, 2000; Joshua Blu Buhs, "Where the Wild Things Are: The Legendary Bigfoot and Boys' Culture in 1970s America," *Western Folklore* (forthcoming).

**35.** Dorothy Gallagher, *How I Came into My Inheritance* (New York: Random House, 2001), 129; Bill Sloan, *"I Watched a Wild Hog EAT my Baby!" A Colorful History of Tabloids and Their Cultural Impact* (New York: Prometheus, 2001), 12–13.

idiosyncratic and incomplete *Big Footnotes*, tabloids printed 106 articles about Sasquatch between 1969 and 1981, an average of about nine stories per year. There were easily as many articles published in men's magazines during the same time period, and more than thirty books. Individual magazines were tossed aside each month or every few months, but they were replaced by new ones, fresh from the presses, and most months there was at least one that featured Sasquatch. More so than the story in *Reader's Digest*, this incessant publication schedule made Bigfoot into a constant feature of American culture in the 1970s. Indeed, the term Bigfoot came to replace Abominable Snowman as the generic term for wildman the world over—just as Abominable Snowman had replaced What-Is-It. In 2006, for example, reports of a wildman reached the international press from the jungles of Malaysia. And in the headlines for those stories, the monster wasn't called by its native name, but by the name some California loggers had bestowed on their own monster: Bigfoot.[36]

## Bozo, the Minnesota Iceman

Ivan Sanderson was on top of the world. He had cornered the market in Forteana and capitalized on what he called "the greatest story in the field of anthropology . . . not just of this century but . . . of all time." Soon, he expected, money would start to fill his coffers: "Now that we've 'proved' . . . one of these ABSMs, even up to the point we have now, my stuff, which is popular enough with the general public, is becoming very valuable." The only problem? *Argosy* paid Patterson more than it did Sanderson—"I feel, and pretty strongly, that, if anybody is to get more, it ought to be me!" he complained—and then Patterson performed his usual vanishing act when the bills came due. "I may be no judge of character," he told his literary agent, "but [my wife] Alma is, and the rest of my gang are not idiots. All of us felt that this Roger Patterson was both a sincere and honest citizen. However, he took a powder on me the minute I got his bloody documentary film made for him, and he has neither paid me my commission nor answered any of my letters since. (He only owes me 75-bucks!) . . . I am getting used to 'people' but anybody who screws me for less than a thousand grand annoys me."[37]

**36.** Perez, *Big Footnotes*; "'Bigfoot' Fever Is Sweeping across Malaysia," *Taipei Times*, January 10, 2006.

**37.** Sanderson to Tom Allen, March 29, 1968, Allen, Tom file (first quotation), and Sanderson to Ollie, February 6, 1968, Agents file (remainder of quotations), both in Ivan T. Sanderson papers.

Sanderson's bitterness at being bilked passed soon enough—and he was on to an even bigger story: the discovery of an Abominable Snowman. In December 1968, Terry Cullen, an animal importer, called Sanderson at his New Jersey office and told him that a man named Frank Hansen was exhibiting a wildman throughout the Midwest—Cullen had seen the creature at the Wisconsin State Fair the previous year, a friend had seen it in Chicago a few days before. As it happened, Bernard Heuvelmans was visiting Sanderson at the time and so together the two hunters of mysterious animals drove to Rollingstone, Minnesota.[38]

Hansen was a fat man with a winning grin. After retiring from the U.S. Air Force, he settled on a farm and showed a restored 1918 John Deere tractor on the carnival and county fair circuit. In 1967, he started to display a man-like creature encased in ice. For 35¢, gawkers could look at what he variously billed as "The Medieval Man" or a relic from "The Ice Age." The creature sounded like a standard-issue carnival gaffe—a fake exhibit to fool the rubes. Hansen, however, was evasive about the creature's origins and his motives for showing it. He told Sanderson that Russian sealers had found the body off the coast of Kamchatka; he also said that Japanese whalers had fished the body from the sea. In either case, it ended up in Hong Kong, where Hansen's benefactor, a California millionaire, had purchased it. The mysterious owner had then leased it to Hansen for display at sideshows. Hansen said that he didn't know what the creature was and didn't want to know. He also said that the wildman had been examined by scientists in Oklahoma, who took hair, tissue, and blood samples. Hansen claimed never to have heard of Sanderson before and to be unaware of the publicity surrounding Patterson's film—striking for someone in the business of displaying a wildman.[39]

If Sanderson entertained any doubts about Hansen or his monster, they disappeared when he saw the Thing—as he inevitably called the wildman at first. "It required but one look at the specimen, on both our . . . parts, to see that we were looking at a genuine cadaver," Sanderson said. He and Heuvelmans spent two days in the monster's trailer, hunched over a block of ice, examining the creature, drawing it, taking photos. Sanderson was certain that he could smell rotting flesh. Heuvelmans thought that "Bozo," as they

**38.** Sanderson, "The Hansen Case," February 28, 1969, Bozo 5 file, Ivan T. Sanderson papers.

**39.** Ibid.; Sanderson, "Brief no. 2," 4 April 1969, Bozo 4 file, Ivan T. Sanderson papers; Phil Casey, "Strange Ice Man Tale," *Washington Post*, March 27, 1969; Tom Hall, "Tracking the Minnesota Monster," *Chicago Tribune Magazine*, October 24, 1971, 18–26, 30, 34, 36; Joe Nickell, *Secrets of the Sideshows* (Lexington: University Press of Kentucky, 2005), 338.

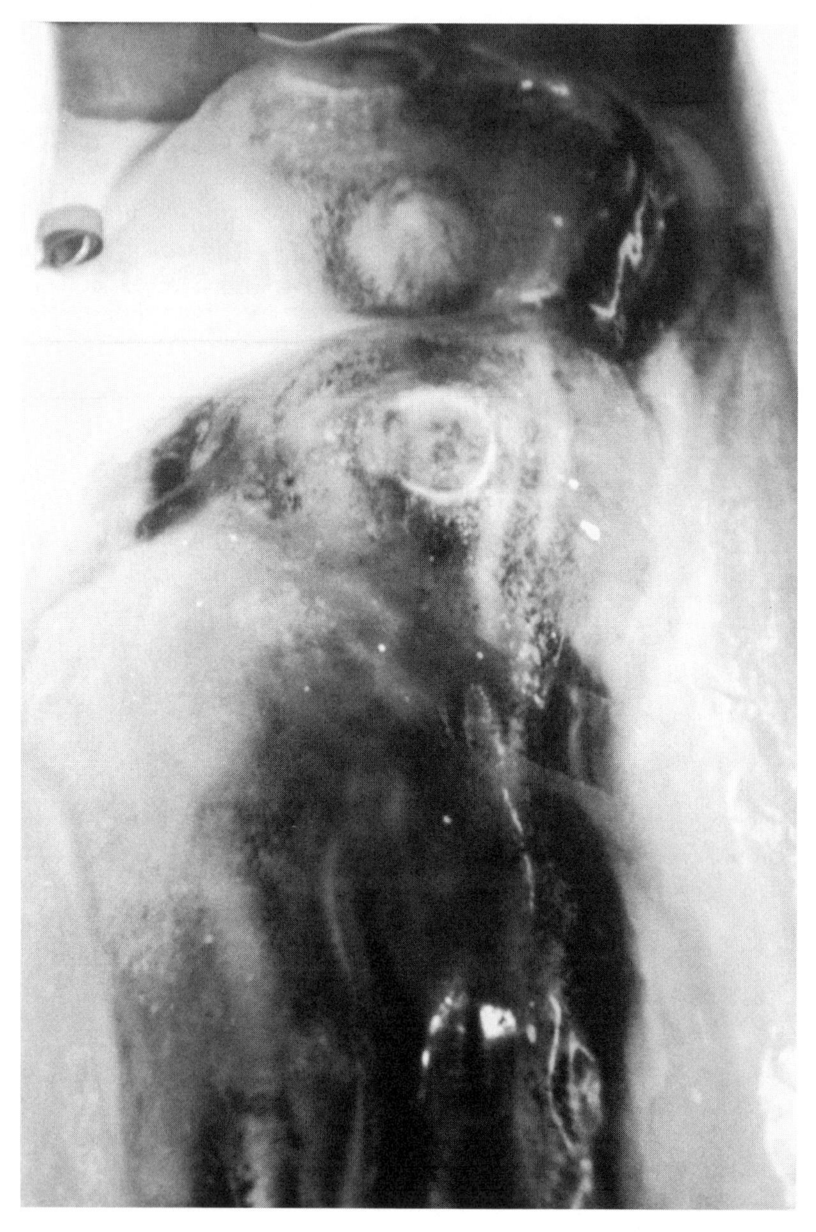

FIGURE 22. About the same time that Roger Patterson was reacquainting America with Bigfoot, Ivan Sanderson and Bernard Heuvelmans stumbled across a carnival exhibit that they thought showed a real wildman encased in ice. This is a picture that Heuvelmans took of Bozo, as they called the creature. (Image 05970. © Musée de Zoologie—Lausanne/Agence Martienne.)

named the creature for some obscure reason, was a Neanderthal. Sanderson disagreed; it was *Homo erectus*, he decided. Eventually, they showed their drawings to many of the same scientists who had studied the Yeti, Carleton Coon and W. C. Osman Hill and George Agogino and John Napier, whipping up a frenzy of excitement. Once back in Europe, Heuvelmans published a paper on the creature in an academic journal, which caught the attention of the international press. Sanderson took to the pages of *Argosy* (and also wrote an article for the Italian journal *Genus*). Coming in the wake of the *Reader's Digest* article about Patterson's movie, these reports further settled Bigfoot into the mass media.[40]

Napier thought that the creature sounded too chimerical to be real but was by this point deeply involved in the study of Patterson's film and so urged the Smithsonian to take up the matter. The Smithsonian's secretary at the time was S. Dillon Ripley, an ornithologist who had worked in Asia beside some of the same people who hunted the Yeti. He approved Napier putting out a press release noting the Smithsonian's interest—albeit a restrained and skeptical interest. Ripley also contacted the FBI on the assumption that Hansen might have broken some law that would force him to forfeit the body to the institute. (The bureau politely refused to investigate the matter.)[41]

As word of the discovery spread, Hansen's behavior became increasingly erratic. His story kept changing—the creature was not found in the ocean, he said, but a river; no, he said, it hadn't been fished from the water at all but bought from a British exporter in Hong Kong. Yes, he admitted, he had heard of Patterson's movie and, with Cullen, had read all of Sanderson's articles on ABSMs. He told Sanderson that Bozo's owner was mad at the

**40.** Sanderson, "The Hansen Case," 28 February 1969, Bozo 5 file, Ivan T. Sanderson papers; Carleton Coon, February 21, 1969; John Napier, "The Iceman Cometh to Britain," March 21, 1969; both in MNH-Dept. V-Zoology Div. Mammals (ICEMAN) file, box 326, RG 99, Office of the Secretary (S. Dillon Ripley), 1964–1971 papers, RU99, Smithsonian Institution Archives, Washington, D.C.; Bernard Heuvelmans, "Note Preliminaire sur un Specimen Conserve dans la glace d'une forme encore inconnu d'Hominide Vivian Homo Pongoides (Sp. Seu Subsp. Nov.)," *Institute Royal des Sciences Naturelles de Belgique Bulletin* 45, no. 4 (1969): 1–24; Ivan T. Sanderson, "Preliminary Description of the External Morphology of What Appeared to Be the Fresh Corpse of a Hitherto Unknown Form of Living Hominid," *Genus* 25 (1969): 249–278; Ivan T. Sanderson, "The Missing Link," *Argosy*, May 1969, 23–31 (quotation, 26).

**41.** Ripley to J. Edgar Hoover, April 10, 1969; J. Edgar Hoover to S Ripley, April 16, 1969; Napier to Ripley, March 27, 1969, all in MNH-Dept. V-Zoology Div. Mammals (ICEMAN) file, box 326, RG 99, Office of the Secretary (S. Dillon Ripley), 1964–1971 papers; S. Dillon Ripley, *Search for the Spiny Babbler* (Boston: Houghton Mifflin, 1952); Napier, *Bigfoot*, 102–3, 106.

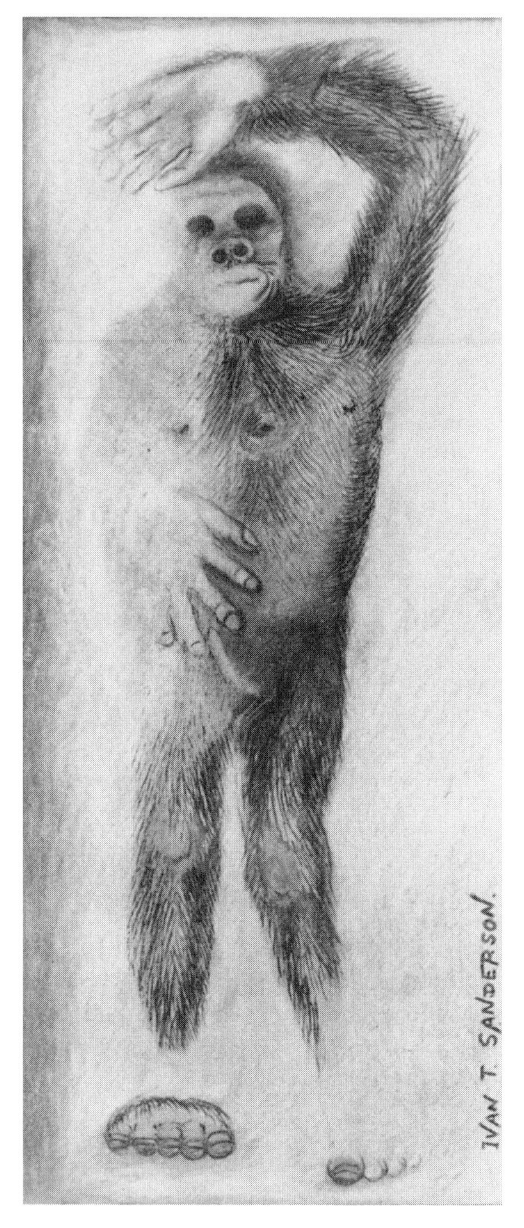


FIGURE 23. Ivan Sanderson drew pictures of Bozo to show more detail. He published several of them in *Argosy* and *Genus*. (Ivan T. Sanderson papers, B Sa3, American Philosophical Society.)

publicity (!) and had ordered that the creature be replaced with something else. He went into hiding. Worried that the body might have disappeared, Napier had the Smithsonian's publicity staff place stories in key newspapers around the country "in the hope that it may smoke out the owner so that a direct appeal" could be made to him. Meanwhile, Cullen told Sanderson that for the price of a stun gun, he could have *another* Bozo—a claim so extravagant that even Sanderson doubted it.[42]

At the end of April 1969, Hansen reappeared and went on tour again, this time with what he acknowledged was a *fabricated* wildman, implying that it was a model of the original. The new display looked different than the old, with more teeth showing, the big toe moved, and other minor alterations. Taking advantage of the publicity, he posted enlarged pages from Sanderson's *Argosy* article on the display boards around the exhibit along with the prominent announcement, "Investigated by the FBI." If it hadn't been obvious before, it was now: the whole thing was a con. The point was proved when the Smithsonian's publicity staff discovered a West Coast company that claimed to have made the creature for Hansen back in 1967.[43]

Napier concluded that Bozo was just another "What-Is-It" and that Hansen deserved the "Barnum Award"—"always one step ahead of the rest of us," he conceded. Napier excused Sanderson and Heuvelmans's lapse in judgment as an understandable reaction: in surreal, almost Gothic conditions, they had seen what for so long they had sought and convinced themselves that the (very good) model was a real beast. But whether Sanderson's actions were justified or not, J. Lawrence Angel, the Smithsonian's curator of physical anthropology, said that his institution would no longer collect Sasquatch reports—there was too much tomfoolery in the subject, which did the Smithsonian's reputation no good.[44]

Sanderson offered a different interpretation of the events. He agreed that Hansen seemed to be acting the part of the con man—but that was just a performance. Hansen and Cullen, he told Napier, had probably really killed a wildman, and they probably did contact him in hopes that he might increase

**42.** Sanderson, "The Hansen Case," February 28, 1969, Bozo 5 file; idem, "Brief no. 2," April 4, 1969, Bozo 4 file, both in Ivan T. Sanderson papers; Magnus Linklater, "Is It a Fake . . . Is It an Ape . . . or Is It Neanderthal Man?" March 23, 1969; John Napier to S. Dillon Ripley, March 27, 1969 (quotation), both in MNH-Dept. V-Zoology Div. Mammals (ICEMAN) file, box 326, RG 99, Office of the Secretary (S. Dillon Ripley), 1964–1971 papers; Phil Casey, "Ice Man Origin Solid Mystery to Smithsonian," *Los Angeles Times*, April 17, 1969; Casey, "Strange Ice Man Tale."

**43.** Sanderson, "The Hansen Case," February 28, 1969, Bozo 5 file; idem, "Brief no. 2," April 4, 1969, Bozo 4 file, both in Ivan T. Sanderson papers; Hall, "Tracking the Minnesota Monster," 18–26, 30, 34, 36; Napier, *Bigfoot*, 107–8; Jerome Clark, "The Iceman Goeth," *Fate*, March 1982, 56–59.

**44.** J. Lawrence Angel to E. H. Gravell, March 12, 1970, box 126, RG 155, Director, National Museum of Natural History, 1948–1970 papers, Smithsonian Institution Archives, Washington, D.C.; Napier, *Bigfoot*, 107–10.

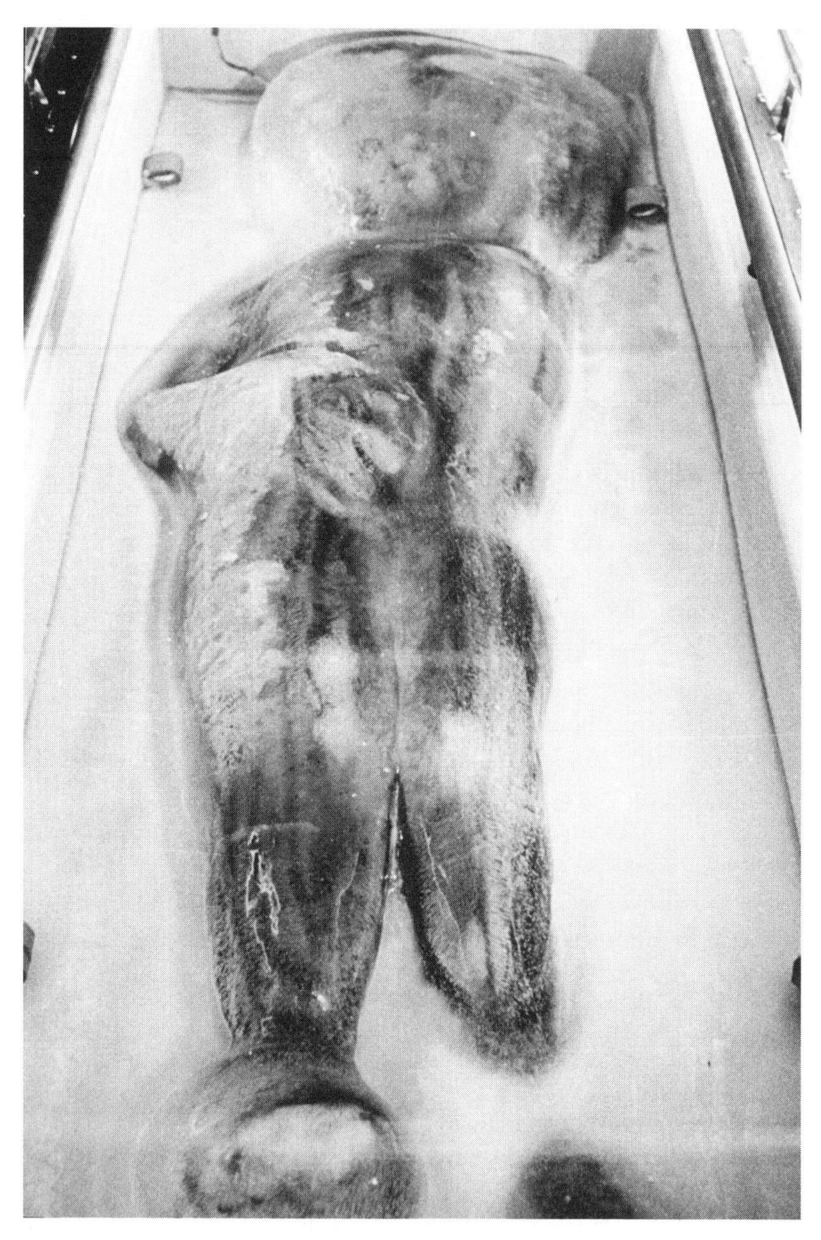

FIGURE 24. In the wake of the publicity that Sanderson generated for Bozo, Frank Hansen, the exhibitor, went into hiding and then returned with what he said was a fabricated version of the beast. Some people took him at his word and thought that he had substituted a fake for the original; others thought that there was only ever one Bozo and that Hansen had dropped out of the public eye long enough to melt its casing, reposition it, and freeze it again. (Image 05990. © Musée de Zoologie—Lausanne/Agence Martienne.)

the publicity surrounding their exhibit. But they hadn't expected quite so much publicity, including the FBI's possible involvement. They became afraid that they might be found out, that they might go to jail—and so they then had tried to make Bozo *look* like a con to throw off the authorities.[45]

Maybe Sanderson even believed this baroque theory, but if so his credulity—as usual—served the bottom line. All of the "cloak-and-daggerish things," Sanderson told his publisher, had "potential value vis-à-vis" the reprint of his "snowman book and potentially much more so for a sequel." The story had "everything, short of . . . straight rape." By maintaining the mystery surrounding Bozo—by refusing to concede that he had been conned—Sanderson could wring a few more nickels out of the subject, as could the men's magazines and tabloids.[46]

That the Minnesota Iceman was a hoax hardly slowed Bigfoot's return to the mass media. A few months after Hansen took the wildman on tour again, the tabloid *National Bulletin* filled in the one part of the story Sanderson felt was missing, running an article titled "I Was Raped by the Abominable Snowman" about a young Minnesota woman who killed the monster after being ravaged. Later, Sanderson arranged for Hansen to publish a story on Bozo for the men's magazine *Saga*—anything to keep the story going. In the article, Hansen offered yet another account of Bozo's origins. He said that he shot the beast while hunting, froze it, and then decided to take it on tour. He had a fake made because he worried that the Air Force would cut off his pension if it found he had killed a man-beast. When the authorities started to investigate, he replaced the real creature with the fake one. The tale ended with the exact ambiguity for which such magazines strived—and which was a boon to Sanderson's mystery-mongering: "There will surely be skeptics that will brand this story a complete fabrication. Possibly it is, I am not under oath and, should the situation dictate, I will deny every word of it."[47]

## Bigfoot on Tour

If Patterson was inclined to laugh at having gotten the better of Sanderson, he probably didn't have the time. He was busy with more legal problems:

**45.** Sanderson to Napier, n.d., Bozo 1 file, Ivan T. Sanderson papers.
**46.** Sanderson to Richard Heller, February 13, 1969, and March 2, 1969 (quotations), Pyramid 1 file, both in Ivan T. Sanderson papers.
**47.** Frank Hansen, "I Killed the Ape-Man Creature of Whiteface," *Saga*, July 1970, 8–11, 55–60 (quotation, 60); Hall, "Tracking the Minnesota Monster," 18–26, 30, 34, 36; Napier, *Bigfoot*, 111–12.

a woman who had invested in his movie was suing him. He started—or helped start—the Northwest Research Association to sell his book and newsletters. And, he was editing the BBC documentary into a film that he could show around the country. A con man had very few opportunities to enjoy the fruits of his labor. It was always on to the next job, the thrill—more than money, more than fame—the real goal.[48]

In the winter of 1968, Patterson finished the feature-length *Bigfoot: America's Abominable Snowman*. Around that time, he and DeAtley met Ron Olson, an Oregon man two years returned from the military and working for his father's film company, American National Enterprises. Olson had seen Sanderson's article in *Argosy* and thought a movie about Bigfoot would be of interest to his audience. Like John Green meeting Ivan Sanderson, this encounter was an important one for Bigfoot's career. Hollywood had rejected Patterson and his film; Olson provided another way to get Sasquatch onto the big screen and the cultural firmament.[49]

At the time, moviegoing was mostly an urban phenomenon, with costly movies playing for months in ornate city theaters. Along with a cadre of competing independents, American National Enterprises was exploiting the holes in this business model and pioneering a new way of marketing and distributing movies. The independents focused on rural areas, renting entire theaters—all four walls, as it were, giving them their nickname, "four-wallers"—to show their movies. (In smaller towns without theaters, they rented high school gymnasiums or Elk's clubs.) The films usually played in the winter, when there was abundant free time and few competing entertainments. In many ways, the shows were an extension of rural popular culture: they were interactive, with hosts and discussion sessions.[50]

But the four-wallers also relied on techniques associated with mass culture. They did intensive market research, surveying rural communities and using computers to discern patterns in the responses. American National Enterprises spent three times more on surveys and marketing than on production, an absurd ratio, unheard of in Hollywood at the time. From their surveys, the companies found that rural communities disapproved of Holly-

**48.** Long, *The Making of Bigfoot*, 142, 260–62, 293–94, 300–11.
**49.** Doug Bates, "The Man Who Chases Bigfoot," *Eugene (OR) Register-Guard Emerald Empire Magazine*, October 21, 1973.
**50.** Lizabeth Cohen, "Encountering Mass Culture at the Grassroots: The Experience of Chicago Workers in the 1920s," *American Quarterly* 41 (1989): 6–33; Frederick Wasser, "Four Walling Exhibition: Regional Resistance to the Hollywood Film Industry," *Cinema Journal* 34 (1995): 51–65.

wood's salaciousness; they wanted nature films and documentaries—there was an intense craving for films that were true, that were authentic. At the same time, the audiences showed a decided interest in the paranormal. One of American National Enterprises's competitors, Sunn, made its name releasing documentaries on the Bermuda Triangle, Noah's Ark, and aliens that gave birth to human civilization. These were not critical examinations of the ideas, but enthusiastic endorsements—mockumentaries, as they were sometimes called. Just like the readers of men's magazines, the four-walling audience did not necessarily consider true and factual to be synonyms.[51]

With their audience's desires carefully measured and films made to match those desires, four-wallers then saturated local television markets with advertisements for their movies—another innovation, since the big Hollywood studios mostly advertised on radio and in newspapers—deploying what they had found in surveys to create a demand for the movie. Crowds came out in droves; the four-wallers took the ticket money, the theater owner received the rent and what was made from concessions—and then the film was gone, in a few days, maybe a week. The quick advertising blitz and quicker exit was necessary. For the most part, these were bad movies, with poor cinematography, dropped sound, muddy tones, no characterization, formulaic plots, and stock footage that the four-wallers recycled again and again. Half of the movie *Toklat*, for example, about a grizzly bear, was filmed before the producers even decided on a story. The four-wallers had to open and close quickly to beat word of mouth. And then it was on to the next town, the next market, sinking the profits into another huge advertising buy.[52]

Fly-by-night though it seemed, four-walling was nonetheless quite lucrative. In January and February of 1972, *Toklat* was the second-highest grossing film in America, behind *The Godfather*. The classic four-wall adventure movie *Billy Jack*, about a Native American martial arts expert, grossed $32,000,000—*after* Warner Brothers had released the film but failed to make

**51.** Michael T. Kaufman, "Unlikely Films Score Success Here," *New York Times*, February 26, 1974; Wasser, "Four Walling Exhibition," 51–65; Michael Kammen, *American Culture, American Tastes: Social Change and the Twentieth Century* (New York: Basic Books, Inc., 1999), 18–26; Jane Roscoe and Craig Hight, *Faking It: Mock-Documentary and the Subversion of Factuality* (Manchester: Manchester University Press, 2002); Frederick Wasser, *Veni, Vidi, Video: The Hollywood Empire and the VCR* (Austin: University of Texas, Austin Press, 2002), 24–44; Lizabeth Cohen, *A Consumers' Republic: The Politics of Mass Consumption in Postwar America* (New York: Vintage Books, 2003).

**52.** Jonathan Kwitny, "In Four-Walling, the Way You Sell the Movie Is More Important Than the Way You Make It," *Wall Street Journal*, September 3, 1974; Wasser, "Four Walling Exhibition," 51–65.

money on it. When American National Enterprises went public in 1970, it grossed $5,000,000 in nine months, netted $605,000, and earned 39¢ per share.[53]

Olson was persuasive enough to convince DeAtley and Patterson to experiment with four-walling. They rented a high school auditorium in Lakeview, Oregon, to show their Bigfoot film. It was a success. After the movie, DeAtley, Patterson, and others working with them retired to their motel room with a trashcan full of money. "We were throwing it on each other on the bed and stuff!" DeAtley remembered. Ones, fives, tens, twenties floated through the air.[54]

"Bullshit on these guys," DeAtley later remembered thinking about American National Enterprises. "We'll just do it ourselves. They'd given me the whole recipe." Patterson's movie was replicated again and again and again until there were thirty-eight prints to show at different venues. Through the early months of 1969, DeAtley, Patterson, and a troupe of employees that DeAtley hired crisscrossed the Pacific Northwest and Midwest, blitzing an area with advertising, showing the film, and leaving with pockets full of cash. According to DeAtley, they ran through the little Oregon towns along Interstate 5; they went to Roosevelt, Idaho, then on to Utah, where they showed the movie at the Salt Palace. "We went over to Denver, Colorado and hit that TV market. Fabulous," DeAtley recalled. "Nebraska! Nebraska feeds both North and South Dakota. It's all one market through there. Fabulous!"[55]

Next, they pushed into Wisconsin and Minnesota, renting theaters, buying commercial slots. As DeAtley remembered years later, about a week before the movie opened there, the Smithsonian announced that Bozo was a fake. "We died!" DeAtley said. They lost $100,000. But they had been up $200,000 for the winter, so they came out all right. And, anyway, by then Patterson and DeAtley were ready to get out of the four-wall business. Patterson was dying from Hodgkin's disease; he was so sickly that he couldn't even introduce his movie at any showings after the Salt Palace. DeAtley missed his family. And his conscience nagged at him. He was selling something that he didn't believe in—he had gotten over his momentary enthusiasm, returned to his senses, and knew that Bigfoot didn't exist. Four-walling the movie wasn't a lie, but was only spitting distance away. Patterson arranged

**53.** Jack Goodman, "Critter Films: Outdoor Movies Post Big Profits," *New York Times*, December 27, 1970; Don Thomas, "Wild-Life Movies Can Reap Fantastic Dividends," *Calgary Herald*, September 16, 1972.

**54.** Long, *The Making of Bigfoot*, 249 (quotation), 263.

**55.** Ibid., 249, 263–64.

for American National Enterprises to buy the film; DeAtley signed off on the deal.[56]

Olson, however, didn't get far with the film. American National Enterprises's board of directors had no interest in Sasquatch. Bigfoot was too strange a subject, the board insisted, too weird. Olson was sure that they were wrong: audiences were interested in science fiction, the paranormal, and Bigfoot—they flocked to see celluloid simians in February 1968 when *Planet of the Apes* debuted (so popular it spawned four sequels). Apparently as a result of a compromise between Olson and the board, American National Enterprises produced a twenty-minute documentary about the creature, which was paired with the feature-length *Cry of the Wild*. A bit later, the company distributed another mockumentary on Bigfoot, this one about the exploits of Sasquatch hunter Robert Morgan. But Olson was unsatisfied, the documentaries neither as lucrative nor as persuasive as he had hoped. There was still a large market to tap, he thought.[57]

Olson was right, too, and in the early 1970s Bigfoot began appearing at suburban drive-ins. These theaters, like rural areas, were largely ignored by Hollywood and so independent filmmakers supplied them with so-called exploitation movies: cheaply made horror flicks and sexually charged films aimed at a mostly teenage audience. Sasquatch starred in *Bigfoot* and *Schlock* and *The Beast and the Vixens* and *Shriek of the Mutilated* and scads more. Most of these films were disposable, seen and forgotten, but there was one breakout hit among them: *The Legend of Boggy Creek*.[58]

Charles Pierce, an aspiring filmmaker, heard tale that a Bigfoot-like creature had been seen in Fouke, Arkansas, and went there to make a movie about the creature. As most four-wall films, *The Legend of Boggy Creek* was made on a shoestring budget, estimates of its cost ranging from $25,000 to $160,000. Area high school students worked as the crew; the Bigfoot was clad in a gorilla costume; and the cast was composed of locals, many of them people whom Pierce met at a gas station. Voiceover narration, interviews, reenactments, and the presence of actual Foukes folks on screen gave the film a documentary feel, that sense intensified by the tagline: "A True Story." Long shots of fog, meandering rivers, and thick swamps, the film's dingy tones

**56.** Bates, "The Man Who Chases Bigfoot," 3–5; Long, *The Making of Bigfoot*, 264–266 (quotation, 264).

**57.** Bates, "The Man Who Chases Bigfoot," 3–5; John Russo and Larry Landsman, *Planet of the Apes Revisited: The Behind-the-Scenes Story of the Classic Science Fiction Saga* (New York: Thomas Dunne Books, 2001); Long, *The Making of Bigfoot*, 263.

**58.** Wasser, *Veni, Vidi, Video*, 39–40; Coleman, *Bigfoot!* 205–10.

and occasional lack of focus, and constant talk—but few glimpses—of the monster gave the movie a creepy aura, a sense of foreboding. Together the verisimilitude and the slowly building tension gave the film its power: a horror story made to seem real by its supposed authenticity; mundane rural life made to seem fantastic by the presence of a monster.[59]

Unable to find a distributor, Pierce chose to four-wall the movie, paying $3,500 to rent the Paramount Theater in Texarkana for a week. Crowds lined up around the building for tickets—one of those moments when rural popular culture and mass culture met, the movie something that many in the audience had themselves helped to make, but when they sat down and darkness fell, it was shhh! no talking. They could not yell to the beast, could not touch it, and could only undress it in their imaginations. Based on the success at the Paramount, Pierce found a distributor, Howco International, which placed movies at drive-ins, and the film became a hit. *The Legend of Boggy Creek* made an estimated $20,000,000 and guaranteed that Bigfoot would continue to walk the path from fringe to center.[60]

The success of *The Legend of Boggy Creek* inspired imitators. During the Thanksgiving holiday of 1974, CBS aired *Mysterious Monsters*, a documentary about the Loch Ness monster and Bigfoot. (It was co-produced by the Smithsonian; apparently Sasquatch's popularity was too much for the institute to ignore, whatever Angel may have said.) The show attracted an estimated sixty million viewers, making it the highest-rated program of the week. Sunn took the documentary on a four-wall tour, bringing the television show to places with limited television access and generating an estimated $24,000,000. Apparently pleased, Sunn later released another pseudo-documentary, *The Legend of Bigfoot*, about the exploits of Ivan Marx. Throughout the decade, the TV series "In Search of . . . ," which adapted mockumentary conventions to television, ran a number of shows on Sasquatch. The beast also guest-starred on a number of TV dramas. Bigfoot was entering its halcyon days, the 1970s, when it was an entertainment icon, object of ardent devotion, and subject of scientific inquiry.[61]

**59.** Smokey Crabtree, *Smokey and the Fouke Monster* (Fouke, AR: Days Creek Production, 1974); Scott Von Doviak, *Hick Flicks: The Rise and Fall of Redneck Cinema* (Jefferson, NC: McFarland & Co., 2005), 146–47, 205.

**60.** Internet Movie Database, Business data for *The Legend of Boggy Creek*, http://imdb.com/title/tt0068837/business (accessed April 10, 2008); Von Doviak, *Hick Flicks*, 146–49.

**61.** Green, *Sasquatch*, 50–51; Scott Forslund, "The Nature of the Beast," *Pacific Northwest*, March 1983, http://www.bigfootencounters.com/articles/forslund.htm (accessed April 10, 2008); Alan Morton, *The Complete Directory to Science Fiction, Fantasy and Horror Television*

## The Secret of Sasquatch

What accounts for Bigfoot's popularity? Why could it entice Peter and Bryan Byrne halfway around the world, disrupt Dahinden's life, and redirect Green's career? Why could *Argosy* sell out an issue in a week? Why could *The Legend of Boggy Creek* make millions? The beast appealed to hunters for the same reason that the Yeti had intrigued British mountaineers: it was evidence that the world was not yet fully explored, that there was still room for a man to test his mettle, to touch the really real behind the false front of consumer goods and scientific arrogance. The hunt gave Byrne and Green and Dahinden and Titmus a chance to be real men—using their skills to seek something vital and alive and precious—in a way few other modern pursuits offered. It gave them an opportunity to leave civilization, to go to out-of-the-way places and live in a world that was not plastic, but was authentic and genuine, a repudiation of the society around them, a society that very often did not value them or their opinions.

Bigfoot's attraction to working-class audiences was different. Writers of Bigfoot stories and producers of Bigfoot films did not always belong to the same class as their audience—indeed, they often chafed at the restrictions of catering to working-class tastes and dismissed the audience as "rednecks" and "dorks"—but through a complex act of cultural ventriloquism they could speak, as cultural historian Michael Denning said of pulp writers, in working-class accents. Four-wallers had a relatively sophisticated system for assuring that their movies appealed to the audience; editors at men's magazines had a more subjective system. The results were the same, though—stories that expressed sentiments familiar to readers, that were understandable to them. Bigfoot was popular among white working-class men because Sasquatchiana reflected their hopes, their fears, and their hidden desires.[62]

In the late 1960s, generally speaking, white working-class men felt besieged from all sides. Black Americans agitated for their constitutionally guaranteed right to be treated equally and laid claims to privileges that had historically been reserved for white men—to the working-class eye, it seemed

*Series: A Comprehensive Guide to the First Fifty Years, 1946–1996* (Peoria, IL: Other World Books, 1997); Hartlaub, "Sasquatch: Kitsch of Death"; Coleman, *Bigfoot!* 210–11; Von Doviak, *Hick Flicks*, 150–52.

**62.** Michael Denning, *Mechanic Accents: Dime Novels and Working-Class Culture in America* (New York: Verso, 1998); Gallagher, *How I Came into My Inheritance*, 28–30; Sloan, *"I Watched a Wild Hog EAT my Baby!"* 173 (first quotation); Josh Alan Friedman, "Throw 'Em a few Hot Words," in Parfrey, *It's a Man's World*, 31, 34, 35 (second quotation).

that blacks were refusing to play by the rules, demanding instead that the fabric of society be changed to accommodate them. Middle-class college students protested a war that mostly claimed as its victims the children of the working class. Gender divisions seemed to be breaking down: women entered the labor force in large numbers while men were being forced to operate in what had traditionally been the women's sphere—shopping, worrying about appearances and clothing. At the same time, the economic position of the white working class was worsening, as the postwar boom came to an end, taxes increased, and real wages stagnated. White working-class men were frustrated and scared about the changing society, the changing economy, worried that they were failing.[63]

Many Bigfoot tales sought to ease these anxieties by affirming the culture of character, the importance of work, skill, and an old-fashioned masculinity. In much of 1970s Sasquatchiana, the beast was made out to be fighting the same battles as white working-class men, standing against America's plastic culture, the vapidity of personality, the femininity of consumerism. A *Saga* article insisted that Bigfoot was reacting to a "civilization [that was] encroaching on their primeval lairs . . . the only way they know how—by fighting back!" Bigfoot did what white working-class men wished that they could do: lived in the forest, far from womanly society. An Illinois man protested the plans to capture Sasquatch: "Fetch poor 'Bigfoot' out to this turpentined rat-race? Hell, no! Let the poor devil stay in his primitive state of happiness." Sanderson's articles tapped into these sentiments, promising that the world was still a place of adventure that could test the mettle of true men.[64]

These hopes, however, were more than fantasies about the restoration of masculinity. They were often elegiac. In Edgar Pangborn's tale "Longtooth," a Yeti stole Leda, the young, modern wife of Harp Ryder, a representative of an old-fashioned masculinity. Ryder tracked down his wife and her abductor, but tragedy was certain: he was old, his attempts to have a baby with Leda unsuccessful—guaranteeing, as the narrator said, the time was soon

**63.** Pete Hamill, "The Revolt of the White Lower Middle Class," *New York*, April 14, 1969, 28–29; Richard Sennett and Jonathan Cobb, *The Hidden Injuries of Class* (New York: Vintage Books, 1972); E. E. LeMasters, *Blue-Collar Aristocrats: Life-Styles at a Working-Class Tavern* (Madison: University of Wisconsin Press, 1975); Lillian Breslow Rubin, *Worlds of Pain: Life in the Working Class Family* (New York: Basic Books, Inc., 1976); Lillian Breslow Rubin, *Families on the Fault Line: America's Working Class Speaks about the Family, the Economy, Race, and Ethnicity* (New York: HarperPerennial, 1994).

**64.** Edward R. Bryant, letter, *True*, March 1960, 4 (second quotation); Al Masters, "America's Mysterious Cave-Man Monsters," *Saga*, November 1970, 22–25, 90–94 (first quotation, 22).

FIGURE 25. Bigfoot was popular in men's adventure magazines for a number of reasons. One of the most obvious was that Bigfoot stood for an old-fashioned masculinity. Here was a creature that could control women, that could live without civilization, that was self-reliant and strong, just as many of the working-class readers of such magazines thought (or wished) themselves to be. (Illustration by John Asaro, from *Saga*, July 1969.)

when "nobody remembered Harp's way of living"—and his word was no longer taken seriously since he started talking about Yetis, his word being the symbol and substance of his character, his values. By the time that he found the beast and his wife, Leda had succumbed to the Yeti: she was wounded by "an inner blindness, a look of a beast wholly centered on its own needs." Disgusted, Harp shot Leda "between the eyes." The act was misogynistic; Leda suffered for making Harp feel inadequate. But there was also an element of self-hatred in Harp's murdering Leda. Harp and the Yeti were the both primitive relics and, in seeing the beast with Leda, Harp saw what he had already done to her.[65]

Even when they ended tragically, though, Sasquatch tales such as "Longtooth" ultimately bred hope. To be among the last to uphold tried, true, yet vanishing, values was to imbue the quotidian with grandness. Repairing a lawn mower, refusing to shop, fishing—these weren't mere everyday happenings, they were instances of a vanishing character, and therefore of mythic importance. Maybe the battle was already lost. But that's when it's imperative to hold onto principles: lost causes are those most worth fighting for.

Not all Sasquatch tales, however, had such hopeful messages; others more directly addressed the fears of working-class men. In these tales, Sasquatch was not an ally, but the thing that attacked working-class men, that revealed the fragility of their masculinity. In the movie *Night of the Demon*, Bigfoot ripped the penis off a motorcyclist. In the novel *Sasquatch: Monster of the Northwest Woods*, a man about to confront Bigfoot worried that the creature would "see this rifle in my hands and want to shove it up my ass." *Saga* reported that Sasquatches might—like blacks and women—insist on receiving privileges that had once only accrued to white men, taking a large bite out of what seemed to be a rapidly shrinking pie. "They would have the same rights as any other citizen. This would include the right to vote, own property, enter into legal contracts and, of course, be responsible for their own acts . . . . The government would undoubtedly decide they were wards of the state . . . . The politicians would create another government bureau to manage their affairs. Some politician would start thinking about the Snowman vote and we would have another poverty program!"[66]

**65.** Edgar Pangborn, "Longtooth," *Magazine of Fantasy and Science Fiction*, January 1970, 5–36 (quotations, 10, 34).

**66.** Richard L. Tierney, "On the Legal Status of Bigfoot," *Bigfoot Bulletin* 26 (April–May–June 1971): 6–7; Warren Smith, "America's Terrifying Woodland Monster-Men," *Saga*, July 1969, 93 (second quotation); Michael E. Knerr, *Sasquatch: Monster of the Northwest Woods* (New York: Belmont Tower Books, 1977), 215 (first quotation); Green, *Sasquatch*, 460.

FIGURE 26. Bigfoot also represented many of the fears that working-class readers of men's magazines felt—about their economic position, their masculinity, and their power. This drawing of Bigfoot by famed sports illustrator Gabe Perillo Jr. for *UFO Report* re-interpreted the classic image from *True* and Patterson's movie, turning the creature into a threat. This time, the man with the gun is far away, and the reader is face-to-face with a Bigfoot that was clearly a killer. (Used by permission of Gabe Perillo Jr.)

Bigfoot was, thus, a nightmare bogey; and Bigfoot was a dream hero. These seem to be simple opposites. But they were not always so. Some Sasquatch fiction made nightmares and dreams the same. To the frequent question, is Bigfoot a man or beast, they answered, both: simultaneously other and self, black and white, woman and man.

The constraints of working-class life were rigid, and obeying them could be stifling. Blacks, in the racist imagination of whites, seemed to live without regard to such strictures; while whites repressed their desires, blacks indulged theirs. Blacks were free, and their freedom brought them in touch with something vital, with their soul—they ate soul food, listened to soul music. Working-class men were envious of this connection. Women were also seen as having a primitive contact with life; they were forces of civilization, domesticating and castrating men, but they were also, in the imagination of white working-class men, slaves to their biology, controlled by cycles of hormones. And this connection to their wildness made them powerful. Women had the mysterious and awesome ability to give birth.[67]

By imagining themselves into the body of Sasquatch, white working-class men could imagine themselves as black, as women, could come in contact with their own souls, their own repressed and forbidden desires. Becoming Bigfoot was a way of regaining a potency that these men worried they lacked, of standing against an enervating civilization—just as residents of Willow Creek sometimes dressed as the monster to touch their essential selves. The female Sasquatch in Walter Sheldon's novel *The Beast*, for example, symbolized modern, castrating womanhood—she was smart, ambitious, unsatisfied by even the best male in her tribe. She also represented "the darker-skinned peoples of the earth, the hungry fighters," who would "emerge as dominant" against "white man's America," a nation grown decadent with its "prosperity and comfort." The reader, though, was almost forced to identify with her. She was called Self, which was supposed to be a way of representing the primitive thought patterns of the creatures—so undeveloped they had not yet invented the concept of naming—but also worked to make her an extension of the reader, his representative in the story, himself, in fact. And she was the hero. The story ended with her outliving the rest of her tribe,

**67.** Sennett and Cobb, *The Hidden Injuries of Class*, 138; Harold Schechter, *The Bosom Serpent: Folklore and Popular Art* (Iowa City: University of Iowa Press, 1988), 107–8; S. Elizabeth Bird, *For Enquiring Minds: A Cultural History of Supermarket Tabloids* (Knoxville: University of Tennessee Press, 1992), 76–77.

walking to find another group where she would be celebrated and where she would give birth, assuring the perpetuation of her kind. "She began to shuffle slowly, in time with the beating of her heart. A keening song rose in her throat, softly at first, and then in a great crescendo, rising to the sky. It was as clear as the waters of the spring; it was as reedy sweet as the lonely cry of the loon. It was a song of triumph and love, and of the exquisite joy of living."[68]

Identifying with blacks and women provided white working-class men with a subversive thrill, a chance to experience emotions that were otherwise repressed, denied, or hidden. They allowed for the discovery of a secret self. But there was also an element of horror in these stories. Masquerading had its dangers. The first risk was that the possession would be too successful. This was the haunting fear that animated so many tales of Bigfoot abducting humans: that the humans would disappear forever into the world of the Sasquatch. Fredric Brown's short story "Abominable" expressed this fear succinctly. Chauncey Atherton, a Brit, brave, knighted, and "a connoisseur of women," fell in lust with a film actress. When she went missing in the Himalayas, Atherton went after her. While on the hunt, he saw a Yeti in the distance and killed it; moments later, he was captured by another Yeti. His captor explained that Yetis were a Sherpa-like people who had developed a drug that adapted them to the high altitudes, giving them fur, making them huge. Their number was small, though, and so they recruited outsiders. The actress had been one of their recruits, had turned into the monster that Atherton had shot. And now Atherton himself was to be her replacement. "Take her place?" he said. "But—I'm a *man*." The Yeti who had captured him replied, "Thank God for that—because I am an Abominable Snowwoman." The story's title, "Abominable," was both a shortened form of the traditional Abominable Snowman to allow for the surprise ending, and also a commentary on Atherton's situation: he who had been white was now and forever no longer; he who had been a man in pursuit of women was now kept, like a woman, by a more powerful mate. Instead of killing the monster and

**68.** Paul Hoch, *White Hero, Black Beast: Racism, Sexism and the Mask of Masculinity* (New York: Pluto Press, 1978); Walter J. Sheldon, *The Beast* (New York: Fawcett, 1980), 203, 288; Carol J. Clover, "Her Body, Himself: Gender in the Slasher Film," *Representations* 20 (1987): 187–228; Eric Lott, *Love & Theft: Blackface Minstrelsy and the American Working Class* (New York: Oxford University Press, 1993); Michael Kimmel, *Manhood in America: A Cultural History* (New York: The Free Press, 1996), 320; W. T. Lhamon Jr., *Raising Cain: Blackface Performance from Jim Crow to Hip Hop* (Cambridge, MA: Harvard University Press, 1998).

FIGURE 27. At times, Bigfoot could simultaneously be an object of fear and desire—a creature that allowed white working-class men to experience taboo emotions. In the 1970s, artist R. Crumb penned a comic called "White Man Meets Bigfoot" that followed the exploits of a straight-laced bureaucrat coming to terms with his innate wildness, a process facilitated by a female Bigfoot that was clearly made out to be a black woman—and thus the white man found his inner black soul. Crumb reprised that female Bigfoot for this 2000 cover of *Fate* magazine. (Courtesy of *Fate* magazine. Used by permission.)

rescuing the girl, he was the monster and he was the girl—it was an unendurable hell, an abominable situation.[69]

As horrible as Atherton's situation was, though, readers probably found pleasure in it, and in similar stories, at least partially accounting for the popularity of Bigfoot among working-class men during the 1970s. The writer Marina Warner noted, "Uttering the fear, describing the phantom, generally scaring oneself and the audience constitutes one way of dealing with the feelings that giants, ogres, child-guzzlers, ghouls, vampires, cannibals, and all their kind inspire. Dreaming of their horrors and desires and crimes, exaggerating them, reinforcing them, repeating them over and over again, works to squeeze pleasure out of the confrontation." Reading these Bigfoot stories was cathartic for the working-class audience; they called forth the unspoken terrors, made them concrete—and then dismissed them, at least temporarily.[70]

The other hazard of inhabiting the body of something else could not be so easily assuaged. The danger was failure: that the soul was not pure but had succumbed to the same corrupting forces that despoiled the world. The journalist Tim Cahill documented this possibility. Cahill saw Bigfoot as "a survivor, a self-reliant primitive in the midst of a vast technocracy: a pleasant reminder that we haven't yet swallowed up all our wilderness." But Cahill was disappointed. He arrived in The Dalles, Oregon, where there had been a rash of Sasquatch sightings; presumably, the beast was following an ancient migration route, but, if so, Bigfoot was confronting a new environment. "There was an aluminum plant nearby, a new shopping center, a Rocket gas station, a new and used car lot, and—strangest of all for [Bigfoot] during the nights—The Dalles Drive-in, which specializes in films like *Deathmaster* and *The Two-Headed Thing*." This landscape was unable to support Cahill's visions. He could not imagine a beast uncorrupted by America's plastic culture surviving here, and ended his article with a tragic fantasy: Bigfoot met with him at a roadside diner. "I need publicity," Sasquatch said. He'd been stopping by the drive-in the past few years and watching those *Planet of the Apes* films, learning English, and now he wanted to go to Hollywood, become an actor—the shallowest, most plastic of America's professions. "A great, inexplicable wave of sadness washed over me," Cahill said. The sadness was, indeed, probably great, but not inexplicable. His fantasy had burst; his hope

**69.** Fredric Brown, "Abominable," in *The Best of Fredric Brown*, ed. Robert Bloch (New York: Ballantine, 1977), 179–81; Lott, *Love & Theft*, 25.

**70.** Marina Warner, *No Go the Bogeyman: Scaring, Lulling & Making Mock* (New York: Farrar, Straus and Giroux, 1998), 6.

for transcendence had been frustrated by a consumer culture so pervasive that it ruined, even, myths.[71]

In the end, Cahill's vision was the truer one. Bigfoot was not a path out of the technocracy, away from the plastic consumer society. Stories about Sasquatch were a weak kind of resistance, offering catharsis and pleasure but no change in society. Real wages remained stagnant throughout the 1970s and 1980s; the status and power of white working-class men declined. Consumer culture did not fall to the giant. Sasquatch was just another example of that culture. Readers of men's magazines, watchers of four-wall films constructed personalities out of these things that they bought. Try as they might, Bigfoot enthusiasts, and those who read about the creature with interest but less attachment, could not find a way out of consumer culture.[72]

The dilemma came to writer Michael Grumley at night:

> I confess to a dream of the now remote and benign giant families growing canny and aggressive on Big Macs and monosodium glutamate. Multiplying at a faster and faster rate, they would turn Arthur Treacher fish cakes and Roy Rogers beef sandwiches to their advantage along the way, growing more dominant with each bite. At last, of course, they would develop a taste for human flesh. As the last scene in my dream tableau I see a smiling Sasquatch franchisee serving finger-lickin' good buckets of Businessman's Buttock and Breast of Starlet to a line of furry teen-agers.[73]

Grumley took pleasure in the inversion, the triumph of Sasquatch and the end of America's soft, plastic culture. But the fantasy is more a nightmare than a dream. The only replacement he can imagine is Bigfoot living the same consumer lifestyle. All the longing, all the searching and hoping, and the beast turned out to have no power to change the world, no path back to a better time, no secret after all.

**71.** Tim Cahill, *A Wolverine Is Eating My Leg* (New York: Vintage Departures, 1990), 205, 206.
**72.** Bird, *For Enquiring Minds*, 205–8.
**73.** Michael Grumley, *There Are Giants in the Earth* (Garden City, NJ: Doubleday, 1974), 126.

CHAPTER EIGHT

# A Contest for Dignity 1969–1977

In the summer of 1968, John Green showed Patterson's film in Willow Creek—this was on the same trip during which he filmed Jim McClarin walking along Bluff Creek. Green also led a small group on an overnight adventure, camping on Bluff Creek and then walking down to the film site. Among those attending both the movie and the field trip was George Haas. Green didn't know it, but Haas was about to become very important to the search for Bigfoot.[1]

Haas was a gardener from Oakland, California, just across the bay from San Francisco; but he was more than that. Haas was a Fortean, a science fiction fan—and a warlock, too, convinced that he had access to supernatural powers. Once, when his television went missing, he cursed the house where he thought the thief resided and a few days later the set was returned; another time, looking to buy a book that was only available in hardback, he nonetheless went into a store that only sold paperbacks . . . and found it! Haas called his home on College Street "The Vaults of Yoh-Vombis," an allusion to a book of fantasy by his favorite author, and populated it with what he and his friends called "eldritch" things.[2]

As a Fortean, Haas was certainly aware of wildmen, but that summer his interest was becoming more active; about the same time that he visited Green, he also advertised in *Fate* for others who shared his fascination and wanted to correspond on the matter. In this, unlike some of his other hobbies, Haas was not unique. During the late 1960s and early 1970s, enthusiasm for Bigfoot was intense and expanding. First Patterson's movie rekindled public

**1.** John Green, *Year of the Sasquatch* (Agassiz, BC: Cheam, 1970), 16; idem, *Sasquatch: The Apes Among Us* (Seattle, WA: Hancock House, 1978), 123.

**2.** Don Herron, *Echoes from the Vaults of Yoh-Vombis: A Compendium of the Life of George F. Haas* (St. Paul, MN: privately printed, 1976); "George Frederick Haas," *Oakland Tribune*, February 23, 1978.

FIGURE 28. Despite Roger Patterson and Ivan Sanderson bringing Bigfoot to the public's attention, and despite the creature's prominence in men's magazines, tabloids, and four-wall films, there was not really a Bigfoot community as much as a collection of individuals interested in the beast, a few of whom knew each other. That is, until George Haas, shown here, started publishing *The Bigfoot Bulletin*. (Photo by Dan Shepard. From Don Herron, *Echoes from the Vaults of Yoh-Vombis*, by permission of the publisher.)

interest—his Northwest Research Association supposedly had about a thousand members—and then *The Legend of Boggy Creek* caught the imagination of a generation; it was probably the most influential piece of Sasquatchiana ever produced. As the audiences for adventure magazines and four-wall films suggested, most Bigfoot enthusiasts were white, male, and working class: apartment managers, aquaria cleaners, college dropouts, commercial fishermen, construction workers, diesel mechanics, ditch diggers, electricians, meat cutters, night club employees, office cleaners, parking lot attendants, policeman, postal workers, teachers.[3]

Haas began corresponding with some of these people, the number growing until he decided that it was easier to maintain contact with other Bigfooters by putting out a newsletter. Jim McClarin designed the masthead, a tracing of an actual Bigfoot print reduced in size, and Haas mailed out the first issue of *The Bigfoot Bulletin* in January 1969—as the Bozo imbroglio un-

**3.** George F. Haas, "Monster Fans, Unite!" *Fate*, June 1968, 134, 145; Ken Castle, "The Search for Bigfoot," *Outdoor Outlook*, February 3, 1974, 8–11; Loren Coleman, *Bigfoot! The True Story of Apes in America* (New York: Pocket Books, 2003), 206–11; Greg Long, *The Making of Bigfoot: The Inside Story* (Amherst, NY: Prometheus, 2004), 140–43.

folded. Using a newsletter to maintain contact with a far-flung network of correspondents probably seemed obvious to Haas, coming out of science fiction fandom as he was. Sci-fi connoisseurs had knitted themselves into a community through fanzines—photocopied newsletters which circulated among enthusiasts (often including writers of the fiction, too); in the zines, fans dissected stories, proposed ideas, and discussed trends in the genre.[4]

Sasquatch fans took to Haas's newsletter readily: with no advertising, the circulation of Haas's newsletter increased to one hundred in nine months and to three hundred in two years. There wasn't any other consistent, focused outlet for their interests. Sanderson's *Pursuit* devoted space to all sort of damned *things*, so was more oriented toward Forteans in general than Bigfooters (indeed, Sanderson hardly reported on Bozo in his own magazine); Patterson was too busy to write the quarterly newsletters for his association, so people tangential to the hunt put them together. The association existed "to make money" only, DeAtley said. By contrast, *The Bigfoot Bulletin* was not a commercial product. It couldn't be bought. The only way to receive a subscription was to submit something in exchange—a citation, an idea, a report from the field. This system meant that each issue was fresh and comprehensive. "For a long time George was getting as many as twelve letters a day filled with clippings from all parts of the country," Haas's biographer said. Haas's noncommercial system also had another, magical effect: formalizing what had been a private, haphazard exchange of information and welding the disparate Bigfoot enthusiasts into a community.[5]

## The Bigfoot Community

A fair amount of the mail that Haas received (and published) recounted the experiences of Sasquatch hunters on expedition. Sometimes such trips seemed like nothing more than excuses for the hunters to spend the weekend with the guys, drinking beer, enjoying the out of doors. Bigfooter Ken Coon, for example, once found tracks while he was in the forest, but he had no tape measure! If he wasn't carrying the basic tools of the trade, then what *was* he doing out there? Certainly male camaraderie and escape from the dreary world of work were some of Bigfoot hunting's great joys, and likely these pleasures

**4.** *Bigfoot Bulletin* 3 (March 20, 1969): 3; Green, *Year of the Sasquatch*, 16–20; Joe Sanders, ed., *Science Fiction Fandom* (Westport, CT: Greenwood Press, 1994).
**5.** *Bigfoot Bulletin* 4 (April 30, 1969): 4; Herron, *Echoes from the Vaults of Yoh-Vombis*, 31 (second quotation); Long, *The Making of Bigfoot*, 140–43, 261 (first quotation).

helped to bond Sasquatch seekers into a community. But there was more to these expeditions than recreation. Expeditions were also (as the examples of Green and Titmus make clear) a chance to develop and deploy skills, to prove masculine competence: to go out to those remote places far away from shopping malls and TV and play at being real men. "I think I became interested in the Bigfoot thing because it gave me an excuse to get out and use my wilderness skills. My life-long love of the wilderness exploration has a purpose beyond just getting there and back," said Bigfoot hunter Thom Powell. Tom Morris, a contractor, said, "Maybe I'm only trying to justify all my trips to the mountains by calling them research. I like wildlife, I like to see anything I can. The more I go, the more I'm amazed at how elusive wildlife can be. I'm happy just to be up there, watching animals move around. I want to come back with the best pictures I can. The ultimate would be that shot of Bigfoot."[6]

Mostly what Haas received, though, and what the newsletter reported on, were references to books and articles about Bigfoot. Sasquatch enthusiasts were readers first and foremost, catalogers and archivists. This focus owed something to the Fortean background of many Bigfooters. Subscribers to the *Bigfoot Bulletin* combed through old newspapers, compiled bibliographies—such as one listing all the wildmen articles published in *Fate*—and assigned reading. "Homework," Haas called it. Swapping citations and articles helped to bind the community; it was also another way to demonstrate skill, in the library rather than the woods. Enthusiasts were to gird their loins and seek out musty tomes, plunge into newspaper morgues, battle with microfilm readers, just as Charles Fort had. Haas boasted, "I have over 2,000 newspaper and magazine clippings and card files on 500 to 700 direct sightings. I've been told I have the largest collection of Bigfoot information available." McClarin kept a briefcase with about 2,500 bibliographical citations on index cards. He valued it at $2,000.[7]

**6.** *Bigfoot Bulletin* 16 (April 30, 1970): 2; Anne Secord, "Science in the Pub: Artisan Botanists in Early Nineteenth-Century Lancashire," *History of Science* 32 (1994): 269–315; Paul McHugh, "Believers Still Pursuing the Legend of Bigfoot," *San Francisco Chronicle*, December 8, 1994 (second quotation); Thom Powell, *The Locals: A Contemporary Investigation of the Bigfoot/Sasquatch Phenomenon* (Surrey, BC: Hancock House, 2003), 109 (first quotation).
**7.** *Bigfoot Bulletin* 4 (April 30, 1969): 4 (first quotation); *Bigfoot Bulletin* 12 (December 31, 1969): 4; Ivan T. Sanderson, *Abominable Snowmen: Legend Come to Life*, rev. abridgment (New York: Pyramid, 1968), 10; Green, *Year of the Sasquatch*, 10–20; John Lawson, "Will the Computer Corner Bigfoot?" *Redding (CA) Record-Searchlight*, May 29, 1970; Castle, "The Search for Bigfoot," 9 (second quotation).

FIGURE 29. In a cartoon from George Haas's *Bigfoot Bulletin*, two Sasquatches are inspecting the tracks of a car's tires. The creatures, like those who hunted them, applied Zadig's method to learn about their quarry from the prints left behind. (*Bigfoot Bulletin* 13 [January 31, 1970]: 4.)

The majority of the references that Haas received, of course, came from men's magazines and tabloids. Bigfoot enthusiasts were devoted consumers of the rags. Sasquatchers probably experienced the same pleasures from the stories in these magazines as other working-class men—the same hopes, fears, and subversive thrills—but they wanted something more from the articles, too. For them, Bigfoot was not only a symbol; the beast actually existed. They wanted the true truth about it, the real reality: they wanted access to what was beyond magazines and stores and TV shows. Maybe they could go into the forest themselves, or maybe all they could do was read about others who went into the forest, but either way they wanted to touch that living, vital thing out there, that Bigfoot. They were not naive about the quality of the magazines, however—or, if they were, quickly found themselves cured of the affliction. Haas said "most" of the stories in men's magazines were "downright lies." Only about 2 percent of what tabloids published was accurate, reckoned another Bigfooter. So, then, why did they collect articles that they knew to be false? Why did they make notes on them, catalog them, and swap them? Why did they value them?[8]

One reason was because of the community's Fortean roots. Charles Fort collected all kinds of odd reports without regard to their accuracy. "I shall be accused of having assembled lies, yarns, hoaxes and superstitions," he wrote in his book *Lo!* "To some degree I think so, myself. To some degree I do not. I offer the data." Haas's guiding philosophy for his newsletter echoed that dictum: "It is the policy of the *Bigfoot Bulletin* to print *news*," Hass declared, "and if the validity of any report is questionable, then let it be determined by subsequent investigation." It may also be that assessing the magazines' veracity offered another arena for demonstrating skill: through diligence, comparison, and careful reading, that 2 percent of truth could be extracted from the slabs of baloney in which it was encased.[9]

Collecting such articles was also a way to call the beast into being. The magazines were there—abundantly there—while Sasquatch was not, and they could be used to create the monster. Bigfooter Danny Perez assiduously collated the citations that he gathered into a published bibliography, which he introduced with a riddle: "If such primates do not exist why the abundance of references? Could it be that both journalist [*sic*] and writers have documented something that isn't out there? If so, that is incredible on the grounds that most of the evidence is contradictions to that premise. This is a reasonable answer, that some-*thing* is behind the literature, ghost or physical animal, uncaught, unclassified and unbottled by twentieth century science." On the cover was an artist's rendition of a Sasquatch face. It's hard to escape the conclusion that the bibliography was not just a sign of life, not just proof of Bigfoot's existence, but was Sasquatch itself, the references its body, the cover its face. The book was the idea made real. And collecting the articles, then was not so different than the Sherpas' penchant for gathering Yeti relics. Both were magical ways of making the beast real, of calling the monster into existence.[10]

This need to imagine Sasquatch into being points to another sentiment

**8.** George Haas, "The Present State of Bigfoot," January 1973 (quotation), in Bigfoot file 1, Andrew Genzoli papers, Humboldt State University, Arcata, CA; Daniel de Vise, "Bigfoot Expert's Scholarly Feat," *Long Beach (CA) Press Telegram*, July 11, 1994.

**9.** *Bigfoot Bulletin* 23 (November 30, 1970): 4 (second quotation; original emphasis); Charles Fort, *Lo!* (New York: Cosimo, Inc., 2004), 27 (first quotation).

**10.** Danny Perez, Big Footnotes: *A Comprehensive Bibliography Concerning Bigfoot, the Abominable Snowmen, and Related Beings* (Norwalk, CA: privately printed, 1988), 10; Susan Stewart,

shared by Bigfooters, a feeling that united them and underlay all of their activities, their expeditions, their reading, their correspondence. In 1974, the Fortean author Michael Grumley wrote, "There is something compelling about the urgency with which [Bigfoot] is now being pursued—compelling and fascinating and indicative of an interest that transcends the hunter's interest in his game or the ethologist's interest in his subject." That transcendent interest was love. Bigfooters *loved* the monster. Sasquatchers referred to the creature in Patterson's film as Patty, incidentally the first name of Roger Patterson's wife but also a term of endearment. Ivan Sanderson dubbed her the "Adorable Woodsman"; Ron Olson called her Harriet. The back cover of Perez's book showed him standing against a Bigfoot carved of redwood, the beast's arm around his waist, his hand on its hand. In 2002, he explained why he was the only one of four children in his family not to have married: "I fell in love at a very early age with the field of Bigfooting. So I guess you could say I got married before any of them." *The Bigfoot Bulletin* was a collective valentine to Sasquatch.[11]

Love can never be fully rationalized or explained: much of it exists beyond the power of words to express, beyond the ability of any representation to re-create, which is one reason why so much art, popular and high, focuses on love. Here is an emotion everyone experiences, everyone knows, that motivates and frustrates, and that still challenges all attempts to describe perfectly, completely. The power of a hairy, stinking, monstrous, probably mythical beast to attract white working-class men, to make them leave families, end friendships, quit school, and devote their lives to what was widely considered a quixotic quest is a mystery and will remain so forever, a twist in the chromosomal material, a combination of personality traits, a concatenation of biographical events mixed with something ineffable and indefinable. But the cross-species love affair is not entirely beyond comprehension, and understanding Sasquatch's pull on these men helps to explain why Bigfoot was so popular all through the 1970s, why the creature mattered. Bigfoot

*On Longing: Narratives of the Miniature, the Gigantic, the Souvenir, the Collection* (Durham, NC: Duke University Press, 1993).

**11.** Ivan T. Sanderson, "First Photos of 'Bigfoot,' California's Legendary 'Abominable Snowman,'" *Argosy*, February 1968, 24; Doug Bates, "Computer (with Human Help) 'Paints' Picture of Elusive Big Foot," *Eugene (OR) Register-Guard*, August 2, 1972; Michael Grumley, *There Are Giants in the Earth* (Garden City, NJ: Doubleday, 1974), 125 (first quotation); Daniel de Vise, "Bigfoot Expert's Scholarly Feat" (second quotation).

reflected their hopes, their fears, their secret dreams, yes, and Bigfoot also held out the promise that the world could be different, that they—white working-class men, boxed in by the civil rights movement, the women's movement, the student movement, the declining economy—could be on top of the social order, *should* be on top of the social order.

Attempts to prove Bigfoot's existence were an example of what sociologist Richard Sennett called a "contest for dignity." Scientists were exemplars of middle-class autonomy, free to study any problem they wanted. And yet they ignored the mystery of Sasquatch, worried about their reputations, their careers, their paychecks. It was a betrayal of their own liberty. "Those clodhoppers," René Dahinden said, echoing Sanderson. "Science is the pursuit of the unknown. Now maybe the scientists think there is nothing unknown, since they know it all, and therefore they don't have to pursue it. I don't know, it looks like the scientists get up every morning and pray, 'Please God, let me go through another day without a new thought.'"[12]

In contrast to the Babbittry of scientists, the hunters applied themselves to the study of Sasquatch and came to know Bigfoot intimately, through their work, their hands, their skill. To their minds, the knowledge was absolutely irrefutable. Their thought was disciplined; they had attended, in Danny Perez's words, "Bigfoot University," and gained a real understanding of the world, not faith or belief, but knowledge. "Let's get this business about belief straight," Green said. "The believers are the scientists, they're the ones who are clinging to a belief. The people who think that there are Sasquatches are the ones who are investigating—the ones who have become convinced on evidence. The scientists are the ones going on pure faith and don't actually know much about it and make darn sure they don't know anything about it."[13]

This knowledge gave Bigfooters a sense of power—it's another reason why men's adventure magazines insisted that they printed the truth: telling readers that they knew the *real* story flattered them as powerful. All they needed was to find Bigfoot, to capture a Sasquatch, kill an Abominable Snowman,

**12.** Richard Sennett and Jonathan Cobb, *The Hidden Injuries of Class* (New York: Vintage Books, 1972), 147 (first quotation); Clive Cocking, "The Magical, Mystical, Mythical Sasquatch," *Weekend Magazine*, May 10, 1975 (second quotation), http://www.bigfootencounters.com/articles/clive.htm (accessed April 10, 2008); Paul Fussell, *Class* (New York: Ballantine, 1983), 43.

**13.** Danny Perez, "Legend Come to Life," February 1989 (first quotation), *North American Bigfoot Information Network Journal*, file 15.14, Vladimir Markotic papers, University of Calgary archives, Calgary, Ontario; Cocking, "The Magical, Mystical, Mythical Sasquatch" (second quotation).

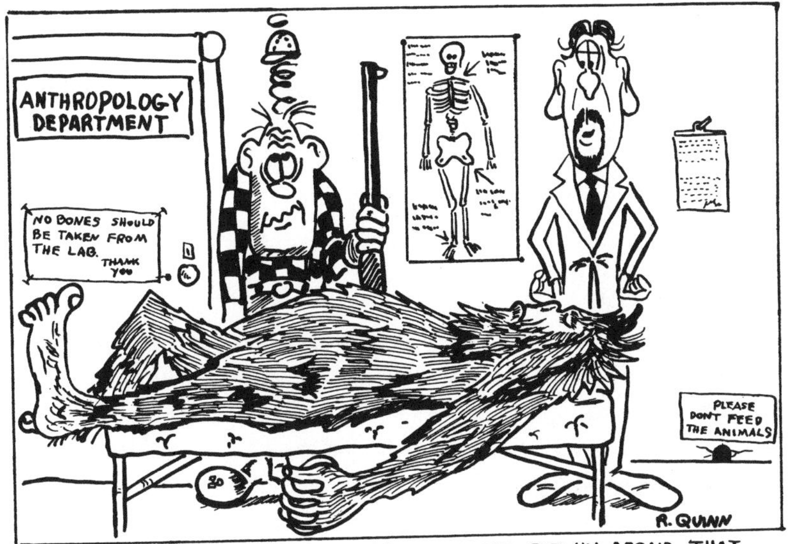


FIGURE 30. Bigfoot enthusiasts saw themselves as hardheaded skeptics grappling with evidence and trying to make sense of the universe's facts. Scientists, on the contrary, hidden away in their ivory towers, were driven by faith in their own theories, ignorant of how the world really worked. Seeing wasn't believing—belief came first, then the ability to see. Because scientists did not believe in Sasquatch, they could not see the evidence that Bigfooters repeatedly laid before them, as this hapless hunter discovered, in a comic strip that circulated among Sasquatch hunters. (*The ISC Newsletter* 1 [Autumn 1982]: 12.)

and the social order would be inverted. Those who had ridiculed them, those who had kept down white working-class men would be proved ignorant. If he ever caught a Sasquatch, Dahinden fantasized, "I'd take the scientists by the scruff of their collective neck and rub their goddamn faces in—actually, I would like to see all the people—the scientists—who have opened their mouths and made their stupid, ignorant statements, fired from their jobs. . . . They should totally, absolutely, right then and there, without pension, without anything, just be taken and thrown out the front door. Then and there."[14]

And when that dream was realized, those who had always known the truth, those who had come to the right conclusion by dint of hard work and the application of skill, would receive the dignity that the world had otherwise

**14.** "Interview: A Candid Conversation with a Prominent Sasquatch Field Worker Who Says Exactly What He Thinks," *The ISC Newsletter* 4, no. 2 (1985): 3.

denied them. "More credibility," one Bigfooter said, should "be given to . . . the common postal worker, the truck driver, the policeman, the housewife, the fisherman, the farmer, the surveyor, the bum off the street, hippies, hitch-hikers, milkmen, shop-janitors, bookkeepers, etc." What Peter Byrne called the "simple genuine honesty of the country people" would, at long last, be celebrated. The world would be put right.[15]

## Cripplefoot

Of course, any object that inspires such love is certain to generate intense jealousy as well. Just as glory went only to those who first climbed Everest, so it would go only to the first person to bag Bigfoot. "There's no second prize," Green said. Thus, even as Haas bound the Bigfooters into a community, antagonism simmered below the surface. The most contentious issue in ABSMery dates to these days. Early in 1969, only a few months after *The Bigfoot Bulletin* debuted, Sasquatch was sighted in Skamania County, Washington. A month later, the county enacted a law making it illegal to kill a Bigfoot there. Likely, the regulation was a grab at publicity—it went into effect on April Fool's Day and the editor of the local paper, which flogged the Bigfoot story, was Roy Craft, a former Hollywood publicist who no doubt understood the commercial value of the news coverage that followed. But even if the law was just an attempt to stimulate tourism, it excited Bigfoot hunters.[16]

Opinions on the ethics of shooting a Bigfoot were sharply divided. To kill Bigfoot, Haas said, was to commit a crime of world historical importance. "We have the opportunity now to avoid the killing of even one individual for the questionable reasons of expediency, fame, financial returns or supposed medical benefits," he said. "Surely we can rise above such fleeting aspirations and do right by one species. We may never get the chance again." A number of other Bigfooters also opposed killing Bigfoot, among them Jim Mc-

**15.** Peter Byrne, "Being Some Notes, in Brief, on the General Findings in Connection with the California Bigfoot," *Genus* 18 (1962): 55–59 (second quotation, 56); Jon Beckjord, "A Reply to 'Comments on Sasquatchery,'" *Western Canadian Journal of Anthropology* 6 (1976): 163–66 (first quotation, 165).

**16.** Roy Craft, "Talking It Over with the Editor," *Skamania County (WA) Pioneer*, March 14, 1969; Don Hunter and René Dahinden, *Sasquatch* (Toronto: McClelland & Stewart, 1973), 134–39; Kay Bartlett, "There's No Second Prize in the Catch-Sasquatch Game," *Idaho Statesman*, February 4, 1979 (quotation); "New County Ordinance Protects Giant Creatures, Public from Wanton Killers," *Skamania County (WA) Pioneer*, April 20, 1984; Don Hannula, "Where Have All the Sasquatch Gone?" *Seattle Times*, June 19, 1991.

Clarin, who encouraged the Sierra Club to take a public stance against shooting Sasquatch. John Green, on the contrary, spoke for a different contingent, which thought that science would only take the subject of Bigfoot seriously if a body were brought in. "A movie won't help," he said. "We already have one. The man with a gun may rightly pause to determine whether he is looking at some idiot masquerading in a fur suit. He may also wisely consider whether the gun he has is adequate to kill a huge animal whose physical capabilities are unknown. But if he is satisfied on these points, he should not hesitate further. Gun it down, cut off a piece you can carry, and get out of there."[17]

But although the views were diametrically opposed and although the matter of killing Bigfoot was directly relevant to proving Bigfoot's existence, claiming the prize, and winning dignity and respect from the world, the debate was not very heated in the wake of Skamania's laws. Haas and his *Bigfoot Bulletin* created a sense of fellowship; everyone was connected, swapping articles and ideas, bound by the certain knowledge that they were working together to change the world. Competitive feelings were hidden—but only for a time. As soon as it looked as though Bigfoot was about to be caught, the jealousies erupted to the surface. Within a year of the *Bigfoot Bulletin*'s first issue, the community was fragmenting.

During the fall of 1969, Joe Rhode, a butcher from Colville, Washington—in the far northwest corner of the state—found Bigfoot tracks around a garbage dump in the nearby town of Bossburg. The tracks looked as though they had been made by a crippled animal, and rumor spread around town that the beast raided the trash because it couldn't compete in the wild. Ivan Marx, who had been on Slick's Pacific Northwest Expedition, lived in the area; word of the discovery passed from him to Green, Dahinden, and Titmus. Green was in eastern Canada, promoting his first book, but Dahinden and Titmus made it out to see the prints.[18]

By the time Dahinden and Titmus arrived, locals and curiosity seekers had trampled all but a few of the tracks. Perhaps, though, they thought, the trackmaker was still in the area. Repeating a trick he had used during the Pacific Northwest Expedition, Titmus hung slabs of meat from trees to attract

**17.** George Haas, "Bigfoot and Tarzan as Folk Heroes," June 1974 (first quotation), in Bigfoot file 1, Andrew Genzoli papers; McClarin to Edgar Wayburn, December 29, 1967, file 23, "Bigfoot" motion pictures, 1967, Box 51, Series 25, Sierra Club Member papers, Bancroft Library, University of California, Berkeley; Green, *The Sasquatch File*, 70–71 (second quotation); Dmitri Bayanov, *Bigfoot: To Kill or to Film? The Problem of Proof* (Burnaby, BC: Pyramid Publications, 2001).

**18.** Hunter and Dahinden, *Sasquatch*, 151–53; Green, *Sasquatch*, 161–62.

the beast. The bait didn't work, however, and after a few days he left the area to follow some other leads. Dahinden wanted to spend more time investigating Cripplefoot, as the Sasquatch was being called, so he rented a trailer and moved it onto Marx's property. (He paid his bills by showing Patterson's movie to service groups.)[19]

A few weeks later, Dahinden and Marx found another run of tracks. "This is the most impressive set of tracks I have ever seen," Dahinden wrote to the *Bigfoot Bulletin*. All told, there were 1,089 prints; whatever made them had stepped over a 43-inch barbed-wire fence four times, leaving eight hairs at one crossing, lain down at one spot, and stepped aside to urinate at another. The news sent a shiver of excitement through the Bigfoot community. When Green finished his promotional tour, he drove the 2,500 miles from Toronto in three days to see the tracks. Patterson and some of his associates flocked to the area, as did some of Dahinden's friends. At some point during the winter, Grover Krantz also came to Bossburg. Krantz was a professor of anthropology at Washington State University, and a Bigfooter. He had been getting his bachelor's and master's degrees at Berkeley when the Yeti was in the news and wondered if the Abominable Snowman wasn't, perhaps, a remnant Neanderthal. When he first heard word of Bigfoot, he thought that it, too, might be a Neanderthal, having come to America across the Bering Strait. He visited Bossburg to see if he was right.[20]

For a while, the hunters, in various constellations, scoured the woods in pick-ups, snowmobiles, and airplanes. They found nothing, which, at least according to his later recollections, made Dahinden reconsider how impressive the set of tracks was. As he thought about them, he realized that for long stretches they ran along a road, which would have been convenient for a hoaxer. Additionally, Ivan Marx was behaving oddly, as though he knew the whole thing was a fraud. An accomplished hunter, he didn't bother taking to the field. Instead, he stayed in his "hovel," as Green called it, and entertained the growing ranks of Sasquatch hunters with tall tales and jokes, such as the one about his chihuahua Finky, "an excitable little creature" that Marx said was his Sasquatch dog. "Once or twice a night," Green reported, Marx put Finky under his left arm, cranked his tail, then released him with the

**19.** Hunter and Dahinden, *Sasquatch*, 151–53.

**20.** "Report from Bossburg," *Bigfoot Bulletin* 12 (December 31, 1969): 2 (quotation); "Sasquatch Book Taken East," *Bigfoot Bulletin* 12 (December 31, 1969): 5; Grover Krantz, "History," n.d., Sasquatch Items file, and Grover Krantz, "Curriculum Vitae," n.d, no folder (item 0001), both in Grover Krantz papers, National Anthropological Archives, Smithsonian Institute, Suitland, MD; Hunter and Dahinden, *Sasquatch*, 153–58; Green, *Sasquatch*, 162.

command, "Kill, Finky! Kill!" The dog darted back and forth, yipping "like a string of firecrackers."[21]

As the new year dawned, the furor calmed. Green left. Patterson did, too. Dahinden took Patterson's movie on a tour of Calgary to earn a little money. It was during this lull that Joe Metlow, a prospector, arrived, and revealed the Bigfoot community's latent jealousies. Late in January, Metlow announced at one of Marx's performances that he had caught a Sasquatch. Immediately, a bidding war broke out. According to Dahinden, the hunters arrayed themselves into two groups, one around him and another around Patterson—who apparently was being funded by an Ohio businessman named Tom Page. Metlow remained coy about what he had, where he had it, and what he would accept for it, but eventually it became common knowledge among the hunters that the supposed Sasquatch was stashed on Frisco Standard. So, they crisscrossed that snow-capped mountain looking for it and staked out one another to make certain that no one cut a secret deal with Metlow. "You couldn't step behind a tree to take a leak without feeling a dozen pair of eyes on you," one hunter said. Far from the action, Dahinden asked Green to go to Bossburg and represent his interests; Green went, but Dahinden felt betrayed—it seemed to him that Green was just out for himself, tacking between the various factions, looking to cut his own deal.[22]

When bidding for the Sasquatch reached $55,000 (offered by Dahinden), Metlow dropped his claim—without admitting whether it was true or false—and replaced it with another: his sister had a Sasquatch leg in her freezer. Metlow's previous shenanigans didn't dissuade any of the Sasquatch hunters. Another bidding war broke out. As Dahinden remembered it, Metlow finally reached an agreement with one faction: Green was going to write the book about the monster, Titmus was going to skin it, and Grover Krantz was going to bring it to science. Dahinden and Patterson were cut out. It was a brutal lesson about the limits of trust. "In a way I'm not sorry it happened. It taught me a lot about the people I had been working with," Dahinden said. "I thought it had been all for one and one for all before this, but that's not the way it worked out."[23]

The leg, however, never appeared, nor did its owner, and Metlow skipped out with a small amount of money. In October, though, while the various hunters were at home, licking their wounds, Marx reignited the passions. He claimed to have made a movie of the crippled Sasquatch. Word spread from

**21.** Hunter and Dahinden, *Sasquatch*, 156, 158; Green, *Sasquatch*, 162–63 (all quotations).
**22.** Hunter and Dahinden, *Sasquatch*, 150–73.
**23.** Ibid., 150–73 (quotations, 160, 165).

the local paper to the international wires and again brought much of the Bigfoot community to Bossburg—never mind that there had already been at least two hoaxes played in that very town, never mind that in the intervening months Bozo had been revealed as a con. After seeing the movie, Green wrote in the *Bigfoot Bulletin*, "I am satisfied . . . that [Marx] could not have faked all he has to show, and that the film is genuine." Another bidding war broke out. Sanderson was reportedly drawn in, as were Patterson and Dahinden and all the rest. According to Dahinden, Tom Page, Patterson's backer, offered Marx $25,000 to confirm or deny the authenticity of the film. Marx refused.[24]

Sometime toward the end of the year, Peter Byrne also came to Bossburg. Byrne had left Asia and founded the International Wildlife Conservation Society in Washington, D.C., shortly before, convinced that the tigers and other big game animals that he once killed now needed his protection. Under the society's aegis, he returned to the hunt for Bigfoot—but armed only with camera and tranquilizer gun. Byrne was funded by a wealthy Washington, D.C., heiress and possibly Tom Page, as well. Relatively flush with funds, Byrne hired Marx as a Sasquatch hunter, paying him $750 each month, and had him put the original film in a bank security vault. Over the next few months—as Marx collected his checks but, according to John Green, refused to hunt—Byrne played copies of the film for his backers and service groups and such. Byrne also set about investigating the circumstances under which the film was taken.[25]

Marx, Byrne learned, had been seen buying a bunch of fur in Spokane not long before he claimed to have made the film. Byrne also found the location where the movie had been made and discovered that Marx had been lying. A branch that Marx claimed was nine feet above the ground, for instance, proved to be less than six. Looking at the landscape, Byrne realized, as well, that Cripplefoot must have stayed in one small area—while the cameraman moved around, getting the beast against different backgrounds to make

**24.** John Green, "Reports from Our Correspondents," *Bigfoot Bulletin* 23 (November 30, 1970): 2 (quotation); "Ivan Marx Takes Color Movies of a Bigfoot," *Bigfoot Bulletin* 23 (November 20, 1970): 1; "More on the Ivan Marx Film," *Bigfoot Bulletin* 24 (December 31, 1970): 1–2; "Regarding the Ivan Marx Film," *Bigfoot Bulletin* 26 (April–May–June 1971): 1; Hunter and Dahinden, *Sasquatch*, 165–68.

**25.** Philip Fradkin, "Sightings of Legendary 'Bigfoot' Stir Wide Search in Northwest," *Idaho Statesman*, December 10, 1972; Hunter and Dahinden, *Sasquatch*, 169; Bill Hazlett, "Tracking Bigfoot Is No Small Feat," *Westways*, March 1974, 52–55, 79; Green, *Sasquatch*, 166–67; Curtis D. MacDougall, *Superstition and the Press* (New York: Prometheus, 1983), 257; Robert Michael Pyle, *Where Bigfoot Walks: Crossing the Dark Divide* (Boston: Houghton Mifflin, 1995), 192.

it look as though the wildman was moving. The film was a hoax—a point proved further when Byrne examined the movie in the vault and found that it was just scrap. A bit gleefully, the local paper, the *Statesman-Examiner*, declared Bigfoot dead.[26]

Like the announcement of Mark Twain's death, however, this one was exaggerated—Bigfoot was on the ascendancy in the 1970s. What did die in Bossburg—or at least drastically waned—was the *esprit de corps* that Haas tried to promote. Feeling betrayed, Dahinden turned on Green: he did "anything for a dollar," Dahinden said, without regard to loyalty or friendship. No doubt, Green felt humiliated by Byrne. He had staked his reputation on the veracity of Marx's film—his investigative prowess, his judgment as a man—and then Byrne, an interloper who had previously usurped control of the Pacific Northwest Expedition, pranced in and proved the whole thing a fraud. The "Great White Hunter," as Green derisively referred to Byrne, was a con man, gifted with a "silver tongue" but without a moral compass.[27]

George Haas stopped publishing *Bigfoot Bulletin* in June 1971—he said that he wanted to spend more time in the field, and that was no doubt true, but there may have been other problems, too. He and his father had both been ill and the *Bulletin* had been forced onto a quarterly schedule at the beginning of that year. (There was also a suggestion, denied by Haas, that he worried some bit of information that he published might lead to Sasquatch being killed.) Jim McClarin briefly tried to resurrect the *Bulletin* as *Manimals Newsletter* in late 1972, but it lasted only two issues, likely done in by the $85-per-month publication costs. Much of the remaining comity within the Bigfoot community went away with it.[28]

The competitive feelings unleashed in Bossburg balkanized the community. "Quite a few of the most active Sasquatch hunters are gambling on being the lucky winner," Green said. "It is in their interest to co-operate to some extent with the others, in order to keep abreast of what is going on, but anything that seems like a really good lead will not be shared." Bigfooters began charging each other for information. (One reason for the high quality of Marian Place's book on Bigfoot was her willingness to pay various groups

**26.** Hunter and Dahinden, *Sasquatch*, 169–70; Green, *Sasquatch*, 166–67.

**27.** Dahinden to Markotic, December 2, 1980 (first quotation), file 15.16, Vladimir Markotic papers; Hunter and Dahinden, *Sasquatch*, 163; Green, *Sasquatch*, 168 (second quotation); Bartlett, "There's No Second Prize in the Catch-Sasquatch Game" (third quotation).

**28.** Editorial, *Bigfoot Bulletin* 24 (December 31, 1970): 5; Editorial, *Bigfoot Bulletin* 25 (Jan–March 1971): 8; "Apologies to George F. Haas," *Manimals Newsletter* vol. 1, no. 2 (September 27, 1972): 3; "Photocopy Service," *Manimals Newsletter* vol. 1, no. 2 (September 27, 1972): 4.

for information.) René Dahinden captured the emerging mean-spiritedness in his pithy way, quipping, "The search for the Sasquatch is a bit like looking for the Holy Grail, except that it is performed by very unholy people."[29]

As the community fragmented, paranoia became common. Differing views on the ethics of killing Bigfoot could now be grounds for dissolving relationships—as could having the wrong acquaintances. Green, for instance, not only refused to have anything to do with Byrne, he also refused to be friendly with people who were friendly with Byrne. When a *True* article about Bigfoot hunters quoted Krantz saying that he hoped Byrne would catch the Sasquatch, both Green and Dahinden told Krantz that if the quote was accurate (they knew enough not to believe that what was written in *True* was actually true), then their associations were ended, full stop. No more sharing information. Dahinden added that if Krantz had allied himself with Byrne, he would withdraw permission for Krantz to use photos he had taken of some Sasquatch tracks, thus stymieing any analysis that Krantz hoped to perform. "If you play with everybody, you will be like a Girl who plays this game," Dahinden lectured in one of his typically ungrammatical letters, "everybody will play with you, but when the chips are down, you will be left standing." Managing the multiple relationships took delicacy and foresight and diplomacy, every utterance having to be measured by how the many different hunters would understand it. The pressure could be great. Krantz, for example, swore that he had said no such thing to *True*, but he still maintained a relationship with Byrne. When, in 1993, Krantz finally split with Byrne, he sighed, "It's a relief not to be walking a tightrope in my relations."[30]

## The Center That Wasn't

As the Bossburg flap ended, Byrne moved to The Dalles, Oregon, where he set up a trailer and opened the Bigfoot Information Center. At the time, he admitted that he wasn't sure whether Bigfoot really existed—he gave it a 95 percent chance—but thought that he could clear up the matter in six to twelve

**29.** Marian Place, undated letter, folder 4, box 11, Marian T. Place papers, Arizona State University, Tempe; Green, *Sasquatch*, 159 (first quotation); Marnie Ko, "Footprints Leading to Nowhere: René Dahinden Spent Much of His Life Believing in Something He Could Never Find," *Report News*, May 28, 2001 (second quotation).

**30.** Green to Krantz, May 11, 1975; Dahinden to Krantz, May 12, 1975; Dahinden to Krantz, May 26, 1975 (first quotation), all in *True* Correspondence file; Krantz to Byrne, September 4, 1993 (second quotation), René Letters file, all in Grover Krantz papers; Al Stump, "The Man Who Tracks 'Bigfoot,'" *True*, May 1975, 28–31, 74–77.

months. When that didn't happen, he settled in for the long haul, cultivating a new source of funding—the Academy of Applied Science in Boston, the brainchild of Robert Rines, a polymathic inventor who was also sponsoring a search for the Loch Ness monster. Those first years were good ones for Byrne. He put together educational kits on Sasquatch for local schools, published a book on the beast, gave a lecture on Bigfoot at Yale. In October 1974, he introduced *Bigfoot News*. According to Byrne, the circulation of *Bigfoot News* grew from five hundred to ten thousand in less than five years.[31]

At the time, press reports portrayed Byrne's Bigfoot Information Center as the center of Sasquatchery, which, superficially, it seemed to be—but only superficially. Byrne maintained friendly relations with some in the community—Haas, for example—but many others agreed with Green: Byrne was seen as egotistical, a media hog. His much-voiced opposition to killing a Sasquatch was not seen as sincere, but another way to get a leg up on the competition: he fully planned to kill a Bigfoot, his detractors said, but wanted to spread the idea that killing one was unethical to make it more likely that he was the only one in the forest armed. Unlike the *Bulletin* or even *Manimals Newsletter*, *Bigfoot News* was not free—it cost $5 per year, adding to the suspicion that Byrne was out to make a fast buck. The "Bigfoot Information Center" was a misnomer. The Sasquatch community had no center.[32]

And so, alone or in small groups, Bigfooters tried to fill this power vacuum, jockeying to make themselves the community's foci, to prove the existence of Bigfoot, and to reap the rewards that they were certain would follow. In the early 1970s, John Green sold his newspaper, built a "Sasquatch Office" onto his house, and established himself as a chronicler of the Bigfoot phenomenon, making a living off his Bigfoot books. He "mailed brochures individually in [his] own handwriting to every university and public library on the North American continent, and . . . got several thousand books into major public libraries and university libraries," overall selling somewhere between one hundred thousand and two hundred thousand books. It was a way of hedging his bets against not being the first to find the animal—he

**31.** "The New Bigfoot Educational and Learning Kits," *Bigfoot News* 17 (February 1976): 2; Bill Hill, "Big Foot: New Effort to Capture Huge Creature," *Sacramento (CA) Union*, March 7, 1971; Fradkin, "Sightings of Legendary 'Bigfoot' Stir Wide Search in Northwest"; David C. Anderson, "Stalking the Sasquatch," *New York Times Magazine*, January 20, 1974, 17–25; Walter Sullivan, "Loch Ness Monster: A Serious View," *New York Times*, April 8, 1976; Pyle, *Where Bigfoot Walks*, 16, 191, 198.

**32.** Barbara Wasson, *Sasquatch Apparitions* (Bend, OR: privately printed, 1979), 42–43, 89–91, 103–6.

made himself the "dean" of Sasquatchery, so a certain amount of fame would be deflected onto him no matter who ultimately found Bigfoot.[33]

"All books about the Sasquatch," he advised in his first book about Sasquatch, "should end with a blast at the scientists who try to sweep the whole matter under the rug." With his second, he followed his own prescription and ended *Year of the Sasquatch* with another criticism of scientists for ignoring the problem—for betraying their own liberty. Shortly after the Cripple-foot incident, he also started doing the scientists' job for them. Somehow, he met Ron Olson, the moviemaker, who was by now deeply invested in the hunt for Sasquatch. The two hatched a plan to computerize all of the sighting reports that Green had collected and use American National Enterprises's surveying software to extract meaningful data about the beast—in particular, migration patterns, so that a hunter could intercept the wildman along its route. For several months, Green, McClarin, and a few others translated Green's notes into terms the computers could understand and re-interviewed witnesses in cases where important information was missing. Perhaps—as newspapers around the country said—the computer, humankind's newest technology, could finally capture Bigfoot, one of the species' oldest relatives.[34]

One interesting correlation did emerge. A preponderance of reports came from areas that received more than twenty inches of rain annually. Green noted acidly, "Why mankind's supposed need to imagine monsters should dry up where it doesn't rain much I will leave to someone else to explain." The discovery gained some support from other interesting findings that Bigfooters were making. Analyses of Sasquatch tracks, for example, showed that the prints varied from north to south in a nonrandom way: tracks from California averaged fifteen inches while those from Canada averaged eighteen-and-a-half inches, with footprints from Oregon and Washington falling in between. Bergmann's law, a rule of thumb known to ecologists, stipulates that animal size varies in just this way, presumably since large body size

**33.** "Notes about People," *Manimals Newsletter* vol. 1, no. 1 (12 August 1972): 5 (first quotation); Green, *The Sasquatch File*, 70, 73; Marjorie Halpin, "Things That Go Bump in the Order," *BC Studies*, no. 39 (1978): 63(third quotation); "Interview: Does the Sasquatch Exist and What Can Be Done about It?" *The ISC Newsletter* 8, no. 2 (1989): 3 (second quotation); "Building Beachfront," *Vancouver Sun*, November 2, 2000.

**34.** Green, *On the Track of Sasquatch*, 78 (quotation); idem, *Year of the Sasquatch*, 35, 80; "Ape Meets Computer," *Humboldt (CA) Times-Standard*, June 28, 1970; Lawson, "Will the Computer Corner Bigfoot?"; "Computers Now Looking for Bigfoot," *San Francisco Chronicle*, August 6, 1972.

helps maintain body heat during cold winters and smaller body size allows animals to cool more quickly in warmer climes. It seemed remarkable that the distribution of the tracks would follow esoteric natural laws unless Sasquatch was real. Imagining that Sasquatch existed also explained the puzzling fact that many places throughout the Pacific Northwest incorporated reference to apes: Snowman's Hill, Monkey Creek, Ape Creek, Ape Glacier, Ape Lake, Ape Mountain. Why might that be unless an actual ape—Sasquatch—was known to inhabit the area? There were no other indigenous simians to inspire the names.[35]

For the most part, though, the computer study was disappointing. Only a thin description emerged from the analyses, nothing useful, nothing that couldn't have been derived from simple hand calculations. And, ultimately, Olson yoked the computer analysis to show business, which undermined whatever credibility it had. In 1972, he commissioned a mural that was clearly a copy of one frame from Patterson's movie; Olson, however, said the image was based on the computer analyses. It was, he said in a moment of hyperbole worthy of a carnival barker, "the most accurate picture anybody has come up with yet." More accurate, apparently, than Patterson's movie even, which Olson said was real. Two years later, he built a Bigfoot trap near Applegate, Oregon, along what he said his computer analyses showed was a migration route. (Green insisted the computer provided nothing so useful.) Olson joked, "I know that the trap works because I've already caught two bears and an Indian."[36]

Meanwhile, Dahinden staked his reputation—and hope for dignity—on Patterson's film. He spent a long time nosing into Patterson's personal affairs, convincing himself that the cowboy lacked the resources to have hoaxed the film. Satisfied, in 1971, he set out on a European tour with the movie, hoping that scientists there would be more receptive than their North American counterparts. He traveled to London, Stockholm (twice), Bern, Geneva,

**35.** *Bigfoot Bulletin*, 4 (April 30, 1969): 4; "Bigfoot on the Maps, the Geographical Record," *Bigfoot News*, 4 (January 1975): 4; Green, *Sasquatch*, 171–72 (quotation); George W. Gill, "Population Clines of the North American Sasquatch as Evidenced by Track Lengths and Estimated Statures," in *Manlike Monsters On Trial*, ed. Marjorie Halpin and Michael M. Ames (Vancouver: University of British Columbia Press, 1980), 265–73.

**36.** Bates, "Computer (with Human Help) 'Paints' Picture of Elusive Big Foot" (first quotation); Green, *Sasquatch*, 154–55; Richard M. Dorson, *Man and Beast in American Comic Legend* (Bloomington: Indiana University Press, 1982), 75 (second quotation); "After 18 Years, Trap for Bigfoot Still Empty," *Eugene (OR) Register-Guard*, November 16, 1992; Paul Fattig, "'I've Always Believed There Is a Bigfoot,'" *Medford (OR) Mail Tribune*, September 3, 2006.

Helsinki, and Moscow. In Russia, where he met the researchers with whom Sanderson had been in contact, Dahinden received the most enthusiastic support—the movie was undeniable proof that wildmen still walked the planet, the researchers said. But they provided little support for their contention, only arguing that the thing in the film walked in a coordinated fashion, which of course was true, but was so whether the creature was an animal or man in a monkey suit.[37]

In England, Dahinden received more qualified support—in both senses of that term. He met with Napier, playing the movie again and showing off all the other evidence that he had collected. The primatologist introduced Dahinden to Donald Grieve, an anatomist whose specialty was the study of human walking. Grieve analyzed the film frame by frame and, like Abbott and Napier, was disturbed by what he saw. "My subjective impressions have oscillated between total acceptance of the Sasquatch on the grounds that the film would be difficult to fake to one of irrational rejection based on an emotional response to the possibility that the Sasquatch actually exists," he said. His analysis was similarly undecided. The creature, he determined, was no more than six feet five inches tall, which meant that the stride length was far too short—suggesting that either the tracks or the creature or both had been faked. But the film could not be so easily dismissed. The camera that Patterson used to film the creature—known because it was specified on the arrest warrant sworn out against him—recorded at various speeds, from sixteen frames per second to sixty-four. Patterson wasn't sure which speed he had used. Most likely it was twenty-four frames per second, which was the best speed for television (and was the speed that Sanderson said Patterson used, for whatever that's worth), but no one could say for certain. So, Grieve analyzed the film at three different speeds, sixteen, eighteen, and twenty-four frames per second. If the film was made at sixteen or eighteen frames per second, he concluded, then the creature could not have been human—the characteristics of the walk were outside the range of what humans could do. If, however, the film was taken at twenty-four frames per second, then the walk was that of a normal human being.[38]

**37.** "René Dahinden's Eurasian Trip," *Manimals Newsletter* vol. 1, no. 1 (August 12, 1972): 1–3; Dmitri Donskoi, "The Patterson Film: An Analysis," *Pursuit*, October 1974, 97–98; Daniel Perez, *Bigfoot at Bluff Creek* (Norwalk, CA: Center for Bigfoot Studies, 2003), 15–16; David J. Daegling, *Bigfoot Exposed: An Anthropologist Examines America's Enduring Legend* (Walnut Creek: Altamira Press, 2004), 40, 111–12.

**38.** John Napier, *Bigfoot: The Yeti and Sasquatch in Myth and Reality* (New York: E. P. Dutton, 1973), 93–95, 123–24, 215–21 (quotation, 220).

FIGURE 31. Although Bigfoot was often seen as nothing more than a hoax or legend, a few scientists did investigate reports of the beast. Most prominent in the late 1960s and early 1970s was John Napier, a world-renowned authority on primate anatomy. (From M. H. Day, "Professor John Russell Napier, M.R.C.S., L.R.C.P., D.Sc.," *Journal of Anatomy* 159 [1988]: 228. Courtesy of Blackwell Publishing.)

Dahinden's persistence, combined with Green's blasting away at scientists, helped to gain Bigfoot a degree of respectability—and the hunters a degree of dignity—despite Ron Olson and his publicity seeking. Napier published a book on the subject of wildmen, the first by a card-carrying scientist, arguing that the Yeti probably didn't exist but Bigfoot might. Roderick Sprague, an anthropologist at the University of Idaho and editor of the scientific journal *Northwest Anthropological Research Notes*, had been trying to convince graduate students to study Native American beliefs in Sasquatch for years with no success when he came across Green's writings. He talked with other editors at the journal and decided to open its pages to articles on Sasquatch. "We are not suggesting the acceptance or rejection of belief in Sasquatch," he qualified in an editorial announcing the journal's new policy, "but rather the unfettered anthropological study of such beliefs either positive or negative."[39]

**39.** Roderick Sprague, Editorial, *Northwest Anthropological Research Notes* 4 (1970): 127–28.

In 1972, Ed Killam, a graduate student at Colorado State University, received over $7,400 in grant money, $2,400 in donated equipment, and $2,100 in discounted material for a proposed project to study the creature. Possibly encouraged by Napier sticking his neck out on the matter, the Smithsonian showed an interest in Bigfoot again; the anthropologist Geoffrey Bourne included a chapter on wildmen in his book about gorillas; and another anthropologist, Myra Shackley, started to investigate the possible existence of wildmen. Articles on the creature appeared in the *Anthropological Journal of Canada*, *Antiquity*, *BC Studies*, *Current Anthropology*, *Explorers Journal*, *Western Canadian Journal of Anthropology*, and *Zetetic Scholar*. By the end of the decade, *Northwest Anthropological Research Notes* had published ten articles on Sasquatch; these were reprinted as a book, *The Scientist Looks at Sasquatch*, which went through two editions and sold about 2,500 copies, fair for an academic text.[40]

But, ultimately, the film proved incapable of commanding respect from science. At the same time that Grieve and Napier offered their tepid endorsements, William Montagna, director of the Oregon National Primate Research Center in Beaverton, disparaged what he called a "few-second-long bit of foolishness" and "blushed for those scientists who spent unconscionable amounts of time analyzing the dynamics, and angulation of the gait and shape of the animal, only to conclude (cautiously, mind you!) that they could not decide what it was! For weal or woe," he said, "I am neither modest about my scientific adroitness nor cautious about my convictions. Stated simply, Patterson and friends perpetrated a hoax." It was the same opinion offered by scientists at the University of British Columbia and the American Museum of Natural History.[41]

The problem for Dahinden and others who wanted to use the movie to prove Bigfoot's existence was the poor quality of the film and the many mysteries that surrounded its taking. Patterson didn't know the speed at which he took the movie. Neither he nor Gimlin could say the exact distance be-

**40.** Dahinden to Markotic, November 6, 1981, file 15.16, Vladimir Markotic papers; "Ed Killam's Project," *Manimals Newsletter* vol. 1, no. 1 (August 12, 1972): 4; Russ Kinne, "The Search Goes on for Bigfoot," *Smithsonian*, January 1974, 68–72; Geoffrey H. Bourne and Maury Cohen, *The Gentle Giants: The Gorilla Story* (New York: G. P. Putnam's Sons, 1975), 281–302; Roderick Sprague and Grover S. Krantz, eds., *The Scientist Looks at the Sasquatch* (Moscow: University of Idaho Press, 1977); Myra Shackley, "The Case for Neanderthal Survival: Fact, Fiction, or Faction?" *Antiquity* 56 (1982): 31–41; idem, *Still Living? Yeti, Sasquatch and the Neanderthal Enigma* (London: Thames & Hudson, 1983).

**41.** William Montagna, From the Director's Desk, *Primate News* 14, no. 8 (1976): 7–9.

tween the creature and Patterson or the angle at which the film was taken. In addition, the image of the beast on the screen is only two millimeters high. This small size and the poor quality of the film severely limit the amount of information that can be extracted from the movie. Over the years, Bigfoot enthusiasts have claimed that certain features of the film prove that there could have been no hoax—the beast's shoulders were too broad for a human, the ratio of its arm length to leg length is outside the range of human possibility, muscles can be seen to move under the fur. As anthropologist David Daegling pointed out, however, none of these claims are true, and certainly not in the strong way enthusiasts expressed them: measuring the film is not just fussy, it's often impossible. If the movie was a hoax, then this inconclusiveness was its genius, a guarantee that no matter the tools, it would always defy analysis.[42]

Stymied from proving Bigfoot's existence with the film, Dahinden found another use for it. Patterson had left ownership of the movie in a vicious tangle, having sold the same rights again and again. Over the years, Dahinden convinced the aggrieved parties—Gimlin, who had seen no money from the four-wall tour, Patterson's financial backers in Yakima, who had been left holding the bag, buyers of rights that were worth less than the paper on which they were printed—to sell him their share of the film, then sued Patterson's estate for the moneys that had not been paid, as well as anyone who had used the film, from American National Enterprises to Peter Byrne to the makers of the Smithsonian documentary. Eventually, he won 51 percent of the rights to the film footage; Patty Patterson, Roger's widow, retained the other 49 percent and also had all of the TV rights. As he had done before with pictures that he took, Dahinden wielded ownership like a weapon, preventing those he disliked from seeing the film, studying it, or publishing stills from it.[43]

Dahinden's lawsuits further fragmented the Bigfoot community. But there was a method to his madness. Like Green's books, Dahinden's lawsuit was a hedge against not being the one to capture Bigfoot. It was a way of making him a focus of power in a community that lacked a center. "I'm a great fan of Mao Tse Tung's writings on protracted war," Dahinden said. "Mao looked 20 years ahead; that's what I'm doing. If somebody else finds the Sasquatch

**42.** Daegling, *Bigfoot Exposed*, 120, 122–49.

**43.** Superior Court of Washington for Yakima County, Case no. 58594, Gimlin *v.* DeAtley and Patterson, Finding of Facts and Conclusions of Law, February 6, 1976; Dahinden to Markotic, December 2, 1980, file 15.16, Vladimir Markotic papers; Bayanov to Beckjord, March 17, 1982, Beckjord's Baby file, Grover Krantz papers; Long, *The Making of Bigfoot*, 315–27.

first I'm just an 'also ran.' But as soon as that happens, this film will become as important as any 24 feet ever taken."[44]

## Hoaxing, the Unconquerable Problem

Even as Byrne and Green and Dahinden and Krantz and other Sasquatchers jockeyed for position within the Bigfoot community, they all had to overcome the same two problems if they hoped to prove Sasquatch's existence and win the dignity that would attend the discovery. The first problem was hoaxing. Byrne estimated that 85 percent of the reports that he investigated were no good, many of them frauds. There's no way to check his numbers, but if they are even close to correct, then hoaxing was rampant. Certainly, a good-sized book could be filled with the pranks pulled in the 1970s alone.[45]

Ivan Marx, for one, continued to shop outrageous stories and blatantly faked films, taking them on television, to the men's adventure magazines, and on four-wall tours. Ray Pickens, a bricklayer from Arden, Washington, admitted to faking footprints and Sasquatch photos for years. In Redding, California, a woman staged her own kidnapping by Bigfoot and gained national notoriety, for a small time anyway. A New York lawyer exhibited an odd-looking chimp as a Bigfoot's son. And Jean Fitzgerald, a woman from Roseburg, Oregon, claimed to have a family of Sasquatches under observation for three years. Marian Place, the children's author, sincerely believed her and was set to write a biography—or ghost write an autobiography—until Fitzgerald learned that she wouldn't become wealthy and lost interest in the matter.[46]

Ray Wallace, by now transplanted to Toledo, Washington, where he ran a roadside zoo, continued to spin yarns about Bigfoot. Every so often, he graced his old hometown newspaper, the *Klamity Kourier*, with a fifty-page-or-so letter about the Sasquatches that he had caught and the gold that they guarded. Sasquatches were people, he said, with their own language—"Yuk!

**44.** Scott Forslund, "The Nature of the Beast," *Pacific Northwest*, March 1983, (quotation), http://www.bigfootencounters.com/articles/forslund.htm (accessed April 10, 2008).
**45.** Letters to the editor, *Bigfoot News* 4 (January 1975): 4.
**46.** Place to Joe Ann, November 30, 1976, folder 4, box 19; Ray Lincoln to Place, March 23, 1977, folder 5, box 20, both in Marian T. Place papers; "Woman's Abduction—'Bigfoot' or Hoax?" *Humboldt (CA) Times-Standard*, May 24, 1976; Mark Sunlin, "The Hoaxers," *Fate*, August 2002, 25–27; 210–11; James Shreeve, "Oliver's Travels," *Atlantic Monthly*, October 2003, 94; Mike Quast, *Big Footage: A History of Claims for the Sasquatch on Film* (Moorhead, MN: privately printed, 2001), 110–21.

Yuk!" they cried—able to call other animals to them. "Cougars, for example. Cougars are their friends, and they help warn them whenever humans get near," Wallace said. He sold records of their calls, casts of their tracks. Rant Mullens, another Washington state resident, tired of Wallace's endless stream of bullshit, the publicity and money that he made from being associated with Bigfoot, admitted that he had supplied false feet to Wallace, apparently after Wallace left Bluff Creek. Mullens also confessed that he and some friends were responsible for the Ape Canyon incident. When they saw Fred Beck and the other miners bed down, they threw rocks at the cabin and left tracks in the area—albeit mostly of the right foot.[47]

Grover Krantz tried to put a good face on Mullens's admission. "If anything," he said, the disclosure "makes the Bigfoot thing a little cleaner because a very deviant story has dropped out. . . . [T]he miners' sighting . . . didn't follow the regular pattern. In other sightings, the Bigfoot was solitary, not in a group. And they don't normally attack or throw objects." But Mullens's confession was actually a big problem for Sasquatchers. Mullens had been making fake tracks for half a century by the 1970s, and he sold many copies—some to Wallace, which then made their way to others in his family, some, no doubt, to others. This wasn't a conspiracy, certainly not in the way that Green imagined when he suggested that hoaxing could not account for all the tracks—no one was coordinating the various pranks—but it does go some way to explaining how tracks, even similar looking tracks, could be found across the West and across the years. Indeed, Krantz was aware that copies of some eighteen-inch prints "were widely distributed in western Washington."[48]

Another hoax came from Robert Morgan. Morgan set the Bigfoot community ablaze in the mid-1970s when announced that he'd received $45,000 to hunt Bigfoot in Washington, $40,000 of it from the Louisa D. Carpenter Foundation in Ft. Lauderdale, Florida, and another $5,000 from an anonymous Floridian. The grants were administered by the National Wildlife Federation, giving the expedition an air of respectability. Morgan was familiar with southern Washington, having hunted Bigfoot there several different

**47.** William Overend, "Bigfoot Legend Engenders a Feud," *Los Angeles Times*, June 4, 1982 (quotations); Mark Chorvinsky, "Bigfoot and the Ray Wallace Connection," *Fate*, November 1993, 22–29; Pyle, *Where Bigfoot Walks*, 195; John Green, *The Best of Sasquatch Bigfoot* (Surrey, BC: Hancock House, 2004), 9–17.

**48.** "Retired Logger 'Fesses up' to Being Bigfoot 60 Years Ago," *Idaho State Journal*, April 14, 1982 (first quotation); Grover S. Krantz, *Big Footprints: A Scientific Inquiry into the Reality of Sasquatch* (Boulder, CO: Johnson Books, 1992), 33 (second quotation); Mark A. Hall, "The Real Bigfoot and Genuine Bigfoot Tracks," *Wonders* 7 (2002): 99–125.

times. At the end of his trek, he—and what he called his team of scientific experts—claimed that they had found 161 tracks, collected fur, seen the beast, and recorded its vocalizations. But all was not as reputable as it seemed. For a time, Morgan's scientific consultants included Peter Byrne, whose Bigfoot Information Center was headquartered nearby. Byrne accused Morgan of faking the tracks—and was promptly fired.[49]

If it was only Byrne's word against Morgan, there might not be much to the story. But Morgan's actions were those of a con artist. He made grandiose claims, saying that he had been behind Skamania's law against killing a Bigfoot. He misrepresented his hunting team: the scientific experts had bachelor's degrees and a smattering of graduate work, but nothing that could be considered real scientific expertise. (One was Ed Killam, who had won the grants to study Bigfoot.) He left behind a load of unpaid bills. The National Wildlife Federation, leery of the whole situation, tried to distance itself from Morgan, but he repeatedly asserted that the federation itself had sponsored him, and continued to do so after the grant was gone and the National Wildlife Federation no longer had any dealings with him. Even the people who Morgan had brought with him to Washington considered him a liar. One said, "I have dropped all relations with the man, and think associating with him in any way is a threat to anyone's professional reputation."[50]

Although he did his part to make ABSMery look more like show business than science, Ron Olson was worried that the pervasiveness of fraud made Sasquatch an unacceptable subject to scientists and thought that he had a way of solving the hoaxing problem. He suggested that prominent Bigfooters establish a board of examiners to review evidence and distinguish what was obviously fraudulent from what had merit. The idea sounded good to Byrne. Immediately after exposing Morgan, he championed Olson's idea in his *Bigfoot News*. Two months later, Byrne announced that five states had ac-

**49.** Morgan to Louisa d'A. Carpenter, February 17, 1974; Morgan to Byrne, July 24, 1974; Peter Byrne, "Confidential Report on Certain Aspects of the 1974 National Wildlife Federation Bigfoot Expedition (The American Yeti Expedition)," n.d., all in American Yeti Expedition papers, National Wildlife Federation Archives, U.S. Fish and Wildlife Service, National Conservation Training Center, Shepherdstown, WV; John Guernsey, "Summer Sasquatch Season Ends with Big Bigfoot Find in Washington," *Oregonian*, October 19, 1974.

**50.** Byrne, "Confidential Report on Certain Aspects of the 1974 National Wildlife Federation Bigfoot Expedition (The American Yeti Expedition)"; Morgan to Byrne, July 24, 1974; Byrne to Morgan, October 15, 1974; Edward W. Killam to recipients, November 5, 1974; Isadore Hanken to Whom it may concern, November 18, 1974; Laymond Hardy to Thomas L. Kimball, February 1, 1975 (quotation); Thomas L. Kimball to Edward J. Lehman, July 24, 1975, all in American Yeti Expedition papers.

credited analysts who could rule on the authenticity of purported Bigfoot evidence. Byrne also invented what he called a credibility scale, which he used to assess tracks and eyewitness reports, assigning each incident a number from one to ten.[51]

The board of examiners, however, failed to insulate ABSMery from hoaxing. Byrne's accredited staff turned out—according to his own account—to be mostly teenagers. Green didn't join, nor did Dahinden or Olson or Krantz. The competitive nature of the hunt militated against such cooperative ventures. Joining the board meant sacrificing one's own advantages to put Byrne in a better position, both by assuring that he saw a steady stream of the most reliable evidence and because he would have been the nominal head of the board and so received—at the very least—reflected glory if someone else found Bigfoot. The whole point of the board, Olson sneered after Byrne ran with his idea, was to "keep [the Bigfoot Information Center] as the center of Bigfoot activity and keep other little organizations anonymous."[52]

But even if the board had gotten off the ground, it's unlikely to have done much to stop the stream of fraudulent evidence. Because of the competitive nature of the hunt, liars prospered. Long after both Marx and Morgan had been revealed as cons, Krantz maintained a relationship with them. This looked like gullibility—René Dahinden thought that Krantz's "credibility . . . reached a [*sic*] all time low"—and Krantz was certainly credulous at times. (For example, he wondered if timber companies were committing hoaxes to discredit Bigfoot—and thus keep the creature off the endangered species list and the mills running.) In fact, though, maintaining close contacts with known hoaxers was a rational strategy. The competition made skepticism—necessary for the success of any board of examiners—too costly for the hunters. There was a chance, however slight, that some hoaxer, at some time, might come across real evidence—and so every lead, however bizarre, needed to be chased down. Krantz told another anthropologist that his relationship with Morgan "puts me in a unique position to evaluate his claims in a direct manner."[53]

**51.** Olson to Green, February 12, 1975, *True* Correspondence file, Grover Krantz papers; "The Bigfoot Board of Examiners," *Bigfoot News* 2 (November 1974): 3; "The Bigfoot Board of Examiners," *Bigfoot News* 4 (January 1975): 1.

**52.** Olson to Jon Green, February 12, 1975, (quotation), *True* Correspondence file, Grover Krantz papers; Editorial, *Bigfoot News* 37 (October 1977): 1; Peter Byrne, "All Quiet on the Western Front: Bigfoot, April, 1974," *Pursuit*, April 1974, 41–42.

**53.** Krantz to Marx, 1978; Krantz to Edward Lehman, September 4, 1975 (second quotation), both in B.F. Letters Misc. file, Grover Krantz papers; Dahinden to Krantz, May 12, 1975 (first quotation), *True* Correspondence file, Grover Krantz papers; Krantz, *Big Footprints*, 202–4.

What was rational for the hunters, though, ultimately undermined any respect Bigfooters might hope to gain. "Hoaxes have permanently and irreparably contaminated Bigfoot research," wrote Benjamin Radford in the *Skeptical Inquirer*. The community was riddled with pranksters and frauds whom the hunters continued to embrace. In addition, the main vehicles for news about the beast throughout the 1970s were the equally fraudulent men's adventure magazines and tabloids. How could any of this be taken seriously? How was this a path to dignity or respectability?[54]

## The Laughter Curtain

The second problem that Bigfooters had to overcome if they hoped to win dignity for themselves and their subject was the close association between Sasquatch and other forms of stigmatized knowledge. To some degree, the problem had long plagued the subject: giant footprints in California had a connection to the occult Lemurians; Crew was a creationist; Fred Beck and Lee Trippett had attributed paranormal powers to Bigfoot; and Forteans had chronicled the travails of the Yeti, Sasquatch, and Bigfoot since the beginning, sometimes when no one else would. But the problem exploded in the 1970s. As the Bigfoot community grew, it attracted all sorts of new people to the hunt, people with different agendas than the original Sasquatchers, and different views of the beast—people with ideas that took the search for Sasquatch far outside the confines of science.

Scores of reports continued to link Bigfoot with UFOs. "UFOnauts and Bigfoot are teaming up," wrote paranormal investigator B. Anne Slate, author of a series of articles suggesting (*arguing* is too strong a word) that Sasquatches were the advance guard of a hostile alien invasion. Men's adventure magazines and other working-class entertainments were quick to explore the link. Articles on such occult topics flattered readers into thinking that, although they were low on the social totem pole, they had access to truth, real truth, as the magazines might say, and therefore were powerful. "Is there anything we can do to thwart the predicted attempt to take over our world?" Slate asked in another article, and then answered herself, "Initially, knowledge about it will be our greatest weapon."[55]

**54.** Benjamin Radford, "Bigfoot at Fifty: Evaluating a Half-Century of Bigfoot Evidence," *Skeptical Inquirer*, March 2002, 29–34 (quotation, 34).

**55.** B. Anne Slate, "Gods from Inner Space," *Saga*, Jan. 1975, 31–33, 73 –75 (first quotation, 30); B. Ann Slate, "Gods from Inner Space," *UFO Report* 1976, 36–38, 51–54 (second quotation, 54);

Some Christians embraced Bigfoot as an example of science's limits and, by implication, faith's powers. Sasquatch was proof that evolution was false, Holy Scripture true. Some Mormons saw Bigfoot as evidence of their theology. Back in the 1830s, a Saint named David Patten claimed to have seen Cain—Abel's brother—while in Tennessee, the first murderer marked by God and made unkillable. This Cain was tall and hairy, Patten said. In the 1970s and 1980s, some Mormons began to argue that Bigfoot was Patten's Cain—not the result of evolution, but divine creation and punishment, the power of God made manifest.[56]

Other reports turned Bigfoot into a nationwide species. Men's magazines carried stories about Sasquatch frolicking in California deserts and Florida swamps. Bigfoot was even sighted near a highway in Maryland—everywhere except Rhode Island and Hawaii. In Montana, Bigfoot was implicated in otherwise unexplained cattle mutilations. Reported Sasquatch tracks came in all sorts of shapes, some with four toes, some only with three. Forteans Loren Coleman and Jerome Clark proposed a theory somewhat similar to Lee Trippett's to explain this diversity. The materialism of the current age—science, technology, consumerism—squashed the spiritual yearnings of humankind, they said. Fortean things—Bigfoot in all its manifestations and other damned objects—were projections of the Jungian collective unconscious, screams of help from repressed yearnings for a more balanced life. There was a truer, better world still out there, mostly invisible, a more humane world, where dignity was divorced from materialism, and Bigfoot was a guide to that world. The monster knew the way.[57]

Those trying to get Bigfoot accepted as a scientific object bristled at the suggestion that it was connected to UFOs, proof of God's majesty, or three-

S. Elizabeth Bird, *For Enquiring Minds: A Cultural History of Supermarket Tabloids* (Knoxville: University of Tennessee Press, 1992), 205–8.

**56.** "Latest Report from Bluff Creek, Northern California," *Bigfoot Bulletin* 5 (May 31, 1969): 1–2; Marion Shepherd, *Sashwa of the Bigfoot* (n.p.: privately printed, 2000); Shane Lester, *Clan of Cain: The Genesis of Bigfoot* (n.p.: privately printed, 2001); Frank Peretti, *Monster* (Nashville, TN: WestBow Press, 2005); Matthew Bowman, "A Mormon Bigfoot: David Patten's Cain and the Concept of Evil in LDS Folklore," *Journal of Mormon History* 33, no. 3 (2007): 62–82.

**57.** Roberta Donovan and Keith Wolverton, *Mystery Stalks the Prairie* (Raynesford, MT: THAR, 1976), 86–94, 106; B. Ann Slate, "The Ex-Marine Being Stalked by a Bigfoot," *Saga* 1976, 52–54, 79–82; idem, "Florida's Rampaging Man-Ape," *UFO Report*, July 1977, 32–35, 64–69; Jerome Clark and Loren Coleman, *Creatures of the Outer Edge* (New York: Warner, 1978); Green, *Sasquatch*, 6, 227; Marian T. Place, *Bigfoot All over the Country* (New York: Dodd, Mead, 1978); Philip R. Rife, *Bigfoot Across America* (San Jose, CA: Writers Club Press, 2000).

toed. Byrne dismissed the possibility of Bigfoot's extraterrestrial connections in his newsletter. Dahinden said, "I wouldn't place the activities of Sasquatch hunters in the same category as UFO sighters. We here are dealing with the fact that something makes tracks which we have recorded in plaster casts and have been seen with great consistency." Green said that even if he saw a Sasquatch walking through the doors of a UFO, he would doubt that the creature was from outer space, assuming instead that Bigfoot was just investigating the spaceship. Krantz refused to take seriously most reports from outside the northwest and rejected out of hand any prints that had fewer than five toes.[58]

In the mid-1970s, Jon-Erik Beckjord, a relative newcomer to the Bigfoot community, proposed that a new organization be founded, like Olson's board of examiners, to separate fact from fiction, the paranormal from the scientific—already he had taken to the pages of scientific journals arguing that Bigfoot should not be lumped with UFOs or aliens. Green was skeptical, knowing that many such groups were just fancy names for someone's living room and worrying that if the organization sounded too official it would convince the public that Bigfoot was already being studied, reducing pressure on scientists to investigate. He also disliked that Beckjord opposed killing Bigfoot. Still, Green was willing to sign on, impressed by Beckjord's boundless energy and idealism, and so Beckjord created Project Grendel, named in honor of the monster from *Beowulf*. He even recruited some scientific advisors, including the naturalist George Schaller.[59]

But just as the board of examiners failed to defend the study of Bigfoot against hoaxers, so too did Beckjord's group fail to firmly separate the study of Sasquatch from the paranormal. Only a little while after he started Project Grendel, Beckjord joined the heretics. Sasquatches, he said, were not corporeal. They were interdimensional beings. That is, they slipped into and out of mundane existence, which was why none could be killed. When the

**58.** Letters to the editor, *Bigfoot News* 1 (October 1974): 3; "Bigfoot in the Cornstalks, A Bigfoot Information Center Commentary and Viewpoint," *Bigfoot News* 38 (November 1977): 2; Letters to the editor, *Bigfoot News* 39 (December 1977): 4; "B.C. Man Planning Hunt for Sasquatch," October 16, 1969 (Dahinden quotation), file 15.16, Vladimir Markotic papers; Green, *Sasquatch*, 256; idem, Review of *The Evidence for Bigfoot and Other Man-Beasts*, *Cryptozoology* 5 (1986): 98–99; Krantz, *Big Footprints*, 41, 197–219.

**59.** John to John [*sic*], July 8, 1976, Green file; Krantz to anonymous, April 22, 1983, René Letters file, both in Grover Krantz papers; "George Schaller Lends His Support to Project Grendel," *The Gigantopithecus Gazette* 1 (April–May 1977): 1, in Judith Merril fonds, R2929-0-4-E, Library and Archives of Canada, Ottawa, Ontario; Beckjord, "A Reply to 'Comments on Sasquatchery,'" 163–66; Jon Beckjord, "A Rebuttal to Krantz's Three Step Approach to Sasquatch Identification," *Northwest Anthropological Research Notes* 12 (1978): 72–74; Green, *Sasquatch*, 152–59.

bullets flew, they simply—if that's the word—exited our reality for another. That's why trackways suddenly stopped and started, too, not because they were made by careless pranksters. Beckjord started showing pictures of trees and rocks and shrubbery that, he claimed, he had taken while only seeing the obvious scenery but which, when developed, showed entire Bigfoot families. The faces were "like a flat-headed baboon crossed with a wolf." This was proof, he said, of his contentions: Bigfoot was not detectable by the human eye but made visible by cameras, which could record reality in far greater detail. He also said that Sasquatches could shape-shift. Nor were Sasquatches the only interdimensional travelers: his theories were meant to explain a panoply of Fortean beings, all visitors from another dimension.[60]

Beckjord promulgated his theories in the scientific journal *Current Anthropology* and was tireless in finding his way into newspapers and onto television. Maybe he was just a guy having fun, a performance artist of sorts, seeing what crazy things he could say, but even if that's true, he understood the Bigfoot community well. Beckjord was on the make, looking for power in the centerless field of Sasquatchery. He cultivated Krantz as an ally—at least for a while—by giving him a copy of Patterson's film against Dahinden's wishes. With Patty Patterson, who still owned the TV rights to the film, he also tried to sell the movie to networks; and he determined that some stills from the movie had entered the public domain—and so couldn't be controlled by Dahinden. Saying outrageous things was just another strategy for gaining power and prestige. Journalists were drawn to Beckjord because, like Dahinden, he was eminently quotable, bizarre, and tendentious, even within the confines of the already-odd Bigfoot community—perfect for a mass media that had an insatiable craving for the novel, the strange, and the controversial. From the late 1970s through the 1990s, he was among the most prominent Bigfooters, and certainly influential. After Beckjord, many other Bigfooters also developed paranormal theories. There was Stan Johnson, who argued he'd been in almost continuous telepathic communication with Sasquatch for years, Lunetta Woods, who was visited by the wildman in her backyard, and Jack Lapseritis, who saw Sasquatch as a New Age shaman connected to a race of alien space brothers.[61]

**60.** Jon Beckjord, "Jon Beckjord Comments on Robin Ridington on 'The Sasquatch Image,'" *Zetetic Scholar*, no. 6 (1980): 133–40 (quotation, 135); idem, "A New Method for Calculating Sasquatch Weight," *Pursuit*, Spring 1980, 67–71; idem, "The Bigfoot Evidence," *Frontiers of Science* 3 (1981): 22–27; idem, "Beckjord on Bigfoot," *Skeptical Inquirer* Spring 1981, 64–68.
**61.** Krantz to anonymous, April 22, 1983, René Letters file, Grover Krantz papers; Beckjord to Merril, January 23, 1978; "The Roger Patterson Film Bigfoot Film Is Available for TV,"

Other members of the Bigfoot fraternity, more inclined to thinking Sasquatch an ape, were enraged. "They think I'm some kind of nut," Beckjord admitted. Heuvelmans called him "crazy." Dahinden dismissed the claims of those who said that they were in telepathic contact with the beast as nonsensical, unreal. "That's just like saying you had 235 sexual encounters but never got laid! You know?" He sent Beckjord menacing letters, threatened to sic the FBI on him, demanded compensation for Beckjord's use of stills from the film—and Beckjord either ignored them or returned them with the spelling errors corrected and a red F-minus scrawled across the top. Krantz cut all connections with Beckjord: he refused to engage him in print, refused to allow his articles to appear in the same book as Beckjord's essays. When he was planning a conference about Sasquatch, Krantz even considered hiring a bouncer to ensure that Beckjord did not sneak in—a rather literal example of the line that he was trying to draw between science and nonsense. How could the subject of Bigfoot be taken seriously if one of its loudest voices proclaimed the beast was not an ape but an interdimensional traveler?[62]

But the problem was bigger than Beckjord. Many Bigfoot enthusiasts were similarly drawn to other scientific heresies—and not just the committed Forteans. "I always have been, and continue to be, interested in any unusual phenomena without having any particular belief in any of them," Krantz said. Peter Byrne not only believed in the Sasquatch and the Yeti, but also devoted a fair amount of space in *Bigfoot News* to a consideration of the Loch Ness monster. Green worked closely with Ken Coon—no relation to Carleton—who thought that there were many species of wildmen inhabiting North America, some with less than five toes on each foot. Green also allied

n.d., both in Judith Merril fonds, R2929-0-4-E; Jon Beckjord, letter, *Current Anthropology* 21 (1980): 414; Stan Johnson and Joshua Shapiro, *The True Story of Bigfoot* (Pinole, CA: J & S Aquarian Networking, 1987); Stan Johnson, *Bigfoot Memoirs: My Life with Sasquatch* (Newberg, OR: Wild Flower Press, 1995); Lunetta Woods, *Story in the Snow: Encounters with the Sasquatch* (Lakeville, MN: Galde Press, Inc., 1997); Jack Lapseritis, *The Psychic Sasquatch and Their UFO Connection* (Mill Spring, NC: Wild Flower Press, 1998).

**62.** Heuvelmans to Krantz, August 13, 1985 (second quotation), Heuvelmans file, Grover Krantz papers; Krantz to Markotic, November 28, 1988, file 15.14, Vladimir Markotic papers; Bartlett, "There's No Second Prize in the Catch-Sasquatch Game"; Daniel Cohen, *Monster Hunting Today* (New York: Dodd, Mead, 1983), 118 (first quotation); *Sasquatch Odyssey: The Hunt for Big Foot*, DVD, directed by Peter von Puttkamer (1999; West Vancouver, BC: Big Hairy Deal Films, 2004), (third quotation).

himself with creationists in Texas who said that they had found humanlike tracks in rock strata that also had dinosaur tracks, thus proving that humans and dinosaurs existed at the same time.[63]

Fringe ideas could not be filtered out of Sasquatchery. Partly, the problem was the competitive nature of the hunt—again, Bigfooters would associate with kooks if they thought that there might be the slightest chance that person knew something about Bigfoot. Mostly, though, the problem stemmed from the antagonism toward science that was prevalent among Bigfooters. Once the line between science and science fiction was erased, it was hard to decide where to redraw it. Writing in *Saga*, Ken Coon said, "One anthropologist privately admits that one type of unknown primate probably inhabits North America, but says it is 'scientifically illogical' to believe that there is more than one type. But isn't 'scientifically illogical' typical of the wording used by anthropologists years ago regarding *any* Bigfoot evidence?" There was power in that critique. Green wrote an introduction to Ann Slate's book, admitting that he saw no reason to connect UFOs and Sasquatch, but "knowing so well the difficulties and prejudices that [Slate and coauthor Alan Berry] face," he could not "do otherwise than recommend that the reader give their findings full and fair consideration."[64] Why not Sasquatches in UFOs, then? That was scientifically illogical, too. Why not wildmen mutilating cattle? Why not Bigfoot on the New Jersey Turnpike? Why not interdimensional space travelers? Beckjord said that his theories were not unscientific at all.

This inability to separate Sasquatchery from fringe fields, combined with the prevalence of hoaxes and the beast's penchant for appearing in the *National Enquirer*, guaranteed that Bigfoot would not be taken seriously by mainstream society, not by scientists, and not by the mass media. Despite all of their work, the Bigfooters were losing the contest for dignity. Roderick Sprague sent copies of *The Scientist Looks at Sasquatch* to some of the country's leading scientific journals, yet the book was so disrespected that it never even showed up on the lists the journals published

**63.** Green, *Sasquatch*, 304–6, 323–31; Krantz, *Big Footprints*, 241 (quotation); Massimo Pigliucci, *Denying Evolution: Creationism, Scientism, and the Nature of Science* (New York: Sinauer, 2002), 246.

**64.** Ken Coon, "Monsters in Our Midst—New Clues to the Growing Bigfoot Mystery," *Saga*, July 1975, 36–38, 45, 50 (first quotation, 38); B. Ann Slate and Alan R. Berry, *Bigfoot* (New York: Bantam, 1976), xv (second quotation); Thomas F. Gieryn, *Cultural Boundaries of Science: Credibility on the Line* (Chicago: University of Chicago Press, 1999).

of books that they had received in the mail—let alone was reviewed—this despite the fact that one of Sprague's old professors was the review editor at *American Anthropologist*. Green also had a terrible time gaining support for his book *Sasquatch: The Apes Among Us*. He worked with Sanderson's one-time literary agent, who was obviously used to getting books on strange phenomena in print, but in a year, twenty-four publishers rejected Green's manuscript.[65]

Newspapers, magazines, TV, and radio dutifully reported Bigfoot sightings and published stories on the eccentric characters that hunted the beast, but almost always with a wink and a nod. In 1980, after an Idaho woman was scared from her house by screeches that some attributed to Bigfoot, a radio station "had a field day with the report," according to a local newspaper. Disc jockeys joked about the screams all morning and held "a 'Big Noise' contest, in which a screech was broadcast at unannounced times and the sixth, ninth, or whatever caller was awarded a prize." Newspapers called Bigfoot enthusiasts smallheads, liars, nuts, confused, simple, gullible. Sasquatchers were said to suffer from "weirdo psychology." They were drunkards, seeing the monster only through beer-bottle glasses. When stories were too respectful, newspapers themselves could expect ribbing. After the *New York Times* published a series of stories sympathetic to the existence of Sasquatches (and its parent corporation cosponsored a search for the Loch Ness monster), for example, an editorial in the *Washington Star* ridiculed its credulous competitor.[66]

This orientation reflected the class bias of journalists, who were overwhelmingly white middle-class men. (One survey found that journalists in 1979 were 95 percent white, 79 percent male, and 93 percent college educated. Seventy-eight percent earned more than $30,000 per year.) These middle-class men reduced the complex relationship between Sasquatch and its audience to base emotions—seeing in Sasquatchiana only sex and violence, nothing more interesting. They maligned working-class striving. The very

**65.** Ollie Collier to Place, October 10, 1977, box 11, folder 7, Marian T. Place papers; Roderick Sprague, Review of *The Sasquatch and Other Unknown Hominoids*, *Cryptozoology* 5 (1986): 99–108.

**66.** Terry Smith, "Screech Forces Woman to Flee Home," *Idaho State Journal*, July 18, 1980 (first quotation); MacDougall, *Superstition and the Press*, 257–66; Ronald Westrum, "The Press and Anomaly Reports: Distortions of Time, Tone, and Place," *Cryptozoology* 2 (1983): 162–66; Robin Yocum, *Dead before Deadline: . . . And Other Tales from the Police Beat* (Akron, OH: University of Akron Press, 2004), 116–18.

idea that Bigfoot might exist was a symptom of the pathology that beset the working class, their naiveté, their ignorance, their enslavement to desire. Borrowing a term from UFOlogists, Bigfoot enthusiasts called this middle-class disdain the "laughter curtain." Their struggles, their skill and ideas won them no dignity, only derisive laughter.[67]

**67.** Slate and Berry, *Bigfoot*, 148 (quotation); Sean McCloud, *Making the American Religious Fringe: Exotics, Subversives, & Journalists, 1955–1993* (Chapel Hill: University of North Carolina Press, 2004), 9.

CHAPTER NINE

# Cryptozoology

## 1978–1990

Grover Krantz was influenced by the competitive nature of the hunt for Bigfoot as much as anyone else—he felt the skeptical eyes and the temptation to believe any story because maybe, just maybe, it held the final clue to finding Sasquatch. He had a developed streak of paranoia, too—perhaps an innate characteristic, perhaps a response to the conditions of the hunt. And he relished the social inversion that the capture of Bigfoot promised: the little guys making good, the official establishment made to look ridiculous. But he seemed to have a way, nonetheless, to arbitrate between fraud and fact—he seemed capable of doing what Byrne hoped and failed to do with his board of examiners. He had the power of science on his side. He had received a PhD, taught at a university, had earned some respect in professional circles. And, he had Zadig's method.

Krantz was born in Utah, started college at the state university (his study there was interrupted for military service), and then moved to Berkeley, where he received his bachelor's and master's degrees in anthropology. According to those who knew him at the time, Krantz was an icon around Berkeley, throwing epic parties—parties that lasted twenty-four, thirty-six hours. "And he had all these women around," an acquaintance remembered. He was imposing, standing well over six feet tall, with the hunch of many tall people, which made him look something like a question mark, and, for much of his life, a beard, which made him look something like a Sasquatch.[1]

**1.** "History," n.d., Sasquatch Items file; and "Curriculum Vitae," n.d, no folder (item 0001), both in Grover Krantz papers, National Anthropological Archives, Smithsonian Institute, Suitland, MD; Grover S. Krantz, *Only a Dog* (Wheat Ridge, CO: Hofflin Publishing, 1998); Peter Carlson, "Using His Cranium: Grover Krantz's Last Wish Was to Remain With His Friends. And He Has." *Washington Post*, July 5, 2006 (quotation).

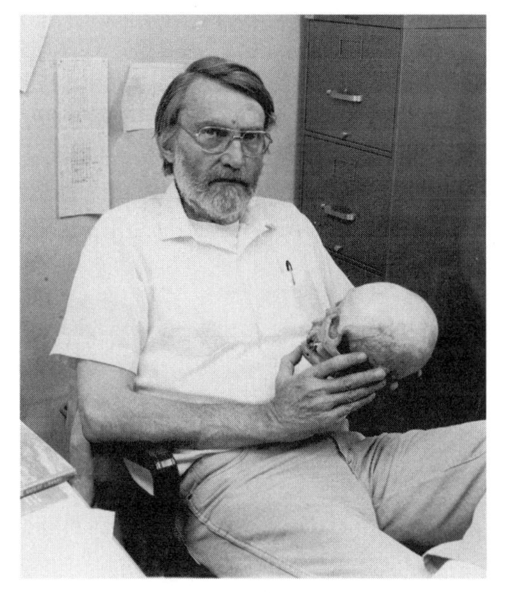

FIGURE 32. The scientist most associated with the study of Sasquatch was Grover Krantz, who started studying the beast in the early 1970s and continued doing so until his death in 2002. Krantz claimed that his interest hurt his scholarly career, but he remained undeterred. (National Anthropological Archives, Smithsonian Institution.)

For all his partying, though, Krantz was irascible, a difficult person. By 1962, when he was thirty-one, he'd already been divorced twice. He was brash and strong willed—characteristics that would later serve him well in the hunt for Bigfoot—arguing with his professors and ultimately dropping out of school in the 1960s because of some disagreement. Afterwards, he said, "It was steadily downhill for me. . . . My life at that time consisted of a part-time job and nearly full-time drinking." Krantz did spare some time to visit Bluff Creek, years after Crew had found those famous tracks and before Patterson made his movie. He also read Sanderson's book *Abominable Snowmen: Legend Come to Life*.[2]

In the mid-1960s, Krantz's life improved. He married again. And he got an Irish wolfhound named Clyde that, he said, turned his life around; the dog taught him love and responsibility. The family moved to Minnesota, where Krantz restarted graduate school, and then to Pullman, Washington, in 1968 when he started teaching at Washington State University, while still completing his PhD. The turmoil did not humble him: he grandly titled his dissertation "The Origin of Man." In one experiment, he wore prosthetic brow ridges for six months to determine their function in *Homo erectus*. And, he continued to investigate Bigfoot, never mind the laughter curtain. His

**2.** Krantz, *Only a Dog*; Carlson, "Using His Cranium" (quotation).

university, he said, supported him only to the extent of not firing him. By his own reckoning, his interest in Bigfoot cost him promotions—and tens of thousands of dollars that would have come with them. (He also thought that timber companies might have tried to discredit him.) Krantz relished the idea of capturing Bigfoot and embarrassing his colleagues: "I want to rub a few faces in the corpse," he said. Like so many others who hunted the beast, or just read about it, Krantz thought that Bigfoot could bestow upon him the dignity that the world otherwise denied.[3]

## Grover Krantz, Sasquatch Scientist

As Wladimir Tschernezky had with the Shipton's Yeti prints, Krantz reconstructed the Sasquatch from its spoor. Krantz's first article for *Northwest Anthropological Research Notes* came only a few months after Sprague opened the journal's pages to consideration of the Sasquatch. In it, Krantz analyzed the handprints that Marx had made, arguing that their size was consistent with the reported dimensions of Sasquatch, that the ratio of hand size to foot size was the same in humans and Bigfoot, and that other subtle correlations pointed to the handprints being real, not fakes: in particular, the hand showed no evidence of an opposable thumb and also lacked a thenar pad, the fleshy part of the hand at the base of the thumb. "It would require someone quite familiar with the anatomy of the human hand to make the connection between a non-opposable thumb and an absence of a thenar eminence," he wrote. "This tends to support the authenticity of these handprints."[4]

In two subsequent papers, Krantz considered the anatomy of a Sasquatch foot based on some Bossburg tracks. He noted that the proportions of Sasquatch feet were different than humans. In Sasquatch, the toes were nearly equal in length, the ball was split in two, there was no arch, the heel was relatively long, and the forefoot short. To Krantz, these differences were telling: they made biological and physical sense. If Bigfoot were as large as everyone reported, then its bones and muscles would have to be in different positions, relative to humans, in order to accommodate the heavier load. Krantz could

**3.** "History," n.d., Sasquatch Items file; and "Curriculum Vitae," n.d, no folder (item 0001), both in Grover Krantz papers; Richard Louv, "The Bigfoot Follies," *Human Behavior* 7, no. 9 (1978): 18–24 (quotation, 20); Krantz, *Only a Dog*; Steve Miller, "Grover Krantz—Mr. Big Foot," *Good Bye! The Journal of Contemporary Obituaries*, January–March 2002, http://www.goodbyemag.com/jan02/krantz.html (accessed April 11, 2008).

**4.** Grover S. Krantz, "Sasquatch Handprints," *Northwest Anthropological Research Notes* 5 (1971): 145–51 (quotation, 149).

calculate those positions according to physical models. The theory matched the measurements of the actual tracks to a great degree of accuracy. Ten years after he finished his reconstruction, Krantz said, "There is no way a faker could have known how far forward to set that ankle. It took me a couple of months to work it out, so you have to figure how much smarter a faker would've had to be. And I don't think there have been any genius anatomists floating around since Leonardo da Vinci. So either the animal is real or its faked by a human being. Of those two choices, the real animal is ridiculous, the fake by the human impossible." More succinctly, he quoted that master of Zadig's method, Sherlock Holmes: "When you have eliminated the impossible, whatever remains, *however improbable*, must be the truth."[5]

There was Krantz's brashness, implicitly comparing himself to Sherlock Holmes and, more grandly, to Leonardo da Vinci. The papers also put on display his paranoia. He worried that some prankster would read his brilliant analysis and use it as the basis of an unsolvable crime, creating a hoax that could not be seen through. Vainly, he hoped that if anyone tried to make a fake Sasquatch handprint based on his description—a handprint that lacked a thenar eminence and had a non-opposable thumb, a handprint with all the right subtleties—experts would recognize the fake because it was too rigid. Oddly, however, he admitted that Marx's handprints were especially rigid, so that seemed to indicate that either Marx had hoaxed Krantz or that rigidity was a poor tool for separating fraud from reality. Perhaps recognizing the inadequacy of that defense, when he analyzed the footprints Krantz kept secret two characteristics that he used to determine the authenticity of a track. That way, he would always know more than those trying to trick him. Krantz was so worried that somebody might discover those two attributes that he never even wrote them down, and, just to be sure no one was snooping through his papers, he also wrote many of his notes about Sasquatch in code.[6]

Krantz's judgments weren't always good. He said, for example, that he was 98 percent certain that the tracks Morgan found were authentic. But his

**5.** Grover S. Krantz, "Additional Notes on Sasquatch Foot Anatomy," *Northwest Anthropological Research Notes* 6 (1972): 230–41; idem, "Anatomy of the Sasquatch Foot," *Northwest Anthropological Research Notes* 6 (1972): 91–104 (second quotation, 103; original emphasis); Patrick Huyghe, "The Search for Bigfoot," *Science Digest* 92, no. 9 (1984): 56–59, 94–96 (first quotation, 94); Grover S. Krantz, *Big Footprints: A Scientific Inquiry into the Reality of Sasquatch* (Boulder, CO: Johnson Books, 1992), 52–68.

**6.** Krantz, "Sasquatch Handprints," 145–51; Roderick Sprague and Grover S. Krantz, eds., *The Scientist Looks at the Sasquatch II* (Moscow: University of Idaho Press, 1979), 21; Krantz, *Big Footprints*, 35.

analysis of the Bossburg track seemed reasonable; Napier, for instance, was also impressed by the Bossburg tracks—they were one of the chief reasons that he refused to dismiss the possibility of Bigfoot, even though he had argued himself out of believing in the Yeti, had doubts about Patterson's movie, and had been conned by Hansen. Napier also reconstructed Sasquatch's feet from some of the tracks plastercasted in Bossburg. The left foot, he found, was humanlike, different from the tracks found in Bluff Creek years earlier—which called into question the authenticity of at least one set of tracks—but made sense according to Zadig's method, structure and function congruent. The right foot was more impressive. It was the crippled foot, and from the print Napier could diagnose the disability, even its etiology: the beast suffered from clubfoot, he wrote in his book, caused by a crushing injury to the foot when the Sasquatch was young. The details were exactly right—the implied bone structure, the bend in toes, the heel impression. "It is very difficult to conceive of a hoaxer so subtle, so knowledgeable—and so sick—who would deliberately fake a footprint of this nature," he said. "I suppose it is possible, but it is so unlikely that I am prepared to discount it." And that left a track that could only be explained by the presence of some outsized, bipedal animal walking through the Pacific Northwest. "I am convinced that the Sasquatch exists, but whether it is all that it is cracked up to be is another matter altogether. There must be something in north-west America that needs explaining, and that something leaves man-like footprints. The evidence I have adduced in favour of the reality of Sasquatch is not hard evidence; few physicists, biologists, or chemists would accept it, but nevertheless it *is* evidence and cannot be ignored."[7] He told Krantz, "I am not convinced about the Sasquatch although I certainly couldn't prove its non-existence. I am not a 'scoffer' but equally I am not a 'believer.'"[8]

Echoing John Green, Krantz took a harder stance, arguing that the Bossburg tracks made skepticism of Bigfoot's existence impossible. "The only alternative explanation for many of these footprints is even more difficult to accept," he wrote in 1979. "The fakery must have been designed by a brilliant human anatomist at least 40 years ago, and one who has ever since been

**7.** John Napier, *Bigfoot: The Yeti and Sasquatch in Myth and Reality* (New York: E. P. Dutton, 1973), 205 (original emphasis).

**8.** Napier to Krantz, August 17, 1971 (final quotation), B.F. Letters Misc. file, Grover Krantz papers; Napier, *Bigfoot*, 120–26 (second quotation, 124), 205 (third quotation); John Guernsey, "Summer Sasquatch Season Ends with Big Bigfoot Find in Washington," *Oregonian*, October 19, 1974 (first quotation).

directing a large group of people placing thousands of these tracks under remarkable circumstances, and all this without ever being seen or having a member expose this colossal hoax." That was Krantz's brashness again, mixed with a paranoid view of the world—either there are hidden beasts, he said, or it is a vast conspiracy. It was also, as the story of Rant Mullens and Ray Wallace proved, wrong: the footprints could be left by a number of related pranksters without requiring much in the way of coordination.[9]

But wrong or not, Krantz's analysis seemed to insulate the Bossburg tracks from the charge of fraud. He used the populist rhetoric that was the *lingua franca* among Bigfooters, but he could seemingly contain its corrosive effects because he was a scientist and his critique came from within the establishment. (He was most decidedly *not* Beckjord, trying to rewrite the laws of physics.) Krantz didn't think that Bigfoot's discovery would be revolutionary, didn't think that it would do more than settle a few scores. Science would still be science; anthropological theory would have a new wrinkle but remain mostly intact. Scientists would move in once America's great ape was captured, and the amateurs would be out. Krantz didn't want to overturn science—just add a new fact to its catalog.[10]

By the middle of the 1970s, it looked as though Krantz had made some progress. In 1976, Marjorie Halpin started organizing the first scientific conference to be held on Sasquatch, at least partly in response to Krantz's work. Halpin was an anthropologist at the University of British Columbia whose interest in Sasquatch had been sparked by her advisor, Wilson Duff, one of those to attend the first showing of Patterson's movie years before. It took a long time, but eventually Halpin convinced officials at the school to host the conference. Apparently, officials saw in Bigfoot a way to increase the university's prestige. Experimental novelist J. Michael Yates had recently been hired by the university's press and was helping to launch a new academic field, the study of monsters, which the conference would initiate. Yates recruited science fiction author Judith Merril to help publicize the conference by writing an article for *Weekend Magazine*, hoping to bring the university's innovation to the wide world's notice. After the conference on Sasquatch would come conferences on monsters in film, theology, and science fiction. The University of British Columbia would become the center for monster studies.[11]

**9.** John Green, *Bigfoot: On the Track of the Sasquatch* (New York: Ballantine, 1973), 46; Sprague and Krantz, *The Scientist Looks at the Sasquatch II*, 11 (quotation).
**10.** Krantz, *Big Footprints*, 174–76, 240–43, 272–73.
**11.** J. Michael Yates to Merril, September 15, 1977, in Judith Merril fonds, R2929-0-4-E, Library and Archives of Canada, Ottawa, Ontario.

With financing set, Halpin announced that the conference would be held in 1978—the same year that Green finally published *Sasquatch: The Apes Among Us*, his summary of two decades of research. (Halpin lauded it in a review.) The number of people who wanted to attend soon overwhelmed her. Some were skeptics: the anthropologist Robin Ridington was there—he had seen Patterson's film with Duff and deemed it a fake. But the outsiders were also given a place at the table. Green and Dahinden and Killam attended. Carleton Coon, still arguing that wildmen existed, gave the plenary address. Krantz gave the conference summary. The laughter curtain seemed to have been split, the contest for dignity about to be won, unexpectedly, by the underdogs. "This conference would not have been possible five years ago," Halpin said. "The academic community is finally opening up to all kinds of strange phenomena." Krantz had much to do with that new openness.[12]

## Anthropology of the Unknown

The conference did not go well. Not at all. Most of the academics gave presentations about the meaning of monsters in various societies—there were talks about man-beast transformations in Inuit culture, monster-making in Mexican villages, the wildman in Gaelic literature. One paper argued that Sasquatch represented all the evil aspects that Canadians saw in themselves but refused to admit and so projected out into the world. Psychologist Wilfrida Ann Mully discussed psychotic patients who invented monsters so that they could control otherwise unmanageable thoughts.[13]

These were within the mainstream of scholarly tradition and fit with the university's hope to become the center of monster studies. Bigfooters, however, thought that the presentations were, as one hunter said, "academic bullshit." They were "irrelevant," Green complained, essays written "just for the sake of giving a paper," rounding out a résumé, and advancing a career in the ivory tower. They didn't engage with the evidence that Sasquatchers had so assiduously collected, did not grapple with the beast's possible existence.

**12.** Marjorie Halpin, "Things That Go Bump in the Order," *BC Studies*, no. 39 (1978): 61–66; Louv, "The Bigfoot Follies," 18–24 (quotation, 18).

**13.** "Anthropology of the Unknown: Sasquatch and Similar Phenomena," May 10–13, 1978, file 15.15, Vladimir Markotic papers, University of Calgary archives, Calgary; E. Wachtel, "Hairy Star, a Hairy Affair: Sasquatch, Bigfoot, Yeti, or What-Have-You Footprints," *Maclean's*, May 29, 1978, 20; Louv, "The Bigfoot Follies," 18–24; Marjorie Halpin and Michael M. Ames, eds., *Manlike Monsters on Trial* (Vancouver: University of British Columbia Press, 1980), 37–46, 86–113.

Arguing that Sasquatch symbolized the evil parts of the Canadian psyche showed a fundamental ignorance of Sasquatchiana—in most reported sightings, Bigfoot just stood among the trees or trotted across the road; the creature didn't act evilly. Presentations such as Mully's were even worse, from the Sasquatch hunters' perspective, not only irrelevant but offensive. They shifted the focus from studying the beast to studying those who thought that it existed. Her paper raised the questions: Were Sasquatch hunters psychotic? Was belief in Bigfoot pathological?[14]

Frustrated and disinterested, most of the Sasquatch hunters, according to the press, spent their time outside the lecture halls, in front of a fair-sized contingent of news reporters, where they sold plaster casts of tracks, books, newsletters—and worked out their vendettas. Beckjord and Dahinden, for instance, manning booths just across the way from one another, shouted taunts and regaled the media with unflattering stories about the other. At least once, their chest puffing nearly came to blows, the fight staved off by a professor who separated them.[15]

Amid this clatter, Peter Byrne arrived with a movie crew in tow—putting on display everything that other Bigfoot hunters detested about him, his showmanship, his publicity seeking. They complained incessantly about his presence until organizers were forced to throw him out. "There'll be no grandstanding here," Dahinden crowed to photographers, according to one report. (At the time, Dahinden was suing Byrne for using images from Patterson's movie without permission.) Byrne's report on the conference for *Bigfoot News* dripped with venomous sarcasm: "What was impressive about the conference and what made a lasting impact on those who attended was the great spirit of friendliness and camaraderie. . . . This was made obvious to all by the generous and unstinted sharing of information . . . and by the fine theme of cooperation that marked the unselfish exchange of ideas. . . . This theme, this aura, is what created the atmosphere of the conference and what made it such a pleasant and meaningful experience, as well as an unforgettable lesson in human relations."[16]

**14.** Green to Markotic, January 4, 1981, file 15.16 (quotation); Gordon Strasenburgh to Markotic, June 2, 1980, file 16.01, both in Vladimir Markotic papers; Dmitri Bayanov, "Letters from Russia," June 1982, *Bigfoot Co-Op*, Bayanov file, Grover Krantz papers; Louv, "The Bigfoot Follies," 18–24 (quotation, 24); Gordon Strasenburgh, letter, *The ISC Newsletter* 2, no. 3 (1983): 10.

**15.** Louv, "The Bigfoot Follies," 18–24; Barbara Wasson, *Sasquatch Apparitions* (Bend, OR: privately printed, 1979), 86–107.

**16.** Editorial, *Bigfoot News* 44 (May 1978): 1 (second quotation); Louv, "The Bigfoot Follies," 18–24 (first quotation, 22); Wasson, *Sasquatch Apparitions*, 42–43, 89–91, 103–6.

Such shenanigans undermined any hope that—since academics had convened a conference on the beast—the news media would now treat the subject respectfully. In fact, for all the reporters in attendance, there was very little coverage of the meeting. (Merril's story was so cut that she had it run under a pseudonym.) And those articles that did appear were unflattering. The magazine *Human Behavior* ran a feature story on the conference titled "The Bigfoot Follies," which mercilessly skewered Dahinden and other believers as lowbrows who romped through the otherwise sedate world of academia. Academic conferences commonly have displays of books and journals for sale, but the article treated the Bigfoot enthusiasts' exhibits as though they were moneychangers in the temple. Using the same psychological language that had run through the presentations, the magazine suggested that Green and Krantz were obsessed, that they were, in the words of another conference attendee, "stuck, like in a forest, they can't get out of it." The conference did not announce a new day. It did not part the laughter curtain: it drew, again, the boundary.[17]

The Bigfoot hunters' antics also seemed to end the university's involvement with Sasquatch—and monsters. Never again, said Marjorie Halpin, would she convene a Bigfoot conference. The politics were too intense, the personalities too difficult. Nothing came of the university's plans to do a series of conferences on monsters, either—whether that was because of institutional politics or the problems that Halpin confronted is unknown, although it seems likely that the Sasquatch conference would have given anyone pause before organizing a similar meeting. If there was even the slightest inclination toward holding a second meeting about monsters, that was probably killed when Halpin tried to compile the presentations into a book, *Manlike Monsters on Trial*.[18]

## Cryptozoology

*Manlike Monsters on Trial* excluded the papers written by Krantz and Coon and Dahinden and Vladimir Markotic, a Harvard-trained anthropologist who

**17.** Merril to J. Michael Yates, April 12, 1978, Judith Merril fonds, R2929-0-4-E; Thorstein, "Close Encounters of the Monstrous Kind," 16–17; Louv, "The Bigfoot Follies," 18–24 (quotation, 24); Wasson, *Sasquatch Apparitions*, 93–94; idem, "Bigfoot Brawl," *Human Behavior* 8 (1979): 8; Ronald Westrum, "The Press and Anomaly Reports: Distortions of Time, Tone, and Place," *Cryptozoology* 2 (1983): 162–66.
**18.** Kay Bartlett, "There's No Second Prize in the Catch-Sasquatch Game," *Idaho Statesman*, February 4, 1979.

had been drawn to the study of Sasquatch; it excluded all of the papers that analyzed the Patterson-Gimlin film; and it excluded a panel discussion that was held on the ethics of killing a Sasquatch. That is not to say all pro-Bigfoot papers were excluded—some did make the final cut. But enough were kept out that Krantz and Markotic were livid, Markotic going so far as complaining to Halpin, the head of the university's anthropology museum, and the University of British Columbia's president.[19]

Halpin shrugged off the griping: rejection was a normal part of the academic process, and she just didn't think that the excluded papers were up to snuff. Although there was some truth to her response—not every paper presented at every academic conference is published—there was more at stake than she allowed. The complaints were only overtly about academic processes. Implied but unexpressed were concerns about power and about class. Halpin explained to Krantz, The point of the conference was not to prove Sasquatch existed but to treat Bigfoot as something that could not be explained, an anomaly. The theme was not "Is there a Sasquatch," but "Why do people believe that there is a Sasquatch?"—a subtle shift, but one which moved the focus from the forests to "the forests of the mind," a phrase that recurred throughout the conference. It was the monster hunters—not the monsters—who were the subject, their credibility, their perceptions, their thoughts that were put under the microscope and analyzed.[20]

Having the right to speak, seeing one's views reflected in the media, these are middle-class prerogatives. The working class does not have the same expectation, the same access to the mass media. When their concerns are expressed, it is usually by middle-class interpreters—ventriloquists, like those who wrote for men's adventure magazines and tabloids. Often, the working class is simply silenced, the subject of analytical discourse, refused the right to speak back. Halpin's snub, the conference's focus, the bad press—all of these served to erase the distinctions that Krantz had tried to erect between his studies and the work of amateurs. It made him the same as the working class, made him into a scientific subject.[21]

After stewing on the problem, Krantz and Markotic decided to put out

**19.** Markotic to Halpin, June 19, 1979; Green to Markotic, July 3, 1979; Krantz to Halpin, July 25, 1979; Krantz to Markotic, July 26, 1979, all in file 15.15, Vladimir Markotic papers; Halpin and Ames, *Manlike Monsters on Trial*.

**20.** Halpin to Krantz, July 3, 1979, file 15.15, Vladimir Markotic papers.

**21.** Richard Sennett and Jonathan Cobb, *The Hidden Injuries of Class* (New York: Vintage Books, 1972); Michael Denning, *Mechanic Accents: Dime Novels and Working-Class Culture in America* (New York: Verso, 1998), 147–48.

their own book in response to Halpin's. Danny Perez advised them to expose the chicanery of the conference organizers, of the press, of mainstream culture, not just speak out but speak back: reverse the microscope and make the scientists into subjects, analyze the conference organizers, and the media. "What we'd all like to see in print is what went on behind the scenes, so to speak," he told Markotic. "We'd like to see condensations of newspaper clippings within the book, transcripts of news-broadcastings so as to see how the press treated the whole thing, and how the general public feels about it. We'd also like to see pictures of the people who wrote the articles." Krantz and Markotic resisted that temptation—pictures, especially, could have been cruel—instead publishing the excluded papers from the conference as well as a few others that took the existence of Sasquatch seriously.[22]

Disappointment about the conference, however, was only part of a more general frustration that Sasquatch enthusiasts were feeling in the late 1970s and into the 1980s. A survey revealed that less than 13 percent of anthropologists thought that Sasquatch existed. By contrast, almost a quarter thought there might be a Loch Ness monster. Something more than a book was needed to address this continued lack of scientific respect for Bigfoot. As Krantz and Markotic were putting together their book, *The Sasquatch and Other Unknown Hominids*, others involved in the search for different mysterious animals had the same thought. Roy Mackal, a biologist at the University of Chicago, and J. Richard Greenwell, research coordinator at the University of Arizona's Office of Arid Land Studies, decided that a new scientific discipline needed to be created, one that took seriously the search for legendary creatures. Mackal had spent much of the 1970s fishing for the Loch Ness monster while Greenwell had been involved with investigations of UFOs. Mackal and Greenwell contacted other scientists whom they knew to be working along similar lines—including Krantz and Markotic—and were encouraged by the positive responses. In January 1982, George Zug, a herpetologist with the Smithsonian, invited Mackal, Greenwell, Krantz, and three others to the institute's National Museum of Natural History. There, they founded the International Society of Cryptozoology.[23]

Literally, cryptozoology means the study of hidden animals—it was a

**22.** Perez to Markotic, June 24, 1981, file 16.01, Vladimir Markotic papers.

**23.** Roy P. Mackal, *The Monsters of Loch Ness* (Chicago: The Swallow Press, 1976); idem, *Searching for Hidden Animals* (Garden City, NJ: Doubleday, 1980); J. Richard Greenwell and James E. King, "Attitudes of Physical Anthropologists toward Reports of Bigfoot and Nessie," *Current Anthropology* 22 (1981): 79–80; "Formation of the Society," *The ISC Newsletter* 1, no. 1 (1982): 1–3.

term that Sanderson and Heuvelmans sometimes used. Those at the meeting defined their discipline as the search for "unexpected" animals: animals that were thought to be extinct but weren't, that inhabited areas where they were not thought to live, or that were a surprising size. Cryptozoology was a sister of paleontology—just as paleontologists sought animals lost to time, crypto-zoologists hunted animals hidden in space, both relying on reconstructions and Zadig's method to create knowledge. The conventioneers also selected a board, named Heuvelmans the first president, and made plans for a new journal, *Cryptozoology*. There was a "real need" for *Cryptozoology* (and cryptozoology), Roderick Sprague wrote. The response from many scientists, he said, had been "as anti-intellectual as the Spanish Inquisition." It was time to resist the powers that be, find a voice, and respond, to bring enlightenment to a world grown decadent and arrogant.[24]

Not everyone, however, was pleased with the International Society of Cryptozoology. Some uncredentialed Bigfoot hunters disdained the society and what it represented. Sasquatch hunting had long been about attacking the scientific establishment, inverting the social order, and now professionals seemed to be horning in, re-asserting their power. "A Sasquatch will never be found by putting out damn learned articles," Danny Perez said. "The Bigfoot researchers don't need you people." Dahinden seemed to have some appreciation for the society, but, as with everything else in his life, he eventually turned on it. He told Greenwell that if he captured a Bigfoot, he would sell it, but only for the right price. "I would say that either I get [the price] or the evidence is going to lay out there and rot away and I wouldn't move a finger. I don't give a damn about science. I couldn't care less about science."[25]

John Green had reservations about the ability of science, as an institution, to solve the problem as well. Throughout his career, he tacked between different truth-making technologies, from journalism's bullshit detectors to science's empiricism to law's adversarial process—as when he had witnesses sign affidavits. In 1989, he recommended that Sasquatchers again look toward the legal profession instead of science. Lawyers, Green noted, were more comfortable than scientists with the kind of evidence that Sasquatch hunters gathered: eyewitness testimony, prints, circumstantial clues. They

**24.** "Formation of the Society," 1–3; Roderick Sprague, Review of *The Sasquatch and Other Unknown Hominoids*, *Cryptozoology* 5 (1986): 101 (first quotation), 103(second quotation).
**25.** Daniel Perez, letter, *The ISC Newsletter* 2, no. 3 (1983): 10–11 (quotation); Gordon Strasenburgh, letter, *The ISC Newsletter* 2, no. 3 (1983): 10; "Interview: A Candid Conversation with a Prominent Sasquatch Field Worker Who Says Exactly What He Thinks," *The ISC Newsletter* 4, no. 2 (1985): 3.

would see the value of what Sasquatchers collected and "would have to conclude that Sasquatches do exist. They could not reach any other conclusion." Certainly, he acknowledged, the results of an official commission would not definitively prove the existence of Sasquatch, but lawyers were often politicians and so controlled the purse strings. Once they were satisfied that Sasquatch existed, they could fund an intense hunt for the beast.[26]

The International Society of Cryptozoology also had trouble pleasing some of its most venerable members. The Russian hominologists thought that the society erred by not focusing more on the study of wildmen, and especially on Patterson's film, which Dmitri Bayanov said should end all debate. They were also upset that Napier was on the society's board, since he was content to accept Bigfoot as a myth if better evidence for its existence could not be found. Bernard Heuvelmans had a very particular—and peculiar—vision for cryptozoology, and he was saddened to see that the International Society of Cryptozoology did not match it. In Heuvelmans's view, cryptozoologists were to gather reports of mysterious animals, sift and analyze them so that mythical elements were filtered out, leaving only a naturalistic description of the beasts, their habits and range. "It is the final result of these activities which will help us locate the relevant animals with the greatest accuracy, to recognize them, and to teach us where, when, and how to approach them." He compared cryptozoology to theoretical astronomy—just as astronomers had predicted the existence of Neptune by studying "a slight perturbation in the elliptic motion of Uranus," cryptozoologists could accurately predict the existence of animals by studying reports of them. "The ambitious aim of cryptozoology is to make one able to describe an animal scientifically before it has been captured or collected," he wrote. But, most cryptozoologists, Heuvelmans noted, just wanted to go into the field and *look*, unarmed with prior study. That, he said, was mere zoology, a less exalted kind of science.[27]

Heuvelmans tried for a while to convince cryptozoologists to accept his view of the field, but became increasingly frustrated. In 1974, he and

**26.** John Green, "The Case for a Legal Inquiry into Sasquatch Evidence," *Cryptozoology* 8 (1989): 37–42 (quotation, 38).

**27.** Bayanov to Markotic, April 17, 1980; Bayanov to Vladimir Markotic, June 26, 1980; Bayanov to Markotic, August 30, 1984, all in file 15.16, Vladimir Markotic papers; Bernard Heuvelmans, "What Is Cryptozoology?" *Cryptozoology* 1 (1982): 1–12; idem, "The Birth and Early History of Cryptozoology," *Cryptozoology* 3 (1984): 1–30 (quotations, 23–24; original emphasis); idem, "The Metamorphosis of Unknown Animals into Fabulous Beasts and of Fabulous Beasts into Known Animals," *Cryptozoology* 9 (1990): 1–12.

the Russian historian Boris Porshnev coauthored a book in French arguing that Neanderthals still existed. Bozo was a centerpiece of the book. But the book was never translated into English and disappeared down a black hole. "Typical American arrogance," he sneered. When Vladimir Markotic wrote in *The Sasquatch and Other Unknown Hominids* that no one had ever seen a wildman—Markotic had visited Hansen's display and proclaimed Bozo a fraud—Heuvelmans took it as a slap in the face and broke off all contact with him. Eventually, he simply withdrew from the society that he had once led. In 1988, the Russians formed their own cryptozoological society, where they could practice the science as they saw fit.[28]

Some scientists outside of cryptozoology also disliked the new discipline. One anthropologist told Greenwell that any scientist associated with the International Society of Crytpozoology was making "an ass of himself." Chico State University anthropologist Turhon Murad used cryptozoology as a case study in pseudoscience and faulty logic. Even former supporters decided that enough was enough. Back in 1959, the paleontologist George Simpson had offered qualified support for the possibility that the Yeti might inhabit the Himalayas. In 1984, he attacked the society as intellectually vapid. Zoologists had already discovered most of the world's large animals, he said, and those who thought otherwise were ignorant or were driven by a "primeval urge" to see monsters where none existed.[29]

Thus, like the conference in British Columbia, cryptozoology never lived up to its promise—the International Society of Cryptozoology could not establish itself as a legitimate scientific society, could barely hold itself together. Throughout its history, the society was troubled by a lack of funds; newsletters and *Cryptozoology* were published *years* late. And the people who most wanted to join were those that the society least wanted: Jon Beckjord threatened to sue to become a member. Other paranormalists flocked to Bigfoot gatherings. By the end of the century, according to some estimates,

**28.** Heuvelmans to Krantz, January 22, 1993 (quotation), Heuvelmans file, Grover Krantz papers; Heuvelmans and Boris F. Porshnev, *L'Homme de Neanderthal Est Tojours Vivant* (Paris: Plon, 1974); "Soviets Form Cryptozoology Society," *The ISC Newsletter* 7, no. 3 (1988): 7–8.
**29.** George Gaylord Simpson, "Creatures Extinct, Living, or Fictional," *Natural History* 48 (1959): 492–94, 544–46; J. Richard Greenwell and James E. King, "Scientists and Anomalous Phenomena: Preliminary Results of a Survey," *Zetetic Scholar*, no. 6 (1980): 23 (first quotation); Robert May, "Science Journals," *Nature* 307 (1984): 687; George Gaylord Simpson, "Mammals and Cryptozoology," *Proceedings of the American Philosophical Society* 128 (1984): 1–19 (quotation, 14); Turhon A. Murad, "Teaching Anthropology and Critical Thinking with the Question "Is There Something Big Afoot?" *Current Anthropology* 29 (1988): 787–89.

about 25 percent of those who attended Bigfoot conferences thought that Bigfoot was from outer space or another dimension. "The zoological idea is a dead end. It is over," Beckjord gloated sometime around 1989. That was an overstatement, but curled inside was a kernel of truth. Who could blame scientists for wanting to stay away from a field so tied to the occult? What good were Green's blasts if even the official society dedicated to the scientific study of Bigfoot could not afford to publish its own journal? What kind of science was this anyway?[30]

## Science Police

At about the same time that the University of British Columbia hosted "Anthropology of the Unknown," a group of scientists and scientific acolytes was worrying over the effect that mass culture was having on American society. For the better part of the twentieth century, interest in the occult, paranormal, and such had been restricted to fringe groups; mass culture changed that, bringing what had been hidden to the mainstream. The late 1960s saw a so-called occult explosion, a seemingly sudden irruption of belief in UFOs, ESP, Eastern religions, alternative medicines . . . and ABSMs. Newspapers started printing astrology columns; mockumentaries appeared in theaters around the country; according to one bibliography, more than 1,800 books on occult topics were published between 1971 and 1975. In that latter year, philosopher Paul Kurtz organized a large and august body of scientists to decry the renaissance of astrology.[31]

The attention generated by that act of criticism encouraged Kurtz and in 1976 he founded the Committee for the Scientific Investigation of Claims of the Paranormal, and abbreviated that mouthful as CSICOP—pronounced "sci-cop." The homonymy was accidental but apposite: the committee was the self-proclaimed science police—upholding the dignity of science against

**30.** Krantz to Markotic, November 28, 1988; Greenwell to Markotic December 20, 1988; Jon Beckjord, n. d., Addendum (quotation), all in file 15.14, Vladimir Markotic papers; Shafer Parker, "Bigfoot: Man, Monkey, or Myth?" *British Columbia Report*, June 23, 1997, 32–34; D. Parvaz, "Arguing the Big Case," *Seattle Post-Intelligencer*, September 28, 1999.

**31.** Nat Freeland, *The Occult Explosion* (New York: G. P. Putnam's Sons, 1972); James Webb, *The Occult Establishment* (La Salle, IL: Open Court Publishing Company, 1976); Kendrick Frazier, "UFOs, Horoscopes, Bigfoot, Psychics and Other Nonsense," *Smithsonian*, March 1978, 54–61; Robert Galbreath, "Explaining Modern Occultism," in *The Occult in America: New Historical Perspectives*, ed. Howard Kerr and Charles L. Crow (Urbana: University of Illinois Press, 1983), 11–37.

pseudoscience, occultism, and New Agery. Members included the science fiction author Isaac Asimov, science writer Martin Gardner, astronomer Carl Sagan, engineer Philip Klass, and magician James Randi.[32]

At first, nobody was sure exactly what CSICOP should do to protect science's prestige against the twin forces of irrationality and mass culture. Early on, there were attempts to turn CSICOP into an academic organization. Sociologist Mario Truzzi edited the committee's first journal, *The Zetetic*, which he wanted open to both the so-called skeptics and the so-called believers—a publication in which the paranormal could be rigorously explored and debated, where results of serious studies could be published. In time, though, CSICOP chose to focus on debunking—not exploring—the paranormal, and Truzzi left, starting a competing journal, *Zetetic Scholar*; the committee then introduced *Skeptical Inquirer*, a popular magazine. A few years later, CSICOP gave up actually testing paranormal claims after a study of astrology became a fiasco. One of the researchers claimed that Kurtz had covered up evidence that showed some astrological influence on athletic ability; the researcher was ousted—after which a few members resigned from the committee.[33]

By the mid-1980s, CSICOP had found its footing. The committee had its own publishing house, Prometheus Books, and *Skeptical Inquirer* had a circulation of about twelve thousand. The skeptics' goal, remembered executive director Lee Nisbet, was "to fight mass-media exploitation of supposedly 'occult' and 'paranormal' phenomena" by debunking whatever pseudoscientific ideas bubbled into the mainstream and criticizing the media for devoting time to such nonsense. Ridicule was the skeptics' primary weapon: a favorite quote was H. L. Mencken's aphorism, "One horse-laugh is worth a thousand syllogisms." Cartoons and illustrations that poked fun at occultists were added to the *Skeptical Inquirer*.[34]

**32.** David J. Hess, *Science in the New Age: The Paranormal, Its Defenders and Debunkers, and American Culture* (Madison: University of Wisconsin Press, 1993), 61–65, 88–89, 158–64; Paul Kurtz, *Skeptical Odysseys: Personal Accounts by the World's Leading Paranormal Inquirers* (Amherst, NY: Prometheus, 2001), 42.

**33.** Trevor Pinch and Harry Collins, "Private Science and Public Knowledge: The Committee for the Scientific Investigation of the Claims of the Paranormal and Its Use of the Literature," *Social Studies of Science* 14 (1984): 521–46; George P. Hansen, "CSICOP and the Skeptics: An Overview," *The Journal of the American Society for Psychical Research* 86 (1992): 19–63; Hess, *Science in the New Age*, 33–34.

**34.** Martin Gardner, *Science: Good, Bad, and Bogus* (Amherst, NY: Prometheus, 1981), vii (second quotation); Hansen, "CSICOP and the Skeptics," 19–63; Hess, *Science in the New Age*, 12, 64–65; Lee Nisbet, "The Origins and Evolution of CSICOP: Science Is Too Important to Be Left to Scientists," *Skeptical Inquirer*, November–December 2001, 50–52 (quotation, 50).

Bigfooters made an especially inviting target; there was nothing inherently unscientific about the topic—"Our position was that the existence of such an animal was within the realm of possibility but had not yet been confirmed by hard science," Kurtz said—but the community's infighting, magnetic pull on pranksters, tolerance for the paranormal, and brash members who insisted that they, and they alone, had infallible judgment were easy to ridicule. Members of the committee turned their attention to Bigfoot sporadically in the 1970s and with increasing frequency during the late 1980s and 1990s when skeptic Michael Dennett made much of the subject. The skeptics were tailors, stitching together the laughter curtain even as Sasquatchers hoped that they would finally slip through to the other side. Articles about the Minnesota Iceman, for instance, dubbed the creature "Sasquatchsickle." And shortly after the University of British Columbia conference, *Skeptical Inquirer* detailed the various feuds between the hunters, including Dahinden and Beckjord's antics at the university, concluding with a short but devastating quip: "Science marches onward."[35]

Bickering among Sasquatch enthusiasts also proved useful to the skeptics, because they could always find one Bigfooter to criticize another. A *Skeptical Inquirer* review of *The Field Guide to Bigfoot*, for instance, which took the Minnesota Iceman seriously, quoted another Bigfooter calling the monster a "frozen dummy." And Cliff Crook, a Washington Bigfoot hunter, told *Skeptical Inquirer*, "Science is about discovering truth. It is evident that Grover Krantz has consistently abused his scientific credentials by his constant failure to acknowledge plain facts." That was the exact point that skeptics wished to make, but it sounded sweeter coming from the mouth of a believer. While this strategy could be persuasive, it did open skeptics to the charge of hypocrisy. Crook himself had been accused of hoaxing—not really a champion of scientific investigation, then. And the Bigfooter who called Bozo a "frozen dummy" was Jon Beckjord, whom skeptics otherwise ridiculed as an ignoramus and fabulist. Why believe anything he said? Were the skeptics praising Beckjord for (finally) getting something right—a blind squirrel stumbling across a nut? Or were they just trying to score debate points without concern for the integrity of the evidence that they put forward?[36]

**35.** "Psychic Vibrations," *Skeptical Inquirer*, 1979, 14 (second quotation); C. Eugene Emery, "Sasquatchsickle: The Monster, the Model, and the Myth," *Skeptical Inquirer*, Winter 1982, 2–4.

**36.** Michael R. Dennett, "Bigfoot Evidence: Are These Tracks Real?" *Skeptical Inquirer*, Fall 1994, 498–509 (second quotation, 507); Robert Sheaffer, "Looking for Mr. Goodape," *Skep-*

Such questions did not arise when skeptics turned to René Dahinden. Dahinden had followed a very different path than his onetime partner John Green. Over the decades, he had become increasingly *unsure* about the monster. He could not prove to his satisfaction that Patterson had faked his film, but he was also not willing to accept Bigfoot's existence either. He derided Green's attempts to create a composite image of the beast from all the sighting reports. "There is no pattern," he said. Over the years, he soured on most other Bigfoot hunters. Too many had proven ignorant or credulous, falling for hoaxes—or perpetrating them—and he was not shy about saying so, frequently and publicly. He ruthlessly revealed the inadequacies of the evidence that other hunters turned up, desperate to find Sasquatch but unable to make himself believe. Since he had never pulled a con himself and was so effective at exposing the chicanery of others, he was an especially valuable resource for CSICOP, a Bigfoot hunter who could be trusted, even lauded, without cognitive dissonance.[37]

Skeptics also went in for meatier, more intelligent criticism of Bigfooters (and paranormalists in general) that highlighted faulty logic and egregious errors. Krantz, for example, argued that those who did not accept Bigfoot's existence had to explain every single Bigfoot track, *including those never seen.* Multiplying the thousand reports of tracks that Green had in his files by arbitrary factors, Krantz concluded that skeptics needed to account for a hundred million prints. "This is rubbish," Robert Boston wrote in the *Skeptical Inquirer*, "and Krantz should be embarrassed for suggesting it." Krantz also tried to prove that Patterson's movie was legitimate by subjecting measurements of the film to algebraic manipulations that showed the film was shot at eighteen frames per second—one of the speeds which Grieve said ruled out the possibility of a hoax. A model argument—except the camera Patterson used didn't have a setting for that speed. Skeptics also pointed out the troubling history of the Cripplefoot tracks, associated as they were with Ivan Marx. Boston compared Krantz's science to creationism—it had all the dressings of science, charts and figures and formulae, "but unfortunately for Krantz, fancy drawings and scholarly sounding afterthoughts do not science make."[38]

........................

*tical Inquirer*, November 1999, 20–21; Benjamin Radford, "The Flawed Guide to Bigfoot," review of *The Field Guide to Bigfoot, Yeti, and Other Mystery Primates Worldwide*, *Skeptical Inquirer*, January 2000, 55 (first quotation).

**37.** Barbara Wasson, *Sasquatch Apparitions* (Bend, OR: privately printed, 1979), 29–36 (quotation, 35).

**38.** Krantz, *Big Footprints*, 19, 94–96; Robert Boston, "Review of *Big Footprints: A Scientific Inquiry into the Reality of Sasquatch*," *Skeptical Inquirer*, Fall 1994, 528–32 (first quotation,

Still, even when engaging with—rather than ridiculing—the Sasquatchers, skeptics were prone to be unfair and ignorant. For instance, two articles in *Skeptical Inquirer* suggested that the Bigfoot myth might have originated in tales Daniel Boone told of confronting a ten-foot ape that he called a Yahoo—stories which in turn were probably borrowed from *Gulliver's Travels*. It was a fine theory but showed absolutely no familiarity with the stories about Bigfoot or their history. In 2004, Prometheus Books published what was supposed to be the definitive exposé of Patterson's movie, Greg Long's *The Making of Bigfoot*. But the book was self-refuting, arguing both that Patterson had made a fake Bigfoot suit and bought one, and offering very different descriptions of the two costumes. To be fair, another skeptic pointed out the flaws in the book, but that the book was published at all suggested that CSICOP's *modus operandi* was, Shoot first, ask questions later.[39]

Unsurprisingly, those ridiculed by CSICOP disliked the committee. The science fiction writer and conspiracist Robert Anton Wilson called CSICOP "the New Inquisition." *Fate* published the charges of the dissident skeptic who thought he found evidence of astrological influence and had been forced out of the committee. Truzzi said that CSICOP did not represent real skepticism but pseudoskepticism because members had an unwavering—unskeptical—belief in science. A true skeptic was agnostic about current scientific theories, he said, accepting them, but always ready to abandon them. Bigfooters made similar charges, building on their image of themselves as more scientific than scientists. Sasquatch hunters were the true skeptics because they were willing to revise scientific dogma and accept that Bigfoot existed.[40]

.........................

528; second, 532); Dennett, "Bigfoot Evidence," 498–509; David J. Daegling, "Cripplefoot Hobbled," *Skeptical Inquirer*, March 2002, 35–38; Benjamin Radford, "Bigfoot at Fifty: Evaluating a Half-Century of Bigfoot Evidence," *Skeptical Inquirer*, March 2002, 29–34; David J. Daegling, *Bigfoot Exposed: An Anthropologist Examines America's Enduring Legend* (Walnut Creek, CA: Altamira Press, 2004), 110.

**39.** Hugh Trotti, "Did Fiction Give Birth to Bigfoot?" *Skeptical Inquirer*, Fall 1994, 541–43; David Matthew Zuefle, "Swift, Boone, and Bigfoot: New Evidence for a Literary Connection," *Skeptical Inquirer*, January–February 1997, 57–58; Greg Long, *The Making of Bigfoot: The Inside Story* (Amherst, NY: Prometheus, 2004); Michael R. Dennett, "Some Reasons for Caution about the Bigfoot Film Exposé," *Skeptical Inquirer*, January–February 2005, 58.

**40.** Dennis Rawlins, "Starbaby," *Fate* October 1981, 67–98; Marcello Truzzi, "On Pseudo-Skepticism," *Zetetic Scholar*, no. 12/13 (1987): 3–4; Bernard Heuvelmans, "The Metamorphosis of Unknown Animals into Fabulous Beasts and of Fabulous Beasts into Known Animals," 1–12; George P. Hansen, "CSICOP and the Skeptics," 19–63 (quotation, 21); Jeff Meldrum, *Sasquatch: Legend Meets Science* (New York: Forge, 2006), 271–73.

Despite these criticisms, despite CSICOP's lack of rigor, and despite the skeptics' worry that the scientific enterprise was beleaguered, teetering on the edge of a precipice, the committee's attempts to gain the mass media's ear was successful. The *New York Times* devoted a column to CSICOP's establishment. Four years after it had published a positive piece on Bigfoot—and four years before the International Society of Cryptozoology was founded—*Smithsonian* magazine printed CSICOP's charter manifesto. *Reader's Digest* reprinted it. These were middle-class publications that shared a worldview with the committee members; they were sympathetic to CSICOP's cause. Science was still held in high esteem by the middle class, and popular publications were loath to contradict the word of experts. Whatever its faults, the committee was given wide latitude in drawing the boundaries around science—in policing what was proper knowledge and what was improper.[41]

## "Definitive Proof"

In 1982, as CSICOP attacked Bigfooters for their gullibility and unscientific attitude, Krantz came across a surefire way to answer the skeptics: "definitive proof," as he said, that Sasquatch existed. In June of that year, Paul Freeman, a forest service employee working in the Walla Walla Watershed, along the border between Oregon and Washington, saw Sasquatch. "I could see the muscles in the legs move when he walked," Freeman said. "I could see the muscles in the arms and shoulders. It just plain scared me, and I've never been scared in the woods before. This thing was real. It was big enough to tear the head right off your shoulders if it wanted to." Six days later, Freeman found a number of gargantuan footprints, the first while he and Bill Epoch, another forest service employee, ate lunch near Elk Wallow, a bog on the Low Creek Trail. Starting in the mud and ending in some brush, eighteen prints covered forty feet. After riding his horse back to headquarters to report the discovery, Freeman returned to Elk Wallow and noticed that a new set of Sasquatch tracks crossed over his horse's previous prints, as though he had been followed.[42]

**41.** Boyce Rensberger, "Panel Fears Vogue for the Paranormal," *New York Times*, August 10, 1977; Kendrick Frazier, "From the Editor's Seat: 25 Years of Science and Skepticism," *Skeptical Inquirer*, May 2001, 46–49.

**42.** Joel Hardin to Wayne Long, July 28, 1982, Fingerprints file, Grover Krantz papers; Krantz to Markotic, December 6, 1982 (first quotation), file 16.01, Vladimir Markotic papers; "Sasquatch in Washington State: New Reports Involve Footprints," *The ISC Newsletter* 1, no. 2 (1982): 2–4 (second quotation, 2–3); René Dahinden, "Whose Dermal Ridges?" *Cryptozoology*

Two days after Freeman found the tracks, the Forest Service tapped Joel Hardin, a tracker who worked for the U.S. Border Patrol, and one of their own wildlife biologists to investigate. Hardin concluded that the tracks were "very clever" but were a hoax. There was no variation in their stride, either going uphill or down, in the pressure that they applied to the ground, or in their swing from side to side. There was no debris in them, as though they had been swept clean. They started and ended abruptly, and for no apparent reason (unless one believed Beckjord's theories). And in the soft mud the tracks did not push down to solid matter—indeed, they did not sink as deeply as the feet of some investigators.[43]

These were important reasons to disbelieve Freeman's discovery, enough for René Dahinden, among others, to declare the tracks fake—another case in which Dahinden came out on the side of the skeptics. They were not the only reasons, though—only the first. Over the years, Freeman displayed an uncanny knack to discover Sasquatch signs. He found many, many more footprints in the same general area. He found a handprint. He found broken tree tips that Bigfoot supposedly damaged. He saw Bigfoot three more times and photographed it twice. He tape-recorded the beast's vocalizations, found an elk skin marked by Sasquatch teeth, and—as seemed *de rigueur*—Sasquatch feces. He found Sasquatch fur. He also found cave paintings made by Sasquatch. The sheer abundance of evidence was hard for Dahinden and other longtime hunters to accept—they had toiled for decades without seeing the beast and here was a man who could find tracks almost at will. The quality of the evidence was suspect as well. Freeman admitted that he had faked tracks before—although he insisted not the ones he found that June. (There was an unconfirmed rumor that at one point he had worked for a shoe company.) In the late 1980s, one of Krantz's graduate students analyzed supposed Sasquatch fur found by Freeman. It turned out to be synthetic, probably from a wig—which meant that almost certainly it was not just an innocent misidentification but a deliberate fraud.[44]

.........................

3 (1984): 128–31; Vance Orchard, *Bigfoot of the Blues* (Walla Walla, WA: privately printed, 1993).

**43.** Joel Hardin to Wayne Long, July 28, 1982 (quotation), Fingerprints file, Grover Krantz papers; Dahinden, "Whose Dermal Ridges?" 128–31.

**44.** "Alleged Sasquatch Prints Are Fakes, Says Dahinden," August 4, 1983, *Vancouver Sun*, Fingerprints file, Grover Krantz papers; Michael R. Dennett, "Evidence for Bigfoot?" *Skeptical Inquirer*, Spring 1989, 264–72; Edward B. Winn, "Physical and Morphological Analysis of Samples of Fiber Purported to Be Sasquatch Hair," *Cryptozoology* 10 (1991): 55–65; Den-

Krantz, however, dismissed the naysayers and brashly proclaimed that Freeman's was "the strongest Sasquatch case on record." He wrote off the Border Patrol's conclusions because the tracker had been heard making disparaging remarks about the prints even before he arrived, which was enough for Krantz to question his objectivity. He admitted that Freeman was unreliable and acknowledged that the sudden appearance and disappearance of the tracks was troubling but defended Freeman anyway, explaining his knack for finding tracks by noting that Freeman spent an inordinate amount of time outdoors. Other uncomfortable evidence—faked fur, claims about cave paintings—he simply ignored. Krantz was certain that he would not become another Ivan Sanderson or Bernard Heuvelmans, however, crouched over a block of ice, not another Tom Slick studying a glass jar full of mysterious shit. The conditions under which the tracks were found didn't matter, he said. The person who found them didn't matter. Krantz was convinced that he had discovered a way to verify the tracks' authenticity from their structure alone—Zadig's ultimate clue. And what he found told him these tracks were real, damn the rest of the world's opinions.[45]

The Walla Walla tracks were different than any others found to that point because they had fingerprints—or, more exactly, dermatoglyphs, the swirls of ridges and troughs that make up fingerprints but are also found on palms, heels, and toes. Krantz could also see what he took to be sweat pores. No one could have thought to fake sweat pores and dermatoglyphs. This was the clue, the undeniable clue, the fingerprints the culprit didn't know were left behind, the evidence that would break the case. Immediately upon examining the tracks, Krantz set about verifying the dermatoglyphs. He asked Heuvelmans and the Russian hominologists to recommend fingerprint experts and submitted casts to American experts for their analysis. Greenwell tried to convince the American Dermatoglyphics Association to study the tracks. Early in 1983, Krantz visited the National Geographic Society, the FBI, and the Smithsonian, where he showed casts of the tracks to his brother (who worked as a photographer at the museum), George Zug, J. Lawrence Angel, T. Dale Stewart (who, with Napier and Angel, had been involved with the Hansen fiasco), and several other scientists. "The evidence of this new

nett, "Bigfoot Evidence," 498–509; idem, "Bigfoot Proponent Comes to the End of the Trail," *Skeptical Briefs*, June 2003, 1–2.

**45.** "Walla Walla Casts Show Dermal Ridges," *The ISC Newsletter* 1, no. 3 (1982): 1–4 (quotation, 3); Grover S. Krantz, "*Et Tu*, René?" *Cryptozoology* 3 (1984): 131–34; Krantz, *Big Footprints*, 82–84.

set of tracks should be treated as conclusive by all authorities," Krantz said.[46]

Convinced that Bigfoot's mark could no longer be denied, Krantz set about forcing Bigfoot into science. He reconstructed one of two known *Gigantopithecus* species, *G. blacki*, based on the teeth and jaw bones—following Zadig's method, the teeth and the jaw proscribing what kind of neck the creature had, its shoulders, and on down. The reconstruction showed that *Gigantopithecus* walked on two legs—not four as gorillas—and made tracks like those of Sasquatch. Krantz was so certain (his brashness again) that he formally proposed Sasquatch be given the scientific name *Gigantopithecus blacki*. It seemed to be a watershed moment in Sasquatch's career and the history of Forteanism. Krantz had expanded the borders of science to include this poor, damned fact without compromising science's authority, and made those who believed in it rational, the skeptics foolish.[47]

Krantz's analysis of the dermatoglyphs did what Patterson's movie could not and convinced Heuvelmans that Bigfoot was real, and an ape. Washington State University started to show signs of interest in Krantz's work, hosting the International Society of Cryptozoology's eighth annual meeting. In 1992, Krantz published *Big Footprints*, a summary of his twenty-year investigation, including his analyses of Patterson's film, the anatomy of Sasquatch feet, and the dermatoglyphs. It looked as though Krantz was going to get the chance to rub a few faces into the Sasquatch—if not the corpse, then the next best thing.[48]

## Arrested by the Science Police

Everything, however, could not be tied into such a neat little bow. Dahinden was certain that the tracks were fake—no matter the microscopic detail. When Greenwell asked Green what evidence he would submit to his proposed legal inquiry, Green only mentioned Patterson's movie, not Krantz's

**46.** Heuvelmans to Krantz, December 13, 1982, Heuvelmans file; Bayanov to Krantz, December 12, 1982, Bayanov file; Krantz, "Washington Impressions Impress Washington," 1983; Greenwell to Alex F. Roche, September 13, 1985; Grover Krantz, form letter, November 2, 1982, all in Fingerprints file, Grover Krantz papers; Grover S. Krantz, "Anatomy and Dermatoglyphics of Three Sasquatch Prints," *Cryptozoology* 2 (1983): 53–81 (quotation, 78).

**47.** Grover S. Krantz, "A Species Named from Footprints," *Northwest Anthropological Research Notes* 19 (1986): 24–39; idem, *Big Footprints*, 177–96.

**48.** Heuvelmans to Krantz, December 13, 1982, Heuvelmans file, Grover Krantz papers; "Pullman Symposium Reviews Sasquatch Evidence," *The ISC Newsletter* 8, no. 4 (1989): 1–5.

"definitive proof." Numerous dermatoglyph experts thought that the ridges meant nothing; some noted that sweat pores never make impressions in the soil. (The American Dermatoglyphic Association ignored Greenwell's request that it study the matter.) Several scientific authorities disagreed with Krantz's attempts to grant Bigfoot a scientific name. Conditionally giving a scientific name to a creature went against the accepted practices of taxonomy. Even Heuvelmans told Krantz that he couldn't offer a name without better evidence. It looked as though Krantz was desperate to be the first to name the creature just in case someone ever caught a Bigfoot, a point that he mostly conceded: "There are only a few rewards that" come with Sasquatch hunting, he wrote in *Big Footprints*, "and one of the most outstanding of these is the right to pick the official scientific name." Krantz's reconstruction of *Gigantopithecus* was also criticized as too speculative. His attempts to have his papers published in major journals failed, and so he was forced to print them in *Northwest Anthropological Research Notes* and *Cryptozoology*.[49]

These, though, were not the biggest failings of Krantz's science. Not even close. In 1990, a construction worker from Indiana sent Krantz a plaster cast that met all his criteria, secret and otherwise—significant, since Krantz usually dismissed reports from as far east as Indiana. According to *Big Footprints*, Krantz showed the cast to Titmus and a fingerprint expert in San Diego, California, both of whom thought that it was made from the track of a real wildman. A few years later, skeptic Michael Dennett, who had been Krantz's constant critic, followed up on the cast, contacting Titmus and the fingerprint expert. Neither of them was quite as confident as Krantz made out. Through his contacts in the Bigfoot community, Dennett also tracked down the man who had submitted the cast to Krantz. To a no doubt surprised and delighted Dennett, the man confessed that he had faked the tracks![50]

J. W. Parker—a pseudonym—had been irritated by Krantz's arrogance and wanted to test his claim that he could not be hoaxed. So he read some of Krantz's anatomical work and then filled a cat-litter box with mud. In about

**49.** Heuvelmans to Krantz, March 1, 1982; Heuvelmans to Krantz, August 13, 1985, both in Heuvelmans file; J. Biegert to Krantz, November 30, 1982; Russel H. Tuttle to Krantz, July 15, 1983; "Evaluation of Krantz's Paper on Dermal Ridges," 1983; Alex F. Roche to Greenwell, September 19, 1985; Leigh M. Van Valen to Krantz, January 21, 1986; J.T. Stern, Review of "A Species Named from Footprints," n.d., all in Fingerprints file, Grover Krantz papers; Dahinden, "Whose Dermal Ridges?" 128–31; Krantz. "*Et Tu*, René?" 131–34; "Interview: Does the Sasquatch Exist and What Can Be Done about It?" *The ISC Newsletter* 8, no. 2 (1989); 6; Krantz, *Big Footprints*, 196 (quotation).

**50.** Krantz, *Big Footprints*, 84–85, 200; Dennett, "Bigfoot Evidence," 498–509.

twenty minutes, he sculpted a print. Parker then made a cast from the print using a plastic resin, giving it a slight arch, shaping the ball to appear as though it bore weight, and adding dermatoglyphs (from his hands) "where the least amount wear and abrasion [*sic*] would occur." He didn't know what Krantz's two secret criteria were but guessed that they might be toenail impressions and scars, both of which he added. He also pressed a walnut shell into the fifth toe and added debris to the cast for verisimilitude. The whole process took an hour or two. As proof that he had actually done what he said, Parker showed Dennett the letter from Krantz acknowledging receipt of the cast; he also had the envelope. Once Dennett nailed down the story's particulars to his satisfaction, he published his discovery in *Skeptical Inquirer*.[51]

Krantz tried to explain away his apparent gullibility, protecting his brashness with his paranoia. He argued that, technically speaking, he had never endorsed the track's authenticity, only reported that it met all of his criteria and that others thought that it was real. Still, he thought that the track *was* legitimate and suggested that claims to the contrary were part of a con to make him look ridiculous. Perhaps, he suggested, Parker knew the track was real but was trying to "goad" Krantz into declaring it a fake, so that he could then "jump all over" him "with the information that it was the real thing all along."[52]

Skeptics dismissed Krantz's response as "nothing less than bizarre." How, exactly, was the plot that Krantz imagined supposed to work? Parker could not independently prove that his track was authentic, so he could not ridicule Krantz for declaring it fake. No, the fact of the matter was that Krantz's scientific judgment had been exposed as shoddy. Krantz was the Bigfoot community's most reputable member, its only scientist, and he had been fooled—not by the incarnation of Sherlock Holmes or the reincarnation of Leonardo da Vinci but by a construction worker. In the end, the three best pieces of evidence Krantz found in support of Bigfoot's existence—Patterson's film, Cripplefoot, and dermatoglyphs—were inextricably bound with pranksters and fraud. A shadow of doubt, deep and dark and wide, settled over Krantz's work.[53]

**51.** J. W. P. to Whom it may concern, February 12, 1992; Zoolooker@aol.com to Ira Walters, April 14, 1996; Zoolooker@aol.com to Ira Walters, April 18, 1996; all in Indiana Foot and SI file, Grover Krantz papers.

**52.** Grover S. Krantz, *Bigfoot/Sasquatch Evidence*, 2nd ed. (Surrey, BC: Hancock House, 1999), 299–300.

**53.** Daegling, *Bigfoot Exposed*, 176 (quotation).

Skeptics had not disproven Bigfoot's existence; they hadn't even altered public opinion—a 1978 poll found that about one in eight people believed in Bigfoot; eighteen years later, the number was almost one in five. But they had won. The science police continued their debunking into the next century, while the International Society of Cryptozoology was only one more in a long line of failures to make ABSMery—and those who studied it—respectable. Struggling during the best of times, the society died a quiet death in the early part of the twenty-first century.[54]

**54.** "A Vote for the Sasquatch," *Newsweek*, June 26, 1978, 32; "American Piety in the Twenty-First Century: New Insights to the Depth and Complexity of Religion in the US: Selected Findings from the Baylor Religion Survey," research pamphlet (Waco, TX: Baylor Institute for Studies of Religion, September 2006), 45.

CHAPTER TEN

# *The Death of Bigfoot?* 1980–2002

While American National Enterprises squandered its early advantage in the Bigfoot movie business, Ron Olson did not give up on the creature. In March 1975, he started raising money for another Bigfoot film, something to match his vision, something persuasive and lucrative. By the end of April, Olson had found 134 backers who invested a combined $300,000 (two-thirds to make the movie, the rest for the initial distribution). According to Olson, the investors were "mostly millworkers and such, and a few little old retired ladies who put their pensions and stuff into this." Filming started in August. The talent was all recruited locally (the director was head of the performing arts department at Lane Community College in Eugene). As with most such films, it was poorly shot, padded with stock footage, and repetitive.[1]

*Sasquatch* opened in January 1976, and then started touring the country. Late in 1977, Olson decided to take his movie to New York City and Washington, D.C. Four-wallers had mostly stayed outside of the major markets because competition was too fierce there and expenses too high, but Olson was feeling especially optimistic. Interest in Bigfoot had proved to be intense. Over the course sixteen days, 1,070 advertisements played on five Manhattan television stations—an average of 13 commercials per day per station—at a cost of $250,000. In January 1978, *Sasquatch* opened at 110 New York City theaters. The movie sold out every theater its first day—then a blizzard hit, shutting down the city. Ticket sales fell precipitously. It was a bitter pill for the producers. There were no second acts, no re-releases. Word of mouth moved too quickly and was too brutal. (A re-release of *Sasquatch* even flopped in

1. Mike O'Brien, "Lights . . . Camera . . . Sasquatch!" *Eugene (OR) Register-Guard*, January 25, 1976.

Olson's hometown.) Still, all was not lost. The movie made $800,000, more than recovering its publicity budget.[2]

Olson's was one of the last great four-wall tours. The industry had bigger problems than the contingencies of weather. Theater owners saw the companies making money and demanded ever-higher rental rates. Inflation dramatically increased the cost of advertising, eating into the four-wallers' small profit margins. And, worst of all for four-wallers, Hollywood learned their tricks. Major studios began doing more market research; they advertised on TV and created demand for summer blockbusters; science fiction and the occult were no longer seen as weird, but became industry bread and butter—from 5 percent of box office in 1971 to 50 percent by 1982. In many ways, the most profitable Hollywood films of the late 1970s and 1980s were four-wall flicks with better production values—*Jaws* and *Star Wars* and *Close Encounters of the Third Kind* and *Raiders of the Lost Ark*.[3]

The death of four-walling was part of a more general trend; as mass culture matured and consolidated, working-class entertainments were increasingly absorbed into the mainstream—and the vehicles that had once carried them went extinct. Men's adventure magazines either went out of business or transformed themselves into pornographic rags, but the kind of stories that they had printed became standard fair on TV, at movie theaters, and on bookshelves. (Stephen King, for instance, began his career writing for adventure magazines before becoming the best-selling novelist in the country.) Tabloid circulation peaked in 1982, after which time mainstream news organizations increasingly adopted the tabloids' methods, focusing on celebrity gossip, tales of the weird and wonderful. "We're being out-tabloided by the mainstream press," an editor of the *Globe* complained as he watched his publication's sales decline.[4]

**2.** Fred Crafts, "How 'Sasquatch' Slickered New York City," *Eugene (OR) Register-Guard*, April 6, 1978; Robert D. McFadden, "Millions Marooned; Blizzard Brings 13 Inches to New York City, Worst Since February '69," *New York Times*, January 21, 1978.

**3.** Justin Wyatt, *High Concept: Movies and Marketing in Hollywood* (Austin: University of Texas Press, 1994); Frederick Wasser, "Four Walling Exhibition: Regional Resistance to the Hollywood Film Industry," *Cinema Journal* 34 (1995): 51–65; Thomas M. Disch, *The Dreams Our Stuff Is Made of: How Science Fiction Conquered the World* (New York: Touchstone, 2000), 208.

**4.** Kevin Glynn, *Tabloid Culture: Trash Taste, Popular Power, and the Transformation of American Television* (Durham, NC: Duke University Press, 2000); Stephen King, *On Writing: A Memoir of the Craft* (New York: Pocket Books, 2000), 69; Bruce J. Schulman, *The Seventies: The Great Shift in American Culture, Society, and Politics* (Cambridge, MA: Da Capo Press, 2001), 102–17; Bill Sloan, *"I Watched a Wild Hog EAT my Baby!" A Colorful History of Tabloids and Their Cultural Impact* (New York: Prometheus, 2001), 15–16, 185, 191, 207–15 (quotation, 211); Adam

The blurring of the boundaries between highbrow and low allowed Bigfoot to escape the confines of working-class entertainment and pass into middle-class culture. Stories about the beast were published in *Harper's* and the *New Yorker*; *Newsweek* and *Science Digest* ran sympathetic reports on Bigfoot hunters. Sasquatch appeared in *New York Times*–approved literature. Bigfoot entered those hallowed middle-class preserves, the theater and art galleries. It appeared in poems. In 1987, the creature starred in the distinctly middle-brow movie *Harry and the Hendersons*. The film made $4,000,000 its opening weekend and spawned a TV show that aired 72 episodes over two years.[5]

At long last, Bigfoot had breeched the laughter curtain. But all was not well for the monster. This was not the same creature that had romped through men's adventure magazines and frolicked in four-wall films. The bourgeois Bigfoot was less challenging, not as violent, not as sexual, not as awesome—"the period of the comedic, gentle, giant, nonsexual Bigfoot had arrived," said Loren Coleman. As well, virtually no middle-class depictions of the beast took its existence seriously. Sasquatch had not so much passed through the laughter curtain as become draped in it, still more than a little ridiculous. Bigfoot was presented as a legend, a self-consciously created myth. Indeed, professing belief in Sasquatch was sometimes considered pathological, as during the trial of murderer Cary Stayner when his interest in Bigfoot was used as evidence of mental instability. Thus, even as Sasquatch reached a huge audience, the beast was seen as increasingly ethereal—ghostly, insubstantial, unbelievable. It was a sign of things to come.[6]

## The Green Man

The promiscuous mixings of high- and lowbrow entertainments created a situation not unlike the one that had made the middle class so anxious in

Parfrey, ed., *It's a Man's World: Men's Adventure Magazines, the Postwar Pulps* (Los Angeles: Feral House, 2003), 22.

**5.** "Weekend Box Office," *Los Angeles Times*, June 9, 1987; Alan Morton, *The Complete Directory to Science Fiction, Fantasy and Horror Television Series: A Comprehensive Guide to the First Fifty Years, 1946–1996* (Peoria, IL: Other World Books, 1997); Joshua Blu Buhs, "Camping with Bigfoot: Sasquatch and the Varieties of Middle-Class Resistance to Consumer Culture in Late Twentieth-Century North America," *Journal of Popular Culture* (forthcoming).

**6.** Eric Bailey, "Jurors Asked to Look Into Mind of a Killer," *Los Angeles Times*, August 19, 2002; Loren Coleman, *Bigfoot! The True Story of Apes in America* (New York: Pocket Books, 2003), 213.

Barnum's America: traditional signs of status could no longer be trusted. Identity was obscured, uncertain. Bigfoot was in the *New Yorker* now! Middle-class Americans were uneasy about these changes; they did not—as so many of the working class did—oppose consumerism and mass culture. On the contrary, they often fostered and celebrated it because their relatively large paychecks already proved that society valued them highly and allowed them to create personalities (buy things) consistent with their own self-image. Still, these cultural changes raised unsettling questions: How could one demonstrate an elite status when mass culture made traditional signs of success—ownership of expensive things, TVs and cars and houses—increasingly available to everyone? Where, amid all the stuff, could one find a space uncolonized by consumer goods, a preserve where individualism might still have free run?[7]

Again not unlike in Barnum's America, debates over wildmen were in part debates over middle-class respectability. The skeptics' attacks on Bigfoot—and the paranormal generally—was a defense of middle-class privileges and scientific elitism against a flood of other, often working-class, forms of knowledge. Science, as CSICOP saw it, was untainted by materialism; it was the application of pure logic, of reason. A scientific attitude was proof of one's elite status. Middle-class Americans could also prove their elite status by ironically celebrating tabloids or developing a cultish devotion to the bad movies of the 1970s. Yes, these were trash, but they could be transformed by middle-class taste into something rarefied, thus affirming the status of those who could see the wonder amid the dross.[8]

Or Bigfoot could also be used to flatter the middle-class for their sensibilities and sensitivities. To believe in Bigfoot was to hope, to dream—to resist the crassness of mass culture and insist that one's individuality remained untouched by consumerism. Robert Bly's *Iron John* and Clarissa Pinkola Estés's *Women Who Run with the Wolves*—best-selling self-help books of the 1990s—both demonstrated this approach (although it was not unique to them), arguing that getting in touch with an inner wildman or wildwoman would allow for the discovery of a true self. "Women's flagging vitality,"

**7.** Paul Fussell, *Class* (New York: Ballantine, 1983), 166; Warren I. Susman, "'Personality' and the Making of Twentieth-Century Culture," in *Culture as History: The Transformation of American Society in the Twentieth Century* (New York: Pantheon, 1984), 271–86; Barbara Ehrenreich, *Fear of Falling: The Inner Life of the Middle Class* (New York: Perennial, 1990); Tom Pendergast, *Creating the Modern Man: American Magazines and Consumer Culture, 1900–1950* (Columbia: University of Missouri Press, 2000), 1–13.

**8.** Buhs, "Camping with Bigfoot."

Estés wrote, "can be restored by extensive 'psychic-archaeological' digs into the ruins of the female underworld. By these methods we are able to recover the ways of the natural instinctive psyche, and through its personification in the Wild Woman archetype we are able to discern the ways and means of woman's deepest nature." This Bigfoot was a nursemaid, an angel, healing the wounds inflicted by mass culture and nurturing those hidden, noble qualities that made one a unique and wonderful individual.[9]

Bigfoot was probably most associated with another middle-class movement, environmentalism, which offered a different solution to the problem of mass culture: escape. Escape now. "If a Sasquatch were ever captured, what would his message to the world be?" asked social critic Jim Goad. "Probably, 'Don't shoot!' Bigfoot doesn't bring a message, he keeps running away from us. That's his message—get out of Babylon before it's too late." Enticing people to leave civilization and get back to nature was a traditional duty (and privilege) of the wildman. This image of Bigfoot as a beneficent spirit of the woods was different than working-class views of the monster—but it wasn't new. It was a resurrection. The wildmen of myth come in many forms. One is the Green Man, the Earth Father—male counterpart of Mother Nature—a woodland spirit. Satyrs are Green Men, Robin Hood, the Green Knight who faced Sir Gawain, and—most familiar to modern readers—the Jolly Green Giant. Sasquatch became a symbol of the environmental movement, a myth created to re-enchant the world and make its preservation a sacred task.[10]

Bigfoot, dressed as the Green Man, had made limited forays into middle-class culture at least since the time of Patterson's movie. In Margaret Atwood's 1970 poem "Oratorio for Sasquatch, Man, and Two Androids," for example, Bigfoot represented the natural world. It had green flesh and "the leaves," the wildman said, "are my ears." Driven away by a robot that could only think to dissect it, analyze it, and reduce its mysteries to cold formulae, the Sasquatch found refuge under a mountain, where it waited "until the killers have been changed to roots, to birds, until the killers have become guardians and have learned our language." A few years later, *National Wild-*

**9.** Clarissa Pinkola Estés, *Women Who Run with the Wolves* (New York: Ballantine, 1994), 3–4 (quotation); Buhs, "Camping with Bigfoot."

**10.** Robert Michael Pyle, *Where Bigfoot Walks: Crossing the Dark Divide* (Boston: Houghton Mifflin, 1995), 155; Jim Goad, *The Redneck Manifesto: How Hillbillies, Hicks, and White Trash Became America's Scapegoats* (New York: Touchstone, 1997), 168; Kathleen Basford, *The Green Man* (Cambridge: D. S. Brewer, 2004); Phyllis Siefker, *Santa Claus, Last of the Wild Men: The Origins and Evolution of Saint Nicholas, Spanning 50,000 Years* (New York: McFarland & Co., 2006).

*life* revisited the Bigfoot mystery—having survived Morgan's exploits—and transformed Sasquatch into an embodiment of nature's wisdom. *Life*, *Audubon*, and hiking guides all offered similar perspectives. Skamania County's law against shooting a Sasquatch also tapped into this sentiment.[11]

These middle-class appropriations of Bigfoot were idiosyncratic and uncoordinated—Atwood, for instance, was trying to create a distinctly Canadian literature. But as the working-class arenas where Bigfoot had once roamed collapsed and were absorbed into mass culture, the view that Bigfoot was a modern manifestation of the Green Man became more prominent. In 1983, David Rains Wallace's book *The Klamath Knot* argued that Bigfoot could be used to start a new kind of mythology. Sasquatch, he said, and similar wildmen, "link us to lakes, rivers, forests, and meadows that are our homes as well as theirs. They lure us into the wilderness . . . not to devour us but to remind us where we are, on a living planet." *The Klamath Knot* won the John Burroughs Medal for Nature Writing and a silver medal from California's Commonwealth Club, and readers of the *San Francisco Chronicle* voted it one of the twentieth century's best nonfiction books about the West. The ideas resonated. Other, later works, such as Robert Michael Pyle's *Where Bigfoot Walks*, Molly Gloss's *Wild Life*, Edward Hoagland's *Seven Rivers West*, and films such as *Little Bigfoot*, *Bigfoot: The Unforgettable Encounter*, and *Harry and the Hendersons* owed much to Rains, all of them imagining a Bigfoot that was, as Pyle wrote, "An icon with shoulders broad enough to accept our mortal dread, and honest enough to promise the earth as long as we both last."[12]

Patterson's film had defined the image of Bigfoot since the late 1960s; but because of this rekindled interest in the Green Man, a new image of Bigfoot took shape. Some illustrations showed Bigfoot draped in foliage, just like the Green Men of old, a benign spirit. *Harry and the Hendersons* probably did the most to make this new image dominant. "For a few years," wrote Loren Coleman, "Harry was what people had in mind when the word *Bigfoot* was

**11.** Shana Alexander, "More Monsters, Please!" *Life*, December 8, 1967, 30; Margaret Atwood, "Oratorio for Sasquatch, Man, and Two Androids," in *Poems for Voices* (Toronto: Canadian Broadcasting Corporation, 1970), 14–28 (first quotation, 15; second quotation, 28); John Hart, *Hiking the Bigfoot Country* (San Francisco: Sierra Club, 1975); John G. Mitchell, "Why We Need Our Monsters," *National Wildlife*, April 1978, 12–15.

**12.** Margaret Atwood, "Canadian Monsters: Some Aspects of the Supernatural in Canadian Fiction," in *The Canadian Imagination: Dimensions of a Literary Culture*, ed. David Staines (Cambridge, MA: Harvard University Press, 1977), 97–122; Pyle, *Where Bigfoot Walks*, 157; "The Best in the West: Top 100 Nonfiction," *San Francisco Chronicle*, November 21, 1999; David Rains Wallace, *The Klamath Knot: Explorations of Myth and Evolution* (Berkeley: University of California, 2003), 138.

FIGURE 33. In the 1980s, Bigfoot caught the attention of the American middle class and became something of an environmental icon—which was both an evolution away from its image in the men's adventure magazines and also a reversion to an earlier type: the Green Man of European legend. Bigfoot as the Green Man appeared throughout American culture, from *Harry and the Hendersons* to this early example, an illustration by Dick Oden for the March 1974 issue of *Westways* magazine. (Courtesy of Automobile Club of Southern California Archives.)

mentioned, replacing the Patterson-Gimlin footage's zaftig Bigfoot." Rick Baker, who had earlier created the Sasquatch for *Schlock*, a trashy Bigfoot flick of the 1970s, did the movie's makeup. The Sasquatch of *Harry and the Hendersons* was quite different. He was clearly male, but was not threatening. Harry had a pronounced, bald forehead and a white beard, making him seem avuncular and wise.[13]

Bigfoot the Green Man was a guide to a different, better place. Unlike the paradise to which Sasquatch led white working-class men, this utopia was a place where leisure was valued over work: where artisanal skill was absent, where walkers and hikers and backpackers knew the world better, survived the world better than those who worked in it, those who felled trees or hunted game or fished rivers. In his book, Robert Michael Pyle daydreamed of Bigfoot tearing into a working-class tavern and devouring the drinkers, leaving the woods to him and others like him who appreciated its beauty without altering it, without working in it. This Arcadia was also a place where there was no consumerism, no mass culture. The titular character in Paul Doyle's novel, *Nioka, Bride of Bigfoot*, for instance, left civilization and had to unlearn

**13.** Coleman, *Bigfoot!* 101, 213–15 (quotation, 213).

consumerism—"the biggest cult of all and the craziest." There was also no longer any trace of the working class or their entertainments—four-wall flicks, men's magazines, tabloids—these were the things that Bigfoot was supposed to lead the middle class away from. Instead, stories about Bigfoot the Green Man focused on the creature's connection to Native Americans. Indians, in the imagination of whites, were careful environmental stewards, in touch with the mysteries and wonders of the natural world, inhabiting an Elysium free from the blemishes of consumerism and mass culture.[14]

And by imagining themselves into the body of Sasquatch, environmentally conscious middle-class Americans could reach this Eden—just as the working class once imagined themselves into Sasquatch to experience a different kind of rapture. "What would we learn of forests if we had the senses of wild animals?" David Rains Wallace asked. "Where there are no clocks or roads, time and distance behave differently, like animals let out of the zoo. Where there are no signs or labels, things seem much less predictable." Writing about his experiences hiking through the Oregon wilderness, Pyle made much of his "mutant" feet that could not withstand confinement in shoes and the "abominable" time they had walking, as the title had it, *Where Bigfoot Walks*. Once, he stripped nude and stood naked in the wilderness. There were other moments when he "left myself behind and climbed inside the great hairy headgear and felt the flapping of my massive feet against the rough pumice paths." At the end of his hike, he heard a whistle that he thought might have been made by a Sasquatch—later finding tracks that seemed to confirm his suspicion—and when that reedy sound whispered through the trees, he whistled back, as though they knew the same language. He had found a place where his special talents were important, where his status was unchallenged, where the cacophony of mass culture and consumerism could not drown out his voice. It was a peculiarly middle-class escape to a peculiarly middle-class paradise.[15]

**14.** Paul Doyle, *Nioka, Bride of Bigfoot: A Novel* (Seattle, WA: Daily Planet Press, 1992), 188; Pyle, *Where Bigfoot Walks*, 119, 241; Richard White, "'Are You an Environmentalist or Do You Work for a Living?': Work and Nature," in *Uncommon Ground: Rethinking the Human Place in Nature*, ed. William Cronon (New York: W. W. Norton, 1996), 171–85; Shari M. Huhndorf, *Going Native: Indians in the American Cultural Imagination* (Ithaca, NY: Cornell University Press, 2001).

**15.** Pyle, *Where Bigfoot Walks*, 35–37, 40, 53, 78, 120–23, 293–99 (remainder of quotations, 136, 137, 153, 121); Robert Baird, "Going Indian: Discovery, Adoption, and Renaming toward a 'True American' from *Deerslayer* to *Dances with Wolves*," in *Dressing in Feathers: The Construction of the Indian in American Popular Culture*, ed. S. Elizabeth Bird (Boulder, CO: Westview

## The Death of Bigfoot

As the middle class tried to domesticate Bigfoot and turn the beast into a symbol of "green spirituality," in Pyle's words, Bigfooters were still trying to prove that the wildman was real—not a legend, but a living, breathing creature. In September 2000, six years after the revelation that Krantz had been fooled by fake dermatoglyphs, Sasquatch hunter Rick Noll was checking fruit traps that had been set along Skookum Meadow, in Washington. Noll had been interested in Bigfoot since the late 1960s; he had apprenticed with Byrne while in college, had attended the British Columbia conference, where he'd been so disappointed in the fighting and egos that he almost quit the search altogether. Noll persisted, though, and in the 1990s became associated with the Bigfoot Field Researchers Organization. It was under the organization's aegis that Noll was in Skookum Meadow.[16]

The Bigfoot Field Researchers Organization was the biggest and probably best known of the Sasquatch groups around the turn of the twenty-first century. (Despite criticisms from skeptics, there remained several hundred people dedicated to the cause of finding Sasquatch. Peter Byrne even returned to the hunt for a while, as did Robert Morgan.) Early in the 1990s, as skeptics focused their attack on Krantz, many Bigfoot hunters moved their organizations onto the Internet, replacing newsletters, correspondence, and phone calls with Usenet groups, e-mail, and Web sites—which could expand on the *Bigfoot Bulletin*'s template by not only providing citations to newspaper and magazine articles but reprinting them entirely, something that would have been cost-prohibitive for earlier Bigfooters, such as George Haas. This new generation of Bigfooters built on John Green's efforts, compiling huge databases of sightings. There was a whole host of competing sites, many of the early ones set up by Henry Franzoni; Matt Moneymaker, an information technology consultant from Orange County, California, took over one and established the Bigfoot Field Researchers Organization, eventually making it the dominant group. The organization not only improved on the work of

Press, 1996), 195–210; Molly Gloss, *Wild Life* (New York: Mariner Books, 2001), 214, 235; Wallace, *The Klamath Knot*, 74 (first quotation).

**16.** Richard Louv, "The Bigfoot Follies," *Human Behavior* 7, no. 9 (1978): 24; Pyle, *Where Bigfoot Walks*, 157 (quotation); Coleman, *Bigfoot!* 17–24; Thom Powell, *The Locals: A Contemporary Investigation of the Bigfoot/Sasquatch Phenomenon* (Surrey, BC: Hancock House, 2003), 109–25; Sali Sheppard-Wolford, *Valley of the Skookum: Four Years of Encounters with Bigfoot* (Enumclaw, WA: Pine Winds Press, 2006).

Haas and Green but also seemed to fulfill Peter Byrne's dream of establishing a board of examiners. The group claimed to have well-trained researchers stationed around the country who would go out and investigate reported sightings, ruling them credible or not. (It still had to deal with Beckjord, though, who also had an online presence.)[17]

During the spring of 2000, evidence suggestive of Sasquatch activity was repeatedly found near Skookum Meadow—fittingly, perhaps, since Skookum may have been another Native American word for Bigfoot. Noll, Moneymaker, and ten or so other organization members visited the area. As with past expeditions, they set up cameras—triggered now not by tripwires but infrared beams; they blasted Bigfoot-like sounds from speakers; and they laid out traps—like Titmus they used a sexual attractant, but instead of soiled sanitary napkins they laced fruit with a mixture of human and gorilla scent. They also brought the media along, a film crew with the TV show "Animal X," to document their findings.

Early one morning, Noll noticed that the mud around one of the traps that he was checking had been disturbed—but not with Bigfoot tracks. As he looked at the troughs and crests, he thought that he saw the impression of Sasquatch's body. Noll pointed out what he saw to a pair of fellow Bigfooters, and they agreed with him. They reasoned that Sasquatch had been afraid of leaving tracks and so instead had lain down in the mud and reached out for the fruit. The hunters spent the day casting the impression and then returned to civilization triumphant.[18]

The discovery repudiated skeptics and seemed to counter attempts to craft Bigfoot into a legend. Using the same techniques—updated with current technology—following the same impulses, Bigfoot hunters had found what they considered to be the best evidence yet of Bigfoot's existence. The Skookum cast, as it came to be called, seemed to foretell the dawning of a new day. The old generation—Byrne and Dahinden and Krantz and Green—was giving way, but their torch was being carried forward. Jimmy Chillicut, a fingerprint expert, defended the importance of dermatoglyphic evidence,

**17.** Matthew Burtch, "In Search of Bigfoot," *Reed Magazine*, August 2000, http://web.reed.edu/reed_magazine/aug2000/a_bigfoot/index.html (accessed April 11, 2008); Bruce Barcott, "'Sasquatch *Is* Real!' Forest Love Slave Tells All!" *Outside*, August 2002, http://outside.away.com/outside/news/200208/200208_sasquatch_200201.adp (accessed April 11, 2008); Powell, *The Locals*, 36–48; Timothy Egan, "Bigfoot Fans Cling to Their Faith—Hoax or Not," *Sacramento (CA) Bee*, January 3, 2003.

**18.** Coleman, *Bigfoot!* 17–24; Powell, *The Locals*, 109–25.

insisting that it all could not have been faked. Jane Goodall offered her support to Bigfoot researchers. So did primatologist Daris Swindler and naturalist George Schaller. Wildlife biologist John Bindernagel published a book on Bigfoot, which compared the creature's reported behavior with known ape behavior to demonstrate that reports of the beast could not have been faked unless all of the witnesses were familiar with studies on ape biology. Jeffrey Meldrum, an anthropologist at Idaho State University, followed in Krantz's large footsteps, studying the many casts that had been made over the years. Like Krantz, he argued that their detail and structure proved them true. Loren Coleman thought that the turn of the millennium was a "historical crossroads," a new beginning for Sasquatch research: "The moment is upon us to stand shoulder to shoulder with skeptics and forge ahead, exploring what is out there, continuing the open-minded but critical search for tomorrow's surprises."[19]

The Skookum cast, however, was only another piece of ambiguous evidence, ultimately proving nothing. Analyses of it found no DNA, no fur, and it raised more questions than it answered. The Sasquatch was afraid of leaving footprints so instead left an impression of its entire body. Where was the sense in that? In 2007, Anton Wroblewski, a geologist, pointed out that the cast looked like an imprint of a laying elk—and noted that there were many elk hoof prints in the surrounding mud (but no Bigfoot tracks).[20]

This was not a new dawn but *an* end. Despite passing through the laughter curtain, Bigfoot was marginalized by the mainstream far more than it had been in the 1950s or 1970s. Encouraging words from scientists were nice but no substitute for research published in academic journals, and there was little of that. Nothing in *Science* or *Nature* or *Current Anthropology* or *Northwest Anthropological Research Notes*. *Cryptozoology* was no more. Jeff Meldrum only published a few summary blurbs of his studies, and a longer paper in the obscure *New Mexico Museum of Natural History and Science Bulletin*; his book on Sasquatch was a companion to the video *Sasquatch: Legend Meets Science*—hardly the usual venue for scientific research. By the time Noll discovered the Skookum cast, more than three decades had passed since Patterson's movie went public. And still, still there was no body. No trucker had ever clipped

**19.** John A. Bindernagel, *North America's Great Ape, the Sasquatch: A Wildlife Biologist Looks at the Continent's Most Misunderstood Large Mammal* (Courtenay, BC: Beachcomber Books, 1998); Loren Coleman, "Bigfoot Evidence Evaluated," *Skeptical Inquirer*, July–August 2002, 63 (quotation); idem, *Bigfoot!* 17–24; Theo Stein, "Bigfoot Believers," *Denver Post*, January 5, 2003; Jeff Meldrum, *Sasquatch: Legend Meets Science* (New York: Forge, 2006), 112–23, 254–58.

**20.** Daniel Perez, "Skookum Hokum?" *Bigfoot Times*, May 2007, 3.

a Sasquatch on the road and dragged it into a hospital or university or museum. No hunter had ever bagged one and shown it to the world.[21]

Bigfooters tried to explain away this embarrassing lack of a body. Green blamed environmentalists—they had turned the public against killing a Sasquatch, he complained. Dahinden blamed a "psychological safety switch": people who saw the beast were so confused that they could not pull the trigger. These were unsatisfying responses, though. Throughout the 1990s, interactions with other large wild mammals—bears, cougars—became common. Where was Sasquatch? Why no compelling new video? Why none shot by animal control officers? Dahinden's preferred explanation, that some psychological safety switch prevented humans from killing Sasquatch, seems more a projection of his own ambiguous feelings about the beast. If the people of Humboldt County could mercilessly slaughter Native Americans, why could they not also find it in themselves to kill a humanlike animal?[22]

Instead of rigorous studies or found Sasquatches, there were only more hoax revelations. In November 2002, amid the hoopla over the Skookum cast, Ray Wallace died. His family told a reporter for the *Seattle Times* that, indeed, Ray had faked those famous tracks in 1958, and continued faking tracks throughout the years. The admission was corroborated by the family of Scoop Beal, who had been editor at the *Humboldt Standard*—Eureka's other paper (although it had the same owner as the *Times*)—when Crew made his discovery. Beal allegedly knew that the tracks were fake from the beginning but kept printing stories because interest was so high. (It seems probable, then, that Genzoli was also in on the joke.) Wallace's children showed the press the fake feet Ray had used—and they had the distinctive double-ball. Quickly, newspapers across the country and around the world, which had once promoted the legends of local wildmen into international sensations, picked up the story and now reported the creature's ignoble passing. As the report was repeated, Wallace's legend grew and soon he was responsible not

**21.** D. Jeffrey Meldrum, "Evaluation of Alleged Sasquatch Footprints and Inferred Functional Morphology," *American Journal of Physical Anthropology (Supplement)* 28 (1999): 200; D. Jeffrey Meldrum and Lonny G. Erikson, "Dermatoglyphics in Casts of Alleged North American Ape Footprints," *Journal of the Idaho Academy of Science* 37 (2001): 36; D. Jeffrey Meldrum, "Ichnotaxonomy of Giant Hominoid Tracks in North America," *New Mexico Museum of Natural History and Sciences Bulletin* 42 (2007): 225–31; Benjamin Radford et al., "The Nonsense and Non-Science of Sasquatch," *Skeptical Inquirer*, May–June 2007, 58–61.

**22.** Don Hunter and René Dahinden, *Sasquatch* (Toronto: McClelland & Stewart, 1973), 85 (quotation); "Interview: Does the Sasquatch Exist and What Can Be Done about It?" *The ISC Newsletter* 8, no. 2 (1989): 6–7.

FIGURE 34. After the death of Ray Wallace in 2002, his family admitted that he had faked the Bluff Creek tracks in 1958—which was no surprise, since Wallace himself had earlier admitted to making other fake Bigfoot prints. The story spread over the newswires, and sometimes was blown out of all proportion, with Wallace accused of with faking seemingly all of the evidence for Bigfoot's existence. The minds of Sasquatch hunters, however, remained unchanged by the news. Here, Wallace's nephew Dale Wallace displays some of the carved feet that Ray used to make Bigfoot tracks. (Photo by Dave Rubert. Used by permission.)

only for the tracks around Bluff Creek in the late 1950s and his Washington home in the 1970s but for all the tracks, for Patterson's movie—putting his wife in a gorilla suit, or a neighbor—for the entire phenomenon. "Bigfoot is dead," the *Seattle Times* said. "Really."[23]

## "The Most Abominable Hoaxer"

The Yeti, too, sometime-cousin of Bigfoot, suffered a declining fortune as the twentieth century gave way to the twenty-first, because the things that had once made it seem real—Everest, the mountaineers—were no longer substantial themselves.

Neither Hillary nor Norgay climbed Everest again; they left tracks, however, a trail that others followed. The Swiss reached the summit the year after Hillary and Norgay; the Chinese reached it from the north in 1960. Throughout the 1970s, increasing numbers of people climbed through the Himalayas—about fifty thousand tourists visiting the area each year by the end of

**23.** Bob Young, "Lovable Trickster Created a Monster with Bigfoot Hoax," *Seattle Times*, December 5, 2002; Coleman, *Bigfoot!* 79–80; John Green, *The Best of Sasquatch Bigfoot* (Surrey, BC: Hancock House, 2004), 9–17; Meldrum, *Sasquatch*, 55–72.

the century (and yet, not one took a good picture of the Abominable Snowman). In the 1980s, commercial operations started offering expeditions: an attempt on the summit for $65,000. Aluminum ladders were laid across crevasses, rope handrails hung. Trash littered the path. Late each season, the snow was gray from boot treads. Hillary complained, "It is hardly mountaineering, more like a conducted tour." Reaching the top was mundane enough that an American paper could report that standing on the roof of the world was an "elusive but relatively trite notch in the ice axe" of a seasoned American climber. By the fiftieth anniversary of Hillary and Norgay's historic climb, more than 1,200 people had reached the summit (and almost two hundred had died trying). Mountaineer John Krakauer wrote, "Traditionalists were offended that the world's highest summit was being sold to rich parvenus—some of whom, if denied the services of guides, would probably have difficulty making it to the top of a peak as modest as Mount Rainier. Everest, the purists sniffed, had been debased and profaned."[24]

Commodified, the mountains seemed less real, less daunting. Commercial companies, anthropologist Catherine Palmer said, "all but erased . . . the notions of 'risk' and 'danger'" from climbing so that they could more easily sell their tours: you, too, could climb Everest, as long as you had good credit. Anyone could. As a result, the men who climbed the mountain, too, lost some of their status. Their word counted for less. In the 1990s, Reinhold Messner went in search of the Yeti. He was the first man to scale Everest alone, the first to do so without supplemental oxygen, and the first to do both at the same time. Like Hillary before him and Smythe before that, Messner concluded that the Yeti was a bear. But unlike Hillary's pronouncement, Messner's opinion received no special weight: stories continued to appear in newspapers about mysterious tracks and strange sightings.[25]

In 1989, the mountaineering writer Peter Gillman argued that Shipton had faked the 1951 track—he was "the most abominable hoaxer." It probably

**24.** Timothy Egan, "As Climbers Die, The Allure of Everest Keeps on Growing," *New York Times*, March 11, 1998; Jon Krakauer, *Into Thin Air: A Personal Account of the Mt. Everest Disaster* (New York: Anchor, 1999), 22 (third quotation); Rich Landers, "Peak of Desire," *Spokesman Review* (Spokane, WA), June 8, 2003 (second quotation); Terri Judd, "Fifty Years after Hillary Triumph, Thousands Follow in his Footsteps," *The Independent* (London), May 24, 2003 (first quotation).

**25.** Reinhold Messner, *My Quest for Yeti*, trans. Peter Constantine (New York: St. Martin's Griffin, 2000); Catherine Palmer, "'Shit happens': The Selling of Risk in Extreme Sport," *Australian Journal of Anthropology* 13 (2002): 323–36 (quotation, 323).

started as a lark, Gillman speculated, a joke that snowballed out of control—the attendant publicity so enormous that Shipton could not admit his ruse without damaging his reputation. And then the Himalayan Committee had snubbed him, costing him the chance to lead the expedition that had conquered Everest, so Shipton maintained the story out of spite, thumbing his nose at an establishment that had betrayed him. Shipton's biographer dismissed Gilman's charges as out of character for the mountaineer, and Michael Ward, who was on the glacier with Shipton, was certain that he hadn't faked the print. But the claim was still telling: the best evidence for the Yeti's existence, the testimony of a man who knew the Himalayas better than any Westerner in the years after World War II, could be dismissed. The world was different than it had been only fifty years before, with different rules for what counted as compelling evidence, different rules for who counted as credible, different rules for what seemed possible.[26]

This process—the commodification of the mountain, its declining stature, and the declining respect for those who climbed it—reached a climax in the years just after the millennium. In April 2003, as the fiftieth anniversary of Everest's conquest approached, and only a few months after Wallace's death, the Walt Disney company announced that it planned to open a new ride at its Wild Animal Park in Lake Buena Vista, Florida. The ride would be called Expedition Everest. "Legend holds that high in the Himalayan Mountains lives an enormous creature that fiercely guards the route to Mount Everest," read the press release. "Now that legend comes dramatically to life at Disney's Animal Kingdom in a new high-speed train adventure that combines coaster-like thrills with the excitement of a close encounter of the hairy kind." The roller coaster opened three years later. The mountain on which this Disney-version of the Abominable Snowman lived was two hundred feet tall, the highest mountain in the state.[27]

The Yeti, once the symbol of a world that existed beyond mass culture, incarnation of an obdurate reality, was now part of mass culture, another consumer good.

**26.** Michael Ward, "Everest 1951: The Footprints Attributed to the Yeti—Myth and Reality," *Wilderness and Environmental Medicine* 8 (1997): 29–32; Peter Steele, *Eric Shipton: Everest & Beyond* (Seattle, WA: Mountaineers Books, 1999), 161.

**27.** "Expedition Everest Construction Archive and News—1998 to 2004" (quotations), http://www.wdwmagic.com/everest_construction_archive2.htm (accessed April 11, 2008); Scott Powers, "Disney's $100 Million Gamble; Expedition Everest Roller Coaster Reaches Sky-High for Fresh Thrills," *Orlando Sentinel*, January 22, 2006.

## Bigfoot Is Dead! Long Live Bigfoot!

The Skookum cast proved less than it seemed; the Yeti did, too. And the utopia that Bigfoot promised to middle-class environmentalists also proved elusive. Historian Samuel Hays argues that environmentalism grew out of the same post–World War II changes in American society and economics that allowed for the creation of mass culture and a consumer society. The huge increase in wages that came in the 1950s and 1960s made it possible for Americans to spend money on "quality of life" goods: Americans wanted clean air, fresh water, and green places where they could play—leisure time and the pursuit of happiness part of the "good life." Nature increasingly became a consumer good and environmentalism was a middle-class form of consumerism—nature was something else that could be used to craft a personality, to develop the self. And the same was true of Bigfoot, the Green Man: Sasquatch, Atwood wrote in her poem, "can never be known: he can teach you only about yourself."[28]

For Robert Michael Pyle, an interest in Bigfoot was a way of distinguishing himself, and the things that he bought in pursuit of the beast testified to his dignity. Pyle disparaged obvious manifestations of consumerism. "The tabloidization of the world seems no different from the general spread of dross in the mass culture today," he wrote. "Examples abound: television almost *in toto*. Wal-marts [*sic*] and malls instead of vital town centers. Vocabulary's decline. Bestseller lists. Lite music, food, and beer—oxymorons all. And architecture: I recently saw a historic photograph of a magnificent hotel, the Louvre, that once stood in Astoria, Oregon. On its site now stands a McDonald's. The barbarians are not at the gates; they're well inside." He made a point of noting that he preferred microbrew ales and Chenin blanc. That he valued—and consumed—quirky music. He spent hundreds of dollars on shoes and, like so many who want to get away from it all, plenty of money on Gore-Tex and tents and backpacks. These are items valued by the middle class because they seem to counter consumerism. But they are consumer items just the same. Bigfoot—and environmentalism—promised an escape from consumerism, but only offered a different kind of consumerism.[29]

**28.** Atwood, "Oratorio for Sasquatch, Man, and Two Androids,", 20 (quotation); Samuel P. Hays, *Beauty, Health, and Permanence: Environmental Politics in the United States, 1955–1985* (Cambridge: Cambridge University Press, 1987).

**29.** Pyle, *Where Bigfoot Walks*, 35–36, 64, 81, 160 (quotation), 194, 217; Thomas Frank, *The Conquest of Cool: Business Culture, Counterculture, and the Rise of Hip Consumerism* (Chicago:

Paul Doyle's novel *Nioka, Bride of Bigfoot* recognized this conundrum: that environmentalism and counterculture values were just another species of consumerism; and Doyle tried to save his heroine from the trap into which Pyle fell. Nioka escaped into the forest, away from the clamor of modern civilization—only to find that she was transformed into a consumer product. "Soon [average citizens] were consuming her like she was the latest extra-strength deodorant, the razor with an even closer shave, the paperback with the raciest storyline yet, the tampon with the baking soda added. She was at a favorite store near you. She became part of the hipwazee." Stores sold Nioka jeans and T-shirts and buttons and peanut butter and books. Nioka was horrified by her commodification, and turned away from it: she was content with nature. But the natural world, in Doyle's description of it, resembled nothing so much as a grocery store, addressing all the same problems that consumer products solved in a mass culture. Nature provided her with food; it ended her periods; it too care of bad breath, body odor, earwax, dandruff, and acid indigestion. The sun offered sexual pleasure. Nature was a one-stop shop, and Nioka just another consumer. Bigfoot was a Green Man, but it wasn't clear whether the green was from leaves twining its body, or dollar bills.[30]

And so there's no paradox, no surprise that while Bigfoot was an incarnation "of nature, the earth, and all that is green and contrary to control"—as Pyle had it—the beast was also an advertising icon, perhaps the most enduring use the middle class found for Sasquatch. The monster hawked books and TV shows and satellite systems and TVs and ketchup and beef jerky and McDonald's and pizza and Disney movies and a host of other things. In the 1970s, Canadian Club whiskey featured Bigfoot in magazine ads; the following decade, Kokanee beer featured the beast in a series of very popular and much-beloved commercials. (Dahinden appeared in one of those, too.)[31]

Bigfoot slipped easily into advertising for the same reason that it once had insinuated itself into men's adventure magazines: the creature was pre-adapted to the demands of production. Sasquatch was instantly recognizable, yet its unbreakable association with 1970s' occultism—its irreducible weirdness—helped it to distinguish one brand from another. Sasquatch also retained an aura of authenticity, a resistance to conformity—which was a sentiment that many advertisers wanted associated with their products. In its

University of Chicago Press, 1997); Joseph Heath and Andrew Potter, *Nation of Rebels: Why Counterculture Became Consumer Culture* (New York: Collins, 2004).

**30.** Doyle, *Nioka, Bride of Bigfoot*, 27, 79, 100–1 (quotation), 108, 135–36, 139.

**31.** Pyle, *When Bigfoot Walks*, 155.

campaign, for example, Canadian Club hid a case of liquor in the forest where Bigfoot was purported to live and invited "a few brave souls" to hunt the product, proof of their indomitable spirit. Bryant, Fulton, and Shee, the advertising company in charge of Kokanee beer's commercials, loved working with Dahinden because he was the very image of genuineness. In the ads, Dahinden stood in front of his own trailer and wore his own clothes "because wardrobe could not make him look more authentic," according to one report.[32]

That Bigfoot never actually satisfied the hope for authenticity—that the beast was widely considered a hoax, the very antithesis of authenticity—did not diminish its appeal. In fact, the suspicion of fraud was an asset. Modern advertisers were aware of the many criticisms that had been leveled against consumerism, against TV, against mass culture—and they used those criticisms to mock themselves, to mock their own commercials. This mockery both insulated advertisers from further attack—after all, they were on the side of angels, making fun of mass culture—and complimented viewers for their acuity, their ability to see through advertising's gimmickry (except, of course, advertisers were still trying to sell their wares). Bigfoot allowed the advertisers to make this double-move, simultaneously associating the use of their products with authenticity and undermining all of the conventions that were used to create a sense of authenticity.[33]

In the Kokanee ad, Dahinden responded to a series of questions from an unseen interviewer about his decades-long hunt. At the end, the interviewer asked if he'd heard that Sasquatch preferred Kokanee beer. "Do you think I'm crazy or something?" Dahinden answered, as behind him Sasquatch left his trailer carrying a case of Kokanee. Of course viewers thought that Dahinden was crazy—for spending a lifetime seeking a mythical beast. Who would take his word on anything? The idea that Bigfoot enjoyed Kokanee beer was equally preposterous. Why take the word of a commercial icon, the spot seemed to ask, if it isn't even real? The commercial ridiculed the very idea of celebrity endorsement. "Essentially," said Rick Kemp, the commercial's creative director, "we wanted to send up [Dahinden's] life's work."[34]

**32.** Patrick Allossery, "Dahinden Left His Own Imprint: Ad People Marveled at His Off-The-Wall Yet Down-to-Earth Aura," *Financial Post*, April 30, 2001.
**33.** David Foster Wallace, "*E Unibus Pluram*: Television and U.S. Fiction," in *A Supposedly Fun Thing I'll Never Do Again: Essays and Arguments* (New York: Back Bay Books, 1997), 21–82; John Leland, *Hip: The History* (New York: HarperCollins, 2004).
**34.** Marnie Ko, "Footprints Leading to Nowhere: René Dahinden Spent Much of His Life Believing in Something He Could Never Find," *Report News*, May 28, 2001 (first quotation); Allossery, "Dahinden Left His Own Imprint" (second quotation).

**Bigfoot's feeding ground: We hid a case of Canadian Club here, then ran like crazy.**

**Here's how you can find the C.C.**

*This scene taken from a priceless 16mm movie film (1967) shows a female Bigfoot on the banks of Bluff Creek in California.*

For more than 165 years tales of Bigfoot, a massive 8-foot-tall, 500-pound humanoid, have haunted the natives and visitors in the rugged Pacific Northwest. Considerable evidence indicates that Bigfoot is in fact now stalking the dark forests and lurking in the dank ravines along the Cascade Mountain Range.

**Keep your ears open! Watch your back!**

Our burial party moved with caution once we learned that thousands of respected people from around the world believed in Bigfoot's existence. So with a 48-pound case of Canadian Club strapped firmly to one man's back, we deployed five other men to cover his front, rear and flank.

Each man was carefully trained to spot the incredible 17½ inch footprints so common in the area. Each man had studied firsthand accounts of the creature's behavior, including most of the confirmed Bigfoot sightings. And yes, at times each man struggled with a special kind of fear.

On November 5, 1976, after hours spent threading our way through this primeval forest, we found Bigfoot's feeding ground, buried the case of C.C. and quickly returned to civilization. But for the rest of our lives we will never know what some of us suspected all along...that Bigfoot himself was watching our every move.

**Directions for a few brave souls.**

The buried C.C. is located almost the same number of miles south of Canada's Good Hope Mountain (elev. 10,617) as it is north of Bluff Creek in northwestern California. You'll know you're on the right track when you stumble on a temporarily dormant volcano. Now proceed somewhere between 6 and 9

*This plaster cast of a Bigfoot print measures 17½ inches in length, 7 inches in width.*

FIGURE 35. Whatever one makes of the claims by the Wallace family—whether one thinks Bigfoot is dead or alive—there can be no doubt that the creature exists. Sasquatch has left indelible tracks all throughout American culture—and will continue to do so. Here, the beast is featured in a print ad for Canadian Club whiskey. The monster's simultaneous ferociousness and silliness made it attractive to advertisers. (Courtesy of Canadian Club.)

The advertising agency Carmichael Lynch sought to evoke a similar feeling in ads for Jack Link's beef jerky. An award-winning series of commercials under the collective title "Messin' with Sasquatch" had hikers playing juvenile tricks on Sasquatch, loosening the lid on his saltshaker, lining the eyepieces of binoculars with black ink, slipping the hand of a sleeping Sasquatch into a pot of warm water, and filling the palm of a once-again sleeping Sasquatch with shaving cream and tickling his face. At the conclusion of each spot, Bigfoot attacked the pranksters. On one level, the commercials made a standard pitch, suggesting that eating jerky was supposed to be a fun but also dangerous endeavor undertaken only by the bold. On another level, however, the series ridiculed the very notion of selling. Much of the advertisement's humor derived from the viewer's knowledge that there was no such thing as a Sasquatch, that the commercial's pitch—the association between jerky and danger—was itself silly, another commercial convention to be exposed. All of these ads, those for Kokanee, those for Jack Link's, flattered viewers for seeing through advertising's conventions, for being in the know—for seeing that the authenticity of Dahinden and Sasquatch were stage managed; for seeing that the commercials aimed to manipulate. Invited to be in on the joke, the viewers were also invited to buy Kokanee beer or Jack Link's beef jerky as proof of their brilliance, authenticity, and distinction.[35]

What *does* seem surprising and paradoxical about Bigfoot's career as a commercial icon is that, while advertising made the creature increasingly vaporous—distant and insubstantial, ethereal and ghostly—commercials also made Sasquatch *more* real. Anthropologist Elizabeth Bird wrote, "Relationships with media figures can be time-consuming and even more 'real' than many daily interactions in the 'real' world." That is to say, commercial objects—as Barnum well knew, as the producers of men's adventure magazines and four-wall flicks like *The Legend of Boggy Creek* learned—can be real—even more real—than scientific objects. It wasn't only Tibetans with their snow leopards and snow dragons, with their bears and Abominable Snowmen who promiscuously mixed the categories of fanciful and the factual. North Americans did, too. Advertising made Bigfoot into a constant presence, part of the cultural firmament, a real and true thing to the inhabitants of North America at the end of the twentieth century, part of their lives, something that they thought they knew, understood. In 1989, a Canadian woman and her grandson were driving along one afternoon when they saw what they thought was a Sasquatch. "It was kind of like a Kokanee commercial," the woman said. "My grandson even joked, 'We should roll down the window and see if he wants a beer.'"[36]

Bigfoot was dead! Long live Bigfoot!

**35.** Wallace, "*E Unibus Pluram*,": 21–82; David Phelps, "Monster Brand: Make-Your-Own Sasquatch Videos and YouTube Have Helped Make Jack Link's the Bestselling Beef Jerky," *Minneapolis Star Tribune*, February 9, 2008.

**36.** S. Elizabeth Bird, *For Enquiring Minds: A Cultural History of Supermarket Tabloids* (Knoxville: University of Tennessee Press, 1992), 3 (quotation); Robert Goldman and Stephen Papson, *Sign Wars: The Cluttered Landscape of Advertising* (New York: The Guilford Press, 1996); Shafer Parker, "Bigfoot: Man, Monkey, or Myth?" *British Columbia Report*, June 23,

## Not *The* End, but *An* End

Probably, there will always be a few iconoclasts who take up Bigfoot's cause, despite Bigfoot's career in advertising, the accumulation of hoaxes, and the criticism of skeptics—indeed, as happened in the nineteenth century, it may be that the antagonism of skeptics encourages some people to accept Sasquatch's existence. The world, as Ivan Sanderson often said, is a big place, full of unusual, ambiguous things—and sometimes those things even point to the existence of wildmen. Sometimes, they make people believe in monsters. In 2001, for example, geneticist Brian Sykes tried to sequence DNA from purported Abominable Snowman fur and, to his surprise, failed (although, to be fair, his admission came only one day after April Fools'): "We normally wouldn't have any trouble at all," Sykes said. "It has all the hallmarks of good material. It's not a human, it's not a bear, nor anything else that we've so far been able to identify. We've never encountered any DNA that we couldn't sequence before. . . . I didn't think this would end in a mystery." There remains plenty of room on this big, blue globe for odd and unusual views.[37]

Thus, the admissions of Wallace's family did little to change the minds of most committed Bigfooters. The beast wasn't dead, they said. Wallace, they noted, had been telling outrageous tales for decades. Why believe him now? The enthusiasts raged against the press for garbling the story. Wallace, they said, had nothing to do with Patterson's movie and could not be responsible for all of the tracks. It was the mass media that was being conned, Bigfooters insisted, and there was some truth to the charge. A New Mexico newspaper, for instance, mocked the Bigfoot enthusiasts as gullible—even while credulously reporting that Wallace had, somehow, been involved with Patterson's movie.[38]

Interest in the creature may even revive at some point: Wallace's death may mark only a hiatus, just as interest in wildmen tailed off in the years after Hillary's hunt then resumed after Patterson's movie. Indeed, it's not impossible that an actual wildman may someday be caught. Or, it may be that

1997, 32–34 (quotation, 33); George Ritzer, *Enchanting a Disenchanted World: Revolutionizing the Means of Consumption* (London: Sage, 2005).

**37.** "'Yeti's hair' defies DNA analysis," *Times* (London), April 2, 2001 (quotation); Richard Conniff and Harry Marshall, "In the Realm of Virtual Reality," in *The Best American Science and Nature Writing 2002*, ed. Natalie Angier (New York: Houghton Mifflin, 2002), 24–33.

**38.** "Bigfoot Is Dead; Gullibility Lives on," *New Mexican* (Santa Fe, NM), December 7, 2002; Coleman, *Bigfoot!* 79–80; Egan, "Bigfoot Fans Cling to Their Faith—Hoax or Not"; Green, *The Best of Sasquatch Bigfoot*, 12–16.

in a few years Bigfoot—in all of its manifestations—is succeeded by some other wildman, existing on some other frontier: even as interest in Bigfoot continued, for example, myths about wildmen in cyberspace took shape, a new kind of geek for a new kind of world. And if humans do ever escape Earth for outer space, settle on some other planet, or at least explore extensively, there will doubtless be reports, earnest and heartfelt, of creatures hiding in the shadows. We have already invented these wildmen, already imagined them into being, and they just need a place to roam. Call them aliens. Call them wildmen. Call them Bigfoot. And the dust will record their tracks, evidence of their passing.[39]

But while the turn of the millennium does not mark *the* end of Bigfoot, it certainly marks *an* end. Bigfoot was born of the mass media, spread on the mass media, and its vitality came from the fear of mass media and consumerism. The creature existed on the frontier edge between mass and popular culture, between a society that was commercial and one that was consumer, between a culture of character and one of personality. It thrived best in working-class entertainments where that fear was most palpable. By the end of the twentieth century, however, that fear drained away, as mass culture and consumerism were accepted, even celebrated. Wallace's death came more than eighty years after newspapers introduced the Abominable Snowman to the world, almost fifty since the debut of Sasquatch and Bigfoot on the international stage. A generation had passed. Vladimir Markotic died in 1994. Titmus died in 1997. Dahinden died in 2001. Heuvelmans died in 2001. Krantz died in 2002. Freeman died in 2003. Beckjord died in 2008.

Bigfoot hunting in the years after 2000 became something of a different enterprise than it had been before, in large part because of the way middle-class interpretations and sentiment had changed the beast. Hunting Bigfoot was now often unapologetically commercial—the Bigfoot Field Researchers Organization charged participants to join their expeditions, for example; Tom Biscardi sold subscriptions to his live webcast from Happy Camp, California, which was supposed to show the capture of a Sasquatch; and most Sasquatch organizations peddled T-shirts or knick-knacks with their names emblazoned on them. As well, Bigfoot's Green Man persona softened the wildman some, altering what it meant to go after the beast. Less and less did the hunt seem to be man confronting the monstrous; more and more did it seem a case of humankind making contact with another sentient being. Sto-

**39.** Erik Davis, *Techgnosis: Myth, Magic + Mysticism in the Age of Information* (New York: Three Rivers Press, 1998).

ries abounded of the connection between sensitive women and Sasquatches: Autumn Williams, for example, claimed that her family had been neighborly with a Bigfoot tribe for years; Janice Coy wrote a book claiming that her family's interactions with wildmen had gone on for generations. This softening of Bigfoot's image allowed for paranormal theories about the creature to persist—Sasquatch was not a terror, but a brother—and also made it increasingly acceptable for women to become involved with Bigfoot hunts. Among others, there were Bobbie Short and Kathy Strain and Autumn Williams and Janice Carter Coy and Lunetta Woods. Women in America are not supposed to be Great White hunters, matching their skills against the savagery of nature, but they *are* expected to be excellent communicators, and that's what was needed in this new hunt: someone to reach out to the primates, to draw them in—just as George Haas had advocated so many years before. Jane Goodall had done so with chimpanzees, Dian Fossey with gorillas, and Birute Galdikas with orangutans. Bigfooters at the beginning of the century were on the look out for the woman who could do the same with Sasquatch.[40]

## Curse of the Sasquatch

Skeptics often disparaged Bigfooters for choosing the easy over the real, the thrilling over the true. Way back in the late 1960s, Daniel Cohen wrote, "It is genuinely exciting to believe in ghosts or flying saucers or the Abominable Snowman or the Lost Continent of Atlantis. Real science is nowhere near so thrilling, no matter how well it is presented. A rigorous logical approach to evidence is hard and restrictive; it destroys the beloved romantic myths and is going to be resented. It is a terrible day for a child when he discovers that Santa Claus does not exist, and adults are not much different." But this was wrong.[41]

Love is never easy, and loving a thing as ambiguous, as unclassifiable, as derided as Bigfoot was especially difficult. To love Bigfoot was to suffer what enthusiasts called the "curse of Sasquatch." Like so many ogres, like any geek, Bigfoot was known by its appetite: its love could be all-consuming, devouring those things an enthusiast once held dear. Families, relationships, careers, all were sacrificed for the beast's love. Certainly the hunt brought its joys, its moments of ecstasy. Over the years, for example, Dahinden won the respect

**40.** Donna J. Haraway, *Primate Visions: Gender, Race, and Nature in the World of Modern Science* (New York: Routledge, 1989); Dan Reed, "Bigfoot Deal Gets Hairy, Legally; Expert Sues Locals over Pay, Collection," *San Jose (CA) Mercury News*, July 19, 2006; Buhs, "Camping with Bigfoot."

**41.** Daniel Cohen, *Myths of the Space Age* (New York: Dodd, Mead, 1967), 5.

of many skeptics for his hardheaded investigations and empiricism. Others also won a sense of freedom, a feeling of being a true man, unbeholdened to the corrupt ways of the modern world, following their own inner lodestar. But much of what followed that first rush of infatuation, Daniel Perez said, was drudgery—a job, a daily grind. "If someone came along and dropped a dead Sasquatch in front of me," Dahinden once quipped, "I'd just stuff my pipe and smoke it." Krantz admitted that he could not imagine a happy resolution to his search: if he ever found Sasquatch, whether he killed the creature or not, he would be haunted—for killing a creature that he had come to respect so much or for letting the object of a lifetime's search get away.[42]

The only certainties in the hunt for Sasquatch were the dilemma, failure, death, the creature's lovers left to pass unfulfilled—their dreams denied. In 1999, Green admitted, "I will almost certainly die without [the Bigfoot mystery] being solved, as has happened to so many of my friends." For all of their hard work, none of the Bigfooters who died around the turn of the twenty-first century had the satisfaction of seeing Bigfoot's existence proved definitively. Markotic, Krantz, Dahinden, and Green never even had the satisfaction of seeing the beast. In fact, Dahinden never once found tracks on his own.[43]

Bigfoot enthusiasts were losers in the contest for dignity. For various reasons, psychological, biographical, historical, and sociological, these mostly working-class men saw in Bigfoot a way to strike against the system that humiliated them, that kept them down. But they were doomed to failure: Bigfoot did not exist. (Another humiliation: the skeptics were right.) It was a tragic situation, an impossible set of choices for those who sought the beast: sacrifice character, skill, identity, or spend a lifetime seeking a creature that could not be found. Believing in Bigfoot was anything but easy, the decision not made lightly, or without fear. "I have my doubts all the time about what I'm doing," Dahinden said. "I've always had them. It's a lonely place to be, on one side of the fence with the rest of the world on the other side. But it's where I have to stay."[44]

**42.** Darlene Bryant, "Bigfoot Believer Tells His Story," *Record-Enterprise* (Chico, CA), April 11, 1980; Paul McHugh, "Believers Still Pursuing the Legend of Bigfoot," *San Francisco Chronicle*, December 8, 1994; Mark McDermott, "Phantom of the Woods, Phantom of the Psyche," *Seattle Times*, July 7, 1996; Ross Crockford, "Looking for Mr. Big," *Outdoor Canada*, March 1998, 28–34; Al Ridenour, "I Married a Monster," *Los Angeles New Times*, September 14, 2000; "Sasquatch Hunter for 45 Years Never Found a Sign of One," *Victoria (BC) Times Colonist*, April 29, 2001 (quotation); Coleman, *Bigfoot!* 169–70.

**43.** Michael Taylor, "Screams in the Night," *San Francisco Chronicle*, January 24, 1999.

**44.** John Colebourn, "Bigfoot Loses Big Fan," *Vancouver Province*, April 20, 2001.

# *Bibliography*

## Archival Collections


Abominable Snowman Clippings File, *Daily Mail*, London, England.

American Religions Collection, University of California, Santa Barbara.

American Yeti Expedition, National Wildlife Federation Archives, U.S. Fish and Wildlife Service, National Conservation Training Center, Shepherdstown, West Virginia.

Anthropology Department Records, CU-23, Bancroft Library, University of California, Berkeley.

Bigfoot file, Columbia Gorge Interpretative Center, Stevenson, Washington.

Carleton Coon papers, National Anthropological Archives, Smithsonian Institute, Suitland, Maryland.

"Correspondence and Memos on the Centennial Mascot, Harrison Bigfoot," 1981–1990, AR154_1_15_10, box 13, Washington State Archives, Olympia.

Director, National Museum of Natural History, 1948–1970, RU 155, Smithsonian Institution Archives, Washington, D.C.

Director, National Zoological Park, ca. 1920–1984, RU 380, Smithsonian Institution Archives, Washington, D.C.

Andrew Genzoli papers, Humboldt State University, Arcata, California (n.b., the library also holds complete runs of *Bigfoot Bulletin* and *Bigfoot News*).

Grover Krantz papers, National Anthropological Archives, Smithsonian Institute, Suitland, Maryland.

Vladimir Markotic papers, University of Calgary archives, Calgary, Alberta.

Judith Merril fonds, R2929-0-4E, Library and Archives of Canada, Ottawa, Ontario.

Office of the Secretary (S. Dillon Ripley), 1964–1971 papers, RU99, Smithsonian Institution, Smithsonian Institution Archives, Washington, D.C.

Marian T. Place papers, Arizona State University, Tempe.

Ivan T. Sanderson papers, B Sa3, American Philosophical Society, Philadelphia, Pennsylvania.

Lawrence Swan file, J. Paul Leonard Library, San Francisco State University, San Francisco, California.

H. W. Tilman papers, HWT/27/1/1–22, Royal Geographical Society, London, England.

Sasquatch clippings file, Columbia Gorge Interpretive Center Museum, Stevenson, Washington.

Sasquatch Collection, University of British Columbia, Vancouver, Canada.
Sasquatch file, Oregon Historical Society Research Library, Portland.
Sasquatch material, M-3 Archeology file, Royal British Columbia Museum Archives, Vancouver, Canada.
Sasquatch vertical file, Idaho State Historical Society, Boise.
Sierra Club Member Papers, MSS 71/295e, Bancroft Library, University of California, Berkeley.
*World Book Encyclopedia* scientific expedition to the Himalaya papers, World Book, Inc., Chicago, Illinois.

## Select Bibliography

Abrahamson, David. *Magazine-Made America: The Cultural Transformation of the Postwar Periodical*. Cresskill, NJ: Hampton Press, 1996.
Agogino, George. "An Overview of the Yeti-Sasquatch Investigations and Some Thoughts on Their Outcome." *Anthropological Journal of Canada* 16 (1978): 11–13.
Arment, Chad. *The Historical Bigfoot*. Landisville, PA: Coachwhip Publications, 2006.
Ashley, Mike. *Transformations: The Story of the Science Fiction Magazines from 1950 to 1970*. Liverpool: Liverpool University Press, 2005.
Atwood, Margaret. "Canadian Monsters: Some Aspects of the Supernatural in Canadian Fiction." In *The Canadian Imagination: Dimensions of a Literary Culture*, edited by David Staines, 97–122. Cambridge, MA: Harvard University Press, 1977.
———. "Oratorio for Sasquatch, Man, and Two Androids." In *Poems for Voices*, by Al Purdy, Margaret Atwood, John Newlove, Phyllis Gotlieb, Tom Marshall, and Alden Nowlan, 14–28. Toronto: Canadian Broadcasting Company, 1970.
Bain, Donald. *Long John Nebel: Radio Talk King, Master Salesman, and Magnificent Charlatan*. New York: Macmillan, 1974.
Baird, Robert. "Going Indian: Discovery, Adoption, and Renaming toward a 'True American' from *Deerslayer* to *Dances with Wolves*." In *Dressing in Feathers: The Construction of the Indian in American Popular Culture*, edited by S. Elizabeth Bird, 195–210. Boulder, CO: Westview Press, 1996.
Baritz, Loren. *The Good Life: The Meaning of Success for the American Middle Class*. New York: Perennial Library, 1990.
Bartra, Roger. *Wild Men in the Looking Glass: The Mythic Origins of European Otherness*. Ann Arbor: University of Michigan Press, 1994.
Basford, Kathleen. *The Green Man*. Cambridge: D. S. Brewer, 2004.
Bayanov, Dmitri. *Bigfoot: To Kill or to Film? The Problem of Proof*. Burnaby, BC: Pyramid Publications, 2001.
———. *In the Footsteps of the Russian Snowman*. Surrey, BC: Hancock House, 1996.
Beck, Fred, and R. A. Beck. *I Fought the Apeman of Mt. St. Helens*. n.p.: privately printed, 1967.
Bendix, Regina. *In Search of Authenticity: The Formation of Folklore Studies*. Madison: University of Wisconsin Press, 1997.
Bernheimer, Richard. *Wild Men in the Middle Ages: A Study in Art, Sentiment, and Demonology*. New York: Octagon, 1970.

Beyer, Margaret W. *The Art People Love: Stories of Richard S. Beyer's Life and His Sculpture*. Pullman: Washington State University Press, 1999.

Bindernagel, John A. *North America's Great Ape, the Sasquatch: A Wildlife Biologist Looks at the Continent's Most Misunderstood Large Mammal*. Courtenay, BC: Beachcomber Books, 1998.

Bird, S. Elizabeth. *For Enquiring Minds: A Cultural History of Supermarket Tabloids*. Knoxville: University of Tennessee Press, 1992.

Boas, Franz. *Kwakiutl Tales*. New York: AMS Press, 1969. Reprint of *Columbia University Contributions to Anthropology*. Vol. 2, *Kwakiutl Tales*, 1910.

Bondeson, Jan. *The Two-Headed Boy, and Other Medical Marvels*. Ithaca, NY: Cornell University Press, 2000.

Bourne, Geoffrey H., and Maury Cohen. *The Gentle Giants: The Gorilla Story*. New York: G. P. Putnam's Sons, 1975.

Bowler, Peter J. *Evolution: The History of an Idea*. Berkeley: University of California Press, 1989.

Bowman, Matthew. "A Mormon Bigfoot: David Patten's Cain and the Concept of Evil in LDS Folklore." *Journal of Mormon History* 33, no. 3 (2007): 62–82.

Bradford, Phillips Verner, and Harvey Blume. *Ota Benga: The Pygmy in the Zoo*. New York: St. Martin's, 1992.

Brunvand, Jan Harold. *The Vanishing Hitchhiker: American Urban Legends and Their Meanings*. New York: W. W. Norton, 1989.

Byrne, Peter. "Being Some Notes, in Brief, on the General Findings in Connection with the California Bigfoot." *Genus* 18 (1962): 55–59.

———. *The Search for Big Foot: Monster, Myth or Man?* New York: Pocket Books, 1976.

Cahill, Tim. *A Wolverine Is Eating My Leg*. New York: Vintage Departures, 1990.

Carranco, Lynwood. "Andrew Genzoli—Journalist, Historian." *Humboldt Historian* 24, no. 3 (1976): 24–25.

———. *Genocide and Vendetta: The Round Valley Wars in Northern California*. Norman: University of Oklahoma Press, 1981.

———. "It Wasn't Bigfoot After All." *Western Folklore* 23, no. 4 (1964): 271–72.

———. "Three Legends of Northwestern California." *Western Folklore* 22, no. 3 (1963): 179–85.

Cherrington, John A. *The Fraser River Valley: A History*. Madeira Park, BC: Harbour Publishing, 1992.

Ciochon, Russell, John Olsen, and Jamie James. *Other Origins: The Search for the Giant Ape in Human Prehistory*. New York: Bantam Books, 1990.

Clark, Jerome, and Loren Coleman. *Creatures of the Outer Edge*. New York: Warner, 1978.

Clarke, Arthur C. *Astounding Days: A Science Fictional Autobiography*. New York: Bantam Books, 1990.

Clover, Carol J. "Her Body, Himself: Gender in the Slasher Film." *Representations* 20 (1987): 187–228.

Cohen, Daniel. *Bigfoot: America's Number One Monster*. New York: Pocket Books, 1982.

———. *Monster Hunting Today*. New York: Dodd, Mead, 1983.

———. *Myths of the Space Age*. New York: Dodd, Mead, 1967.

Cohen, Lizabeth. *A Consumers' Republic: The Politics of Mass Consumption in Postwar America.* New York: Vintage Books, 2003.

———. "Encountering Mass Culture at the Grassroots: The Experience of Chicago Workers in the 1920s." *American Quarterly* 41, no. 1 (1989): 6–33.

Coleman, Loren. *Bigfoot! The True Story of Apes in America.* New York: Pocket Books, 2003.

———. *Tom Slick and the Search for Yeti.* Boston: Faber & Faber, 1989. Reprinted as *Tom Slick: True Life Encounters in Cryptozoology.* Fresno, CA: Craven Street Books, 2002.

———. "Was the First 'Bigfoot' a Hoax? Cryptozoology's Original Sin." *The Anomalist* 2 (1995): 8–27.

Conniff, Richard, and Harry Marshall. "In the Realm of Virtual Reality." In *The Best American Science and Nature Writing 2002*, edited by Natalie Angier, 24–33. New York: Houghton Mifflin, 2002.

Cook, James W. *The Arts of Deception: Playing with Fraud in the Age of Barnum.* Cambridge, MA: Harvard University Press, 2001.

Coon, Carleton. *Adventures and Discoveries: The Autobiography of Carleton S. Coon.* Englewood Cliffs, NJ: Prentice-Hall, 1981.

Crabtree, Smokey. *Smokey and the Fouke Monster.* Fouke, AR: Days Creek Production, 1974.

Cushman, Philip. *Constructing the Self, Constructing America: A Cultural History of Psychotherapy.* Reading, MA: Addison-Wesley, 1995.

Daegling, David J. *Bigfoot Exposed: An Anthropologist Examines America's Enduring Legend.* Walnut Creek, CA: Altamira Press, 2004.

Daston, Lorraine, and Katherine Park. *Wonders and the Order of Nature, 1150–1750.* New York: Zone Books, 2001.

Davis, Erik. *Techgnosis: Myth, Magic + Mysticism in the Age of Information.* New York: Three Rivers Press, 1998.

Deagh, Linda. *American Folklore and the Mass Media.* Bloomington: Indiana University Press, 1994.

Dennet, Andrea Stulman. *Weird and Wonderful: The Dime Museum in America.* New York: New York University Press, 1997.

Denning, Michael. *Mechanic Accents: Dime Novels and Working-Class Culture in America.* New York: Verso, 1998.

Derks, Scott. *Working Americans, 1880–1999.* Vol. 1, *The Working Class.* Lakeville, CT: Grey House Publishing, 2000.

Disch, Thomas M. *The Dreams Our Stuff Is Made of: How Science Fiction Conquered the World.* New York: Touchstone, 2000.

Donovan, Roberta, and Keith Wolverton. *Mystery Stalks the Prairie.* Raynesford, MT: THAR, 1976.

Dorfman, Ariel. *The Empire's Old Clothes: What the Lone Ranger, Babar, and Other Innocent Heroes Do to our Minds.* New York: Pantheon Books, 1983.

Dorson, Richard M. *American Folklore.* Chicago: University of Chicago Press, 1959.

———. "Folklore and Fakelore." *American Mercury*, March 1950, 335–43.

———. *Man and Beast in American Comic Legend.* Bloomington: Indiana University Press, 1982.

Douthwaite, Julia V. *The Wild Girl, Natural Man, and the Monster: Dangerous Experiments in the Age of Enlightenment*. Chicago: University of Chicago Press, 2002.
Doyle, Paul. *Nioka, Bride of Bigfoot: A Novel*. Seattle, WA: Daily Planet Press, 1992.
Edwards, Frank. *Strange World*. New York: Lyle Stuart, 1964.
Ehrenreich, Barbara. *Fear of Falling: The Inner Life of the Middle Class*. New York: Perennial, 1990.
———. *The Hearts of Men: American Dreams and the Flight from Commitment*. New York: Anchor Books, 1983.
Ellenberger, Henri F. *The Discovery of the Unconscious*. New York: Basic Books, 1981.
Estés, Clarissa Pinkola. *Women Who Run with the Wolves*. New York: Ballantine, 1994.
Fiedler, Leslie. *Freaks: Myths and Images of the Secret Self*. New York: Anchor, 1993.
Fort, Charles. *The Books of Charles Fort*. New York: Henry Holt, 1941.
Frank, Thomas. *The Conquest of Cool: Business Culture, Counterculture, and the Rise of Hip Consumerism*. Chicago: University of Chicago Press, 1997.
Fraser, Matthew. *Weapons of Mass Distraction: Soft Power and American Empire*. New York: Thomas Dunne Books, 2003.
Freeland, Nat. *The Occult Explosion*. New York: G. P. Putnam's Sons, 1972.
Freeman, Joshua B. "Hardhats: Construction Workers, Manliness, and the 1970 Pro-War Demonstrations." *Journal of Social History* 26, no. 4 (1993): 725–45.
Freud, Sigmund. *Totem and Taboo: Some Points of Agreement Between the Mental Lives of Savages and Neurotics*. New York: W. W. Norton, 1962.
Fussell, Paul. *Class*. New York: Ballantine, 1983.
Galbreath, Robert. "Explaining Modern Occultism." In *The Occult in America: New Historical Perspectives*, edited by Howard Kerr and Charles L. Crow, 11–37. Urbana: University of Illinois Press,, 1983.
Gardner, Martin. *Science: Good, Bad, and Bogus*. Amherst, NY: Prometheus, 1981.
Gieryn, Thomas F. *Cultural Boundaries of Science: Credibility on the Line*. Chicago: University of Chicago Press, 1999.
Ginzburg, Carlo. "Morelli, Freud and Sherlock Holmes: Clues and Scientific Method." *History Workshop*, no. 9–10 (1980): 5–36.
Gloss, Molly. *Wild Life*. New York: Mariner Books, 2001.
Glynn, Kevin. *Tabloid Culture: Trash Taste, Popular Power, and the Transformation of American Television*. Durham, NC: Duke University Press, 2000.
Goad, Jim. *The Redneck Manifesto: How Hillbillies, Hicks, and White Trash Became America's Scapegoats*. New York: Touchstone, 1997.
Goldman, Robert, and Stephen Papson. *Sign Wars: The Cluttered Landscape of Advertising*. New York: The Guilford Press, 1996.
Green, John. *Bigfoot: On the Track of the Sasquatch*. New York: Ballantine, 1973.
———. "The Case for a Legal Inquiry into Sasquatch Evidence." *Cryptozoology* 8 (1989): 37–42.
———. *On the Track of Sasquatch*. Agassiz, BC: Cheam, 1968.
———. *On the Track of the Sasquatch*. Surrey, BC: Hancock House, 1980.
———. *Sasquatch: The Apes among Us*. Seattle, WA: Hancock House, 1978.
———. *The Sasquatch File*. Agassiz, BC: Cheam, 1973.

———. *Year of the Sasquatch*. Agassiz, BC: Cheam, 1970.

———. *The Best of Sasquatch Bigfoot*. Surrey, BC: Hancock House, 2004.

Greenwell, J. Richard, and James E. King. "Scientists and Anomalous Phenomena. Preliminary Results of a Survey." *Zetetic Scholar*, no. 6 (1980): 17–25.

———. "Attitudes of Physical Anthropologists toward Reports of Bigfoot and Nessie." *Current Anthropology* 22, no. 1 (1981): 79–80.

Grumley, Michael. *There Are Giants in the Earth*. Garden City, NJ: Doubleday, 1974.

Hall, Mark A. "The Real Bigfoot and Genuine Bigfoot Tracks." *Wonders*, 7 (2002): 99–125.

Halle, David. *America's Working Man: Work, Home, and Politics among Blue-Collar Property Owners*. Chicago: University of Chicago Press, 1984.

Halpin, Marjorie, and Michael M. Ames, eds. *Manlike Monsters on Trial*. Vancouver: University of British Columbia Press, 1980.

Halttunen, Karen. *Confidence Men and Painted Women: A Study of Middle-Class Culture in America, 1830–1870*. New Haven, CT: Yale University Press, 1986.

Hansen, George P. "CSICOP and the Skeptics: An Overview." *The Journal of the American Society for Psychical Research* 86 (1992): 19–63.

Haraway, Donna J. *Primate Visions: Gender, Race, and Nature in the World of Modern Science*. New York: Routledge, 1989.

Harris, Neil. *Humbug: The Art of P.T. Barnum*. Boston: Little Brown, 1973.

Hays, Samuel P. *Beauty, Health, and Permanence: Environmental Politics in the United States, 1955–1985*. Cambridge: Cambridge University Press, 1987.

Heath, Joseph, and Andrew Potter. *Nation of Rebels: Why Counterculture Became Consumer Culture*. New York: Collins, 2004.

Herron, Don. *Echoes from the Vaults of Yoh-Vombis: A Compendium of the Life of George F. Haas*. St. Paul, MN: privately printed, 1976.

Hess, David J. *Science in the New Age: The Paranormal, Its Defenders and Debunkers, and American Culture*. Madison: University of Wisconsin Press, 1993.

Heuvelmans, Bernard. "The Birth and Early History of Cryptozoology." *Cryptozoology* 3 (1984): 1–30.

———. "The Metamorphosis of Unknown Animals into Fabulous Beasts and of Fabulous Beasts into Known Animals." *Cryptozoology* 9 (1990): 1–12.

———. "Note Preliminaire sur un Specimen Conserve dans la glace d'une forme encore inconnu d'Hominide Vivian Homo Pongoides (Sp. Seu Subsp. Nov.)." *Institute Royal des Sciences Naturelles de Belgique Bulletin* 45, no. 4 (1969): 1–24.

———. *On the Track of Unknown Animals*. Translated by Richard Garnett. New York: Hill & Wang, 1958.

———. "What Is Cryptozoology?" *Cryptozoology* 1 (1982): 1–12.

Heuvelmans, Bernard, and Boris F. Porshnev. *L'Homme de Neanderthal Est Tojours Vivant*. Paris: Plon, 1974.

Hevly, Bruce. "The Heroic Science of Glacier Motion." *Osiris* 11 (1996): 66–86.

Hillary, Edmund. *Nothing Venture, Nothing Win*. New York: Coronet, 1977.

Hillary, Edmund, and Desmond Doig. *High in the Thin Cold Air*. Garden City, NJ: Doubleday, 1962.

Hoffman, Daniel. *Paul Bunyan: Last of the Frontier Demigods*. Lincoln: University of Nebraska Press, 1983.

Hopkirk, Peter. *The Great Game: The Struggle for Empire in Central Asia*. New York: Kodansha America, 1994.

Howard-Bury, C. K. *Mount Everest: The Reconnaissance, 1921*. London: Edward Arnold, 1922.

Huhndorf, Shari M. *Going Native: Indians in the American Cultural Imagination*. Ithaca, NY: Cornell University Press, 2001.

Hunter, Don, and René Dahinden. *Sasquatch*. Toronto: McClelland & Stewart, 1973.

Huxley, Thomas. *Science and Culture*. London, 1881.

Izzard, Ralph. *The Abominable Snowman*. New York: Doubleday & Company, 1955.

———. *The Hunt for the Buru*. London: Hodder & Stoughton, 1951.

Johnson, Stan. *Bigfoot Memoirs: My Life with Sasquatch*. Newberg, OR: Wild Flower Press, 1995.

Johnson, Stan, and Joshua Shapiro. *The True Story of Bigfoot*. Pinole, CA: J & S Aquarian Networking, 1987.

Judd, Denis. *Empire: The British Imperial Experience, from 1765 to the Present*. New York: Basic Books, 1998.

Kammen, Michael. *American Culture, American Tastes: Social Change and the Twentieth Century*. New York: Basic Books, 1999.

Kimmel, Michael. *Manhood in America: A Cultural History*. New York: The Free Press, 1996.

Kirtley, Bacil F. “Unknown Hominids and New World Legends.” *Western Folklore* 23 (1964): 77–90.

Knerr, Michael E. *Sasquatch: Monster of the Northwest Woods*. New York: Belmont Tower Books, 1977.

Krakauer, Jon. *Into Thin Air: A Personal Account of the Mt. Everest Disaster*. New York: Anchor, 1999.

Krantz, Grover S. “Additional Notes on Sasquatch Foot Anatomy.” *Northwest Anthropological Research Notes* 6 (1972): 230–41.

———. “Anatomy and Dermatoglyphics of Three Sasquatch Prints.” *Cryptozoology* 2 (1983): 53–81.

———. “Anatomy of the Sasquatch Foot.” *Northwest Anthropological Research Notes* 6 (1972): 91–104.

———. *Big Footprints: A Scientific Inquiry into the Reality of Sasquatch*. Boulder, CO: Johnson Books, 1992.

———. *Bigfoot/Sasquatch Evidence*. 2nd ed. Surrey, BC: Hancock House, 1999.

———. *Only a Dog*. Wheat Ridge, CO: Hofflin Publishing, 1998.

———. “Sasquatch Handprints.” *Northwest Anthropological Research Notes* 5 (1971): 145–51.

———. “A Species Named from Footprints.” *Northwest Anthropological Research Notes* 19, no. 1 (1986): 24–39.

Kuklick, Henrika. *The Savage Within: The Social History of British Anthropology, 1885–1945*. Cambridge: Cambridge University Press, 1993.

Kurtz, Paul. *Skeptical Odysseys: Personal Accounts by the World’s Leading Paranormal Inquirers*. Amherst, NY: Prometheus, 2001.

Lapseritis, Jack. *The Psychic Sasquatch and Their UFO Connection*. Mill Spring, NC: Wild Flower Press, 1998.

Lears, Jackson. *Fables of Abundance: A Cultural History of Advertising in America*. New York: Basic Books, 1995.

Leland, John. *Hip: The History*. New York: HarperCollins, 2004.

LeMasters, E. E. *Blue-Collar Aristocrats: Life-Styles at a Working-Class Tavern*. Madison: University of Wisconsin Press, 1975.

Lester, Shane. *Clan of Cain: The Genesis of Bigfoot*. n.p.: privately printed, 2001.

Lévi-Strauss, Claude. *The Way of the Masks*. Vancouver: Douglas & McIntyre Ltd., 1982.

Lhamon, W. T. Jr. *Raising Cain: Blackface Performance from Jim Crow to Hip Hop*. Cambridge, MA: Harvard University Press, 1998.

Liebenberg, Louis. *The Art of Tracking: The Origin of Science*. Cape Town: New Africa Books, 1995.

Long, Greg. *The Making of Bigfoot: The Inside Story*. Amherst, NY: Prometheus, 2004.

Lopez, Donald S. Jr. *Prisoners of Shangri-La: Tibetan Buddhism and the West*. Chicago: University of Chicago Press, 1999.

Lott, Eric. *Love & Theft: Blackface Minstrelsy and the American Working Class*. New York: Oxford University Press, 1993.

MacDougall, Curtis D. *Hoaxes*. New York: Dover, 1958.

———. *Superstition and the Press*. New York: Prometheus, 1983.

Mackal, Roy P. *A Living Dinosaur: In Search of Mokele-Mbembe*. Leiden: Brill Academic, 1987.

———. *The Monsters of Loch Ness*. Chicago: The Swallow Press, 1976.

———. *Searching for Hidden Animals*. Garden City, NJ: Doubleday, 1980.

Marchand, Roland. *Advertising the American Dream: Making Way for Modernity, 1920–1940*. Berkeley: University of California Press, 1986.

McCloud, Sean. *Making the American Religious Fringe: Exotics, Subversives, & Journalists, 1955–1993*. Chapel Hill: University of North Carolina Press, 2004.

Meldrum, Jeff. *Sasquatch: Legend Meets Science*. New York: Forge, 2006.

Mencken, H. L. *Newspaper Days, 1899–1906*. New York: Alfred A. Knopf, 1941.

Messner, Reinhold. *My Quest for Yeti*. Translated by Peter Constantine. New York: St. Martin's Griffin, 2000.

Meyer, Karl Ernest, and Shareen Blair Brysac. *Tournament of Shadows: The Great Game and the Race for Empire in Central Asia*. New York: Counterpoint Press, 2000.

Milligan, Linda. "The 'Truth' about the Bigfoot Legend." *Western Folklore* 49 (1990): 83–98.

Mitman, Gregg. *Reel Nature: America's Romance With Wildlife on Film*. Cambridge, MA: Harvard University Press, 1999.

Morton, Alan. *The Complete Directory to Science Fiction, Fantasy and Horror Television Series: A Comprehensive Guide to the First Fifty Years, 1946–1996*. Peoria, IL: Other World Books, 1997.

Murad, Turhon A. "Teaching Anthropology and Critical Thinking with the Question "Is There Something Big Afoot?"" *Current Anthropology* 29 (1988): 787–89.

Murphy, Christopher L. *The Bigfoot Film Controversy*. Surrey, BC: Hancock House, 2005.

———. *Meet the Sasquatch*. Surrey, BC: Hancock House, 2004.

Murray, W. H. *W. H. Murray: Evidence of Things Not Seen: A Mountaineer's Tale*. London: Baton Wicks, 2002.

Murray, William. *Adventures in the People Business: The Story of World Book*. Chicago: Field Enterprises Educational Corporation, 1966.

Nadis, Fred. *Wonder Shows: Performing Science, Magic, and Religion in America*. New Brunswick, NJ: Rutgers University Press, 2005.

Napier, John. *Bigfoot: The Yeti and Sasquatch in Myth and Reality*. New York: E. P. Dutton, 1973.

Nash, Richard. *Wild Enlightenment: The Borders of Human Identity in the Eighteenth Century*. Charlottesville: University of Virginia Press, 2003.

Newton, Michael. *Savage Girls and Wild Boys: A History of Feral Children*. New York: Picador, 2004.

Nickell, Joe. *Secrets of the Sideshows*. Lexington: University Press of Kentucky, 2005.

Orchard, Vance. *Bigfoot of the Blues*. Walla Walla, WA: privately printed, 1993.

Ortner, Sherry. *Life and Death on Mt. Everest: Sherpas and Himalayan Mountaineering*. Princeton, NJ: Princeton University Press, 2001.

———. *Sherpas through Their Rituals*. Cambridge: Cambridge University Press, 1978.

Orvell, Miles. *The Real Thing: Imitation and Authenticity in American Culture, 1880–1940*. Chapel Hill: University of North Carolina Press, 1989.

Osman Hill, William C. "Abominable Snowmen: The Present Position." *Oryx* 6 (1961): 86–98.

Palmer, Catherine. "'Shit happens': The Selling of Risk in Extreme Sport." *Australian Journal of Anthropology* 13, no. 3 (2002): 323–36.

Pangborn, Edgar. "Longtooth." *Magazine of Fantasy and Science Fiction*, January 1970, 5–36.

Parfrey, Adam, ed. *It's a Man's World: Men's Adventure Magazines, the Postwar Pulps*. Los Angeles: Feral House, 2003.

Pendergast, Tom. *Creating the Modern Man: American Magazines and Consumer Culture, 1900–1950*. Columbia: University of Missouri Press, 2000.

Perez, Daniel. *Bigfoot at Bluff Creek*. Norwalk, CA: Center for Bigfoot Studies, 2003.

Perez, Danny. *Big Footnotes: A Comprehensive Bibliography Concerning Bigfoot, the Abominable Snowmen, and Related Beings*. Norwalk, CA: privately printed, 1988.

Perkins, Marlin. *My Wild Kingdom: An Autobiography*. New York: E. P. Dutton, 1982.

Pettit, Michael. "'The Joy in Believing': The Cardiff Giant, Commercial Deceptions, and Styles of Observation in Gilded Age America." *Isis* 97 (2006): 659–77.

Pinch, Trevor, and Harry Collins. "Private Science and Public Knowledge: The Committee for the Scientific Investigation of the Claims of the Paranormal and Its Use of the Literature." *Social Studies of Science* 14 (1984): 521–46.

Place, Marian T. *Bigfoot All over the Country*. New York: Dodd, Mead, 1978.

———. *On the Track of Bigfoot*. New York: Pocket Books, 1978.

Powell, Thom. *The Locals: A Contemporary Investigation of the Bigfoot/Sasquatch Phenomenon*. Surrey, BC: Hancock House, 2003.

Pranavananda, Swami. "Abominable Snowman." *Indian Geographical Journal* 30 (1955): 99–104.

Price, David. "Interlopers and Invited Guests: On Anthropology's Witting and Unwitting Links to Intelligence Agencies." *Anthropology Today* 18, no. 6 (2002): 16–21.

Pyle, Robert Michael. *Where Bigfoot Walks: Crossing the Dark Divide*. Boston: Houghton Mifflin, 1995.

Quast, Mike. *Big Footage: A History of Claims for the Sasquatch on Film*. Moorhead, MN: privately printed, 2001.

Radford, Benjamin. "Bigfoot at Fifty: Evaluating a Half-Century of Bigfoot Evidence." *Skeptical Inquirer*, March 2002, 29–34.

Ramaswamy, Sumathi. *The Lost Land of Lemuria: Fabulous Geographies, Catastrophic Histories.* Berkeley: University of California Press, 2005.

Rawicz, Slavomir. *The Long Walk: The True Story of a Trek to Freedom.* Guilford, CT: Lyons Press, 1997.

Reiss, Benjamin. *The Showman and the Slave: Race, Death, and Memory in Barnum's America.* Cambridge, MA: Harvard University Press, 2001.

Rife, Philip R. *Bigfoot across America.* San Jose, CA: Writers Club Press, 2000.

Rigsby, Bruce. "Some Pacific Northwest Native Language Names for the Sasquatch Phenomenon." *Northwest Anthropological Research Notes* 5 (1971): 153–56.

Riley, Sam G. "A Search for the Cultural Bigfoot: Folklore or Fakelore?" *Journal of Popular Culture* 10 (1976): 377–87.

Ripley, S. Dillon. *Search for the Spiny Babbler.* Boston: Houghton Mifflin, 1952.

Ritvo, Harriet. *The Platypus and the Mermaid and Other Figments of the Classifying Imagination.* Cambridge, MA: Harvard University Press, 1997.

Ritzer, George. *Enchanting a Disenchanted World: Revolutionizing the Means of Consumption.* London: Sage, 2005.

Roscoe, Jane, and Craig Hight. *Faking It: Mock-Documentary and the Subversion of Factuality.* Manchester: Manchester University Press, 2002.

Rourke, Constance. *American Humor: A Study of the National Character.* New York: Harcourt, Brace, Jovanovich, 1959.

Rubin, Lillian Breslow. *Families on the Fault Line: America's Working Class Speaks about the Family, the Economy, Race, and Ethnicity.* New York: HarperPerennial, 1994.

———. *Worlds of Pain: Life in the Working Class Family.* New York: Basic Books, 1976.

Sanders, Joe, ed. *Science Fiction Fandom.* Westport, CT: Greenwood Press, 1994.

Sanderson, Ivan T. "The Abominable Snowman." *Fantastic Universe*, October 1959, 58–64.

———. "Abominable Snowmen Are Here!" *True*, November 1961, 40–41, 86–92.

———. *Abominable Snowmen: Legend Come to Life.* Philadelphia, PA: Chilton, 1961. Revised abridgment, New York: Pyramid, 1968.

———. *Animal Treasure.* New York: Pyramid, 1972.

———. *The Continent We Live on.* New York: Random House, 1961.

———. "First Photos of 'Bigfoot,' California's Legendary 'Abominable Snowman.'" *Argosy*, February 1968, 23–31, 127–28.

———. *Green Silence: The Story of the Making of a Naturalist.* Edited by Sabina W. Sanderson. New York: David McKay Company, Inc., 1974.

———. "Hairy Primitives or Relic Submen in South America." *Genus* 18 (1962): 60–74.

———. *Investigating the Unexplained.* Englewood Cliffs, NJ: Prentice-Hall, 1972.

———. *Invisible Residents: A Disquisition upon Certain Matters Maritime, and the Possibility of Intelligent Life Under the Waters of This Earth.* New York: Ty Crowell Company, 1970.

———. "The Missing Link." *Argosy*, May 1969, 23–31.

———. "More About the Abominable Snowman." *Fantastic Universe*, December 1959, 27–37.

———. "More Evidence That Bigfoot Exists." *Argosy*, April 1968, 72–73.

———. "A New Look at America's Mystery Giant." *True*, March 1960, 44–45, 101–2, 115.

———. "The Patterson Affair." *Pursuit*, June 1968, 8–10.

———. "Preliminary Description of the External Morphology of What Appeared to Be the Fresh Corpse of a Hitherto Unknown Form of Living Hominid." *Genus* 25 (1969): 249–78.

———. "Some Preliminary Traditions of Submen in Arctic and Subarctic North America." *Genus* 19 (1963): 145–62.

———. "The Strange Story of America's Abominable Snowman." *True*, December 1959, 40–43, 122–26.

———. "There Is an Abominable Snowman." *Fantastic Universe*, February 1960, 16–26.

———. "The Ultimate Hunt." *Sports Afield*, April 1961, 66–69, 113–18.

———. "Wisconsin's Abominable Snowman." *Argosy*, April 1969, 27–29, 70.

———. "Wudewasa: The Furry Men of Europe." *Fate*, November 1962, 25–32.

*Sasquatch Odyssey: The Hunt for Big Foot*. Peter von Puttkamer, dir. West Vancouver, BC: Big Hairy Deal Films, 1999; DVD, 2004.

Schechter, Harold. *The Bosom Serpent: Folklore and Popular Art*. Iowa City: University of Iowa Press, 1988.

Schulman, Bruce J. *The Seventies: The Great Shift in American Culture, Society, and Politics*. Cambridge, MA: Da Capo Press, 2001.

Seabrook, John. *Nobrow: The Culture of Marketing, the Marketing of Culture*. New York: Vintage Books, 2001.

Secord, Anne. "Science in the Pub: Artisan Botanists in Early Nineteenth-Century Lancashire." *History of Science* 32 (1994): 269–315.

Sennett, Richard, and Jonathan Cobb. *The Hidden Injuries of Class*. New York: Vintage Books, 1972.

Shackley, Myra. *Still Living? Yeti, Sasquatch and the Neanderthal Enigma*. London: Thames & Hudson, 1983.

Sheldon, Walter J. *The Beast*. New York: Fawcett, 1980.

Sheppard-Wolford, Sali. *Valley of the Skookum: Four Years of Encounters With Bigfoot*. Enumclaw, WA: Pine Winds Press, 2006.

Shipton, Eric. *Eric Shipton: The Six Mountain-Travel Books*. Seattle, WA: Mountaineers Books, 1997.

———. *That Untravelled World: An Autobiography*. New York: Charles Scribner's Sons, 1969.

Siefker, Phyllis. *Santa Claus, Last of the Wild Men: The Origins and Evolution of Saint Nicholas, Spanning 50,000 Years*. New York: McFarland & Co., 2006.

Sloan, Bill. *"I Watched a Wild Hog EAT my Baby!" A Colorful History of Tabloids and Their Cultural Impact*. New York: Prometheus, 2001.

Soule, Gardner. *Trail of the Abominable Snowman*. New York: G. P. Putnam's Sons, 1966.

Sprague, Roderick, and Grover S. Krantz, eds. *The Scientist Looks at the Sasquatch*. Moscow: University of Idaho Press, 1977.

———. *The Scientist Looks at the Sasquatch II*. Moscow: University of Idaho Press, 1979.

Steele, Peter. *Eric Shipton: Everest & Beyond*. Seattle, WA: Mountaineers Books, 1999.

Steinmeyer, Jim. *Charles Fort: The Man Who Invented the Supernatural*. New York: Jeremy P. Tarcher/Penguin, 2008.

Stephens, Walter. *Giants in Those Days: Folklore, Ancient History, and Nationalism*. Lincoln: University of Nebraska Press, 1989.

Stewart, Susan. *On Longing: Narratives of the Miniature, the Gigantic, the Souvenir, the Collection*. Durham, NC: Duke University Press, 1993.

Stonor, Charles. *The Sherpa and the Snowman*. London: Hollis & Carter, 1955.

Storey, John. *Cultural Studies and the Study of Popular Culture*. Athens: University of Georgia Press, 2003.

Streeby, Shelley. *American Sensations: Class, Empire, and the Production of Popular Culture*. Berkeley: University of California Press, 2002.

Susman, Warren I. "'Personality' and the Making of Twentieth-Century Culture." In *Culture as History: The Transformation of American Society in the Twentieth Century*, 271–86. New York: Pantheon, 1984.

Suttles, Wayne. "On the Cultural Track of the Sasquatch." *Northwest Anthropological Research Notes* 6 (1972): 65–90.

Swan, Lawrence W. *Tales of the Himalaya: Adventures of a Naturalist*. La Crescenta, CA: Mountain Air Books, 2000.

Tilman, H. W. *Mount Everest 1938*. Cambridge: Cambridge University Press,, 1938.

Trow, George S. *Within the Context of No Context*. Boston: Atlantic Monthly Press, 1997.

Tschernezky, W. "A Reconstruction of the Foot of the 'Abominable Snowman.'" *Nature* 186 (1960): 496–97.

Ullman, James Ramsey. *The Age of Mountaineering*. Philadelphia, PA: J. B. Lippincott Company, 1954.

Unsworth, Walt. *Everest: The Mountaineering History*. Seattle, WA: Mountaineers Books, 2000.

Von Doviak, Scott. *Hick Flicks: The Rise and Fall of Redneck Cinema*. Jefferson, NC: McFarland & Co., 2005.

Wallace, David Foster. "*E Unibus Pluram*: Television and U.S. Fiction." In *A Supposedly Fun Thing I'll Never Do Again: Essays and Arguments*, 21–82. New York: Back Bay Books, 1997.

Wallace, David Rains. *The Klamath Knot: Explorations of Myth and Evolution*. Berkeley: University of California Press, 2003. Reprint of the 1983 edition by Sierra Club Books.

Ward, Porter Morgan. "Bigfoot: Man or Myth?" In *Montana: The Magazine of Western History*, April 1957, 19–23.

Warner, Marina. *No Go the Bogeyman: Scaring, Lulling & Making Mock*. New York: Farrar, Straus and Giroux, 1998.

Wasser, Frederick. "Four Walling Exhibition: Regional Resistance to the Hollywood Film Industry." *Cinema Journal* 34 (1995): 51–65.

Wasson, Barbara. *Sasquatch Apparitions*. Bend, OR: privately printed, 1979.

Webb, James. *The Occult Establishment*. La Salle, IL: Open Court Publishing Company, 1976.

Westrum, Ronald. "Social Intelligence About Anomalies: The Case of Meteorites." *Social Studies of Science* 8 (1978): 461–93.

White, Hayden. "The Forms of Wildness: Archeology of an Idea." In *The Wild Man Within: An Image in Western Thought from the Renaissance to Romanticism*, edited by Edward Dudley and Maximillan E. Novak, 3–38. Pittsburgh, PA: University of Pittsburgh Press, 1972.

White, Richard. "'Are You an Environmentalist or Do You Work for a Living?': Work and Nature." In *Uncommon Ground: Rethinking the Human Place in Nature*, edited by William Cronon, 171–85. New York: W. W. Norton, 1996.

Williams, Bennett L. *The Legend of Big Foot: The Shoshone Indian Chief Nampuh*. Homedale, ID: Owyhea Publishing Company, 1992.

Woods, Lunetta. *Story in the Snow: Encounters with the Sasquatch*. Lakeville, MN: Galde Press, Inc., 1997.

Wyatt, Justin. *High Concept: Movies and Marketing in Hollywood*. Austin: University of Texas Press, 1994.

Wylie, Kenneth. *Bigfoot: A Personal Inquiry into a Phenomenon*. New York: Viking, 1980.

# Index

*Page number in italics refers to figures*